Advanced Structures: Materials and Technology

Advanced Structures: Materials and Technology

Edited by
Seth Royal

WILLFORD PRESS

www.willfordpress.com

Published by Willford Press,
118-35 Queens Blvd., Suite 400,
Forest Hills, NY 11375, USA

ISBN: 978-1-68285-338-2

Cataloging-in-publication Data

Advanced structures : materials and technology / edited by Seth Royal.
   p. cm.
Includes bibliographical references and index.
ISBN 978-1-68285-338-2
1. Reinforced concrete. 2. Reinforced concrete construction. 3. Concrete construction. 4. Structural engineering.
I. Royal, Seth.
TA683 .A38 2017
624.183 41--dc23

For information on all Willford Press publications
visit our website at www.willfordpress.com

WILLFORD PRESS

Printed in the United States of America.

Contents

Preface...VII

Chapter 1   **Modeling of RC Frame Buildings for Progressive Collapse Analysis**...1
Floriana Petrone, Li Shan and Sashi K. Kunnath

Chapter 2   **Seismic Response Analysis of Reinforced Concrete Wall Structure using
Macro Model**...14
Dong-Kwan Kim

Chapter 3   **Flexural Strength of RC Beam Strengthened by Partially De-bonded near
Surface-Mounted FRP Strip**..28
Soo-yeon Seo, Ki-bong Choi, Young-sun Kwon and Kang-seok Lee

Chapter 4   **Investigation on the Effectiveness of Aqueous Carbonated Lime in Producing
an Alternative Cementitious Material**...41
Byung-Wan Jo, Sumit Chakraborty, Ji Sun Choi and Jun Ho Jo

Chapter 5   **Utilization of Waste Glass Micro-particles in Producing Self-Consolidating
Concrete Mixtures**...55
Yasser Sharifi, Iman Afshoon, Zeinab Firoozjaei and Amin Momeni

Chapter 6   **Determination of Double-*K* Fracture Parameters of Concrete using Split-Tension
Cube: A Revised Procedure**...72
Shashi Ranjan Pandey, Shailendra Kumar and A. K. L. Srivastava

Chapter 7   **Laboratory Simulation of Corrosion Damage in Reinforced Concrete**........................85
S. Altoubat, M. Maalej and F. U. A. Shaikh

Chapter 8   **The use of Advanced Optical Measurement Methods for the Mechanical
Analysis of Shear Deficient Prestressed Concrete Members**...................................94
K. De Wilder, G. De Roeck and L. Vandewalle

Chapter 9   **Seismic Behavior Factors of RC Staggered Wall Buildings**.......................................109
Jinkoo Kim, Yong Jun and Hyunkoo Kang

Chapter 10   **Computing the Refined Compression Field Theory**..126
A. M. Hernández-Díaz and M. D. García-Román

Chapter 11   **Simplified Design Procedure for Reinforced Concrete Columns Based on
Equivalent Column Concept**...131
Hamdy M. Afefy and El-Tony M. El-Tony

Chapter 12   **Seismic Analysis on Recycled Aggregate Concrete Frame Considering Strain
Rate Effect**...145
Changqing Wang, Jianzhuang Xiao and Zhenping Sun

Chapter 13   **Shear Tests for Ultra-High Performance Fiber Reinforced Concrete (UHPFRC) Beams with Shear Reinforcement**...162
Woo-Young Lim and Sung-Gul Hong

Chapter 14   **A Review on Structural Behavior, Design, and Application of Ultra-High-Performance Fiber-Reinforced Concrete**...174
Doo-Yeol Yoo and Young-Soo Yoon

Chapter 15   **Review of Design Flexural Strengths of Steel–Concrete Composite Beams for Building Structures**...192
Lan Chung, Jong-Jin Lim, Hyeon-Jong Hwang and Tae-Sung Eom

Chapter 16   **Comparison of Strength–Maturity Models Accounting for Hydration Heat in Massive Walls**...205
Keun-Hyeok Yang, Jae-Sung Mun, Do-Gyeum Kim and Myung-Sug Cho

Permissions

List of Contributors

Index

Preface

This book elucidates the concepts and innovative models around prospective developments with respect to concrete structures. In order to build better structure it is important to include stability and strength in the designing of concrete structures. This text presents researches and studies performed by experts across the globe. It attempts to understand the multiple branches that fall under this discipline and how such concepts have practical applications. This book will serve as a valuable source of reference for graduate and post graduate students. A number of latest researches have been included to keep the readers up-to-date with the global concepts in this area of study.

This book was inspired by the evolution of our times; to answer the curiosity of inquisitive minds. Many developments have occurred across the globe in the recent past which has transformed the progress in the field.

This book was developed from a mere concept to drafts to chapters and finally compiled together as a complete text to benefit the readers across all nations. To ensure the quality of the content we instilled two significant steps in our procedure. The first was to appoint an editorial team that would verify the data and statistics provided in the book and also select the most appropriate and valuable contributions from the plentiful contributions we received from authors worldwide. The next step was to appoint an expert of the topic as the Editor-in-Chief, who would head the project and finally make the necessary amendments and modifications to make the text reader-friendly. I was then commissioned to examine all the material to present the topics in the most comprehensible and productive format.

I would like to take this opportunity to thank all the contributing authors who were supportive enough to contribute their time and knowledge to this project. I also wish to convey my regards to my family who have been extremely supportive during the entire project.

Editor

Modeling of RC Frame Buildings for Progressive Collapse Analysis

Floriana Petrone, Li Shan, and Sashi K. Kunnath*

Abstract: The progressive collapse analysis of reinforced concrete (RC) moment-frame buildings under extreme loads is discussed from the perspective of modeling issues. A threat-independent approach or the alternate path method forms the basis of the simulations wherein the extreme event is modeled via column removal scenarios. Using a prototype RC frame building, issues and considerations in constitutive modeling of materials, options in modeling the structural elements and specification of gravity loads are discussed with the goal of achieving consistent models that can be used in collapse scenarios involving successive loss of load-bearing columns at the lowest level of the building. The role of the floor slabs in mobilizing catenary action and influencing the progressive collapse response is also highlighted. Finally, an energy-based approach for identifying the proximity to collapse of regular multi-story buildings is proposed.

Keywords: collapse, frame structure, modeling, reinforced concrete, simulation.

1. Introduction

The systematic development of numerical models so as to gain an in-depth understanding of the behavior of concrete frame buildings subjected to extreme loading conditions, such as the sequential removal of load carrying vertical elements, is presented in this paper. This preliminary study is part of a more comprehensive research directed towards the formulation of robustness indices to assess the resistance of reinforced concrete (RC) structures to disproportionate collapse. Assessing the probability of building failure during either an ordinary or exceptional event deserves particular attention because of its relevance on the safety of human communities and the consequent economic impact on society. These studies also encourage the development of new structural design procedures and assessment criteria to prevent failure or minimize damage due to unexpected extreme events.

Within the above framework, most of the research effort during the last decade has been devoted to the study of disproportionate collapse of multi-story buildings. Progressive or disproportionate collapse occurs when a structure has its load pattern or boundary conditions altered in a manner such that some structural elements are loaded beyond their capacity and fail (Krauthammer et al. 2003). As well-documented in El-Tawil et al. (2014), several studies have been carried out in this field especially after the partial collapse of Ronan Point tower in England in 1968 (Pearson and Delatte 2005), with a significant increase in published papers in the field in the last 10 years. Early studies, including the papers by Lewicki and Olesen (1974), Arora et al. (1980), Gross and McGuire (1983), and McConnel and Kelly (1983), while reviewing alternative design methods to prevent progressive collapse and proposing the first computational approach for performing simulations, already highlight the need for unified analytical procedures for evaluating the resistance of structures to progressive collapse. In subsequent years, aided by the increasing availability of powerful computational tools, a large number of numerical studies have been carried out on different types of structures and with different modeling approaches, leading to a wide database of results and a range of research findings.

The simulation models used in previous studies differ from each other in many ways and can be classified into distinct groups based on the model features: (1) the type of analysis: linear or nonlinear—that can significantly affect the response of the structure as documented in Marjanishvili and Agnew (2006), where a comparison between different analyses shows that the predicted responses can vary significantly when performing static/dynamic and linear/nonlinear analyses, (2) the typology of elements: continuum, beam–column elements or a combination of both have been successfully used for modeling local and global phenomena of progressive collapse: examples of micro-models can be found in Khandelwal and El-Tawil (2007), Sasani and Kropelnicki (2008), Kwasniewski (2010), and Bao et al. (2008), while examples of macro-models are those utilized in Kaewkulchai and Williamson (2004), Bao et al. (2008), and Bao and Kunnath (2010) and an example of the use of hybrid models is reported in the work of Alashker et al. (2011), (3) the dimension of the model: planar or three dimensional—that is crucial in capturing spatial effects: most of the published work were conducted on two-

Department of Civil and Environmental Engineering, University of California, Davis, CA 95616, USA.
*Corresponding Author; E-mail: skkunnath@ucdavis.edu

dimensional structures (Khandelwal and El-Tawil 2007; Bao et al. 2008; Kim et al. 2009) and very few on three-dimensional models (Ruth et al. 2006; Alashker et al. 2011) making the comparison of results very challenging, (4) the floor system, modeled using a collection of beam–column, shell or brick elements (El-Tawil et al. 2014), which plays a key role in determining the response of three-dimensional structures, as shown by Yu et al. (2010), Alashker et al. (2011) and Li and El-Tawil (2014), (5) the loads applied on the structure, which can include only the self-weight or the self-weight and a fraction of the design dead/live loads.

Despite the availability of a large number of numerical studies, there still remains the need for procedures and numerical modeling guidelines to carry out simple yet reliable progressive collapse studies of buildings. The development and use of unified modeling criteria also derives from the necessity of having consistent models so that the results of different simulations can be effectively compared. In the long term this could lead to the development of a database of uniform data, useful to define and validate robustness indices for different typologies of structures. In this context, this study aims at identifying basic modeling features that have to be properly taken into account when analyzing the large-deformation behavior of RC buildings in response to extreme loading conditions. The simulations are based on the alternate path method (APM) wherein the extreme event itself is not simulated but the consequence of the event, i.e., the removal (due to failure) of a critical element is considered. Sensitivity analyses are carried out to assess the adequacy of element and section discretization and the efficacy of alternate options for modeling the floor slab are investigated. An energy-based approach to define proximity to partial or total collapse of the structure is also proposed.

2. Prototype Model, Materials and Elements

The study was conducted on prototype concrete moment-resisting building frames of varying height with and without the incorporation of floor slabs, using explicit time integration in LS-DYNA (Hallquist 2007), a general-purpose finite element code. Primary beams and columns were modeled using Hughes–Liu elements with multiple integration points along the length and cross section integration by means of fiber discretization at integration points, while slabs were represented as layered shells with smeared reinforcement. Both material and geometrical nonlinearities, including damage and fracture, were considered. Center-to-center dimensions were used to define beam and column lengths and joint shear deformations were ignored because separate studies carried out by the authors indicated that the overall displacement responses were not influenced by incorporating special joint elements for the moment frame configurations considered in this study whose behavior is controlled by flexure. Considering joint deformations will be important for the case of shear-critical members (Jeong and Kim 2014).

A prototype building, having plan dimensions 45 m × 30 m and story height 4.6 m for the 1st floor and 3.7 m for the remaining floors (Fig. 1), was designed to meet the requirements of seismic design category C for a site in Atlanta, as per ASCE-7 (2010) with detailing conforming to ACI 318 specifications (2014). Design details are summarized in Table 1. The building is characterized by two missing lines of columns—along B and E—to maximize floor space for practical considerations. Transverse reinforcement in the beams consists of single closed ties whereas for columns the closed ties are augmented with a central cross-tie in each direction (as required by ACI 318).

Nonlinearity of steel and concrete in tension and compression was modelled with isotropic elastic–plastic model (material model 124 in LS-DYNA), by defining constitutive laws expressed through effective stress versus plastic strain curves. The basic properties of the adopted materials are: unconfined compressive strength of concrete = 27.6 MPa, yield strength of reinforcement = 413.8 MPa, Young's modulus of concrete = 24.8 GPa, and for steel = 200 GPa. The peak stress and corresponding strain due to confinement as well as the ultimate stress and strain are obtained using the Mander et al. (1988) model. The Scott et al. (1982) model is used to represent the shape of the stress–strain curve up to peak compressive stress based on the material Concrete02 in OpenSees (2015) and a linear representation is used for the post-peak response. Overall three different types of confined concrete and one unconfined concrete were defined, as indicated in Table 2. The stress–strain curve for concrete is shown in Fig. 2a and that for steel in Fig. 2b. The assigned stress–strain curves are different in compression and tension, as shown in Fig. 2. The specified yield stress was the same in tension and compression. The stress–strain relationship of steel in compression accounts for buckling by following the model proposed by Gomes and Appleton (1997). The complete stress–strain curves in tension and compression, respectively were obtained using the ReinforcingSteel material in OpenSees (2015). It was observed that though confinement has an effect on the strength and deformation capacity of RC members, failure at large deformations is more influenced by the ductility of the steel reinforcement.

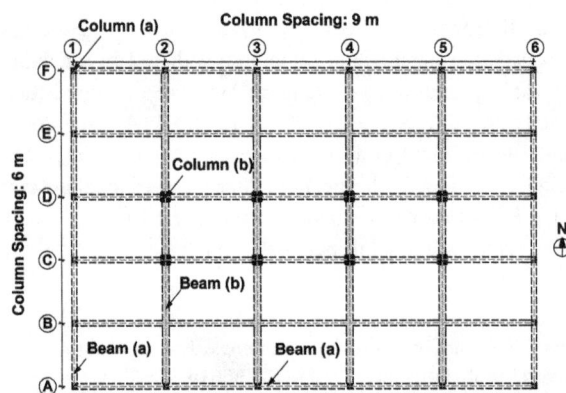

Fig. 1 Plan view of the building considered in study.

Table 1 Summary of design data.

Types	Beam (a)		Beam (b)		Column (a)	Column (b)
Dimensions: $b \times$ h (mm)	500×360		600×430		500×500	600×600
	Mid-span	Support	Mid-span	Support		
1 Story						
Longitudinal reinforcement	3: $\phi16$ (top) 4: $\phi16$ (bot)	4: $\phi22$ (top) 3: $\phi22$ (bot)	3: $\phi19$ (top) 4: $\phi19$ (bot)	4: $\phi19$ (top) 3: $\phi19$ (bot)	16: $\phi25$	16: $\phi22$
Shear reinforcement	2: $\phi12$ @152 mm	2: $\phi12$ @76 mm	2: $\phi12$ @178 mm	2: $\phi12$ @76 mm	3: $\phi12$ @203 mm	3: $\phi12$ @178 mm
3 Story						
Longitudinal reinforcement	3: $\phi16$ (top) 4: $\phi16$ (bot)	4: $\phi22$ (top) 3: $\phi22$ (bot)	3: $\phi19$ (top) 4: $\phi19$ (bot)	4: $\phi19$ (top) 3: $\phi19$ (bot)	16: $\phi25$	16: $\phi22$
Shear reinforcement	2: $\phi12$ @152 mm	2: $\phi12$ @76 mm	2: $\phi12$ @178 mm	2: $\phi12$ @76 mm	3: $\phi12$ @203 mm	3: $\phi12$ @178 mm
6 Story						
Dimensions: $b \times$ h (mm)	550×350		650×450		550×550	650×650
	Mid-span	Support	Mid-span	Support		
Longitudinal reinforcement	3: $\phi19$ (top) 4: $\phi19$ (bot)	4: $\phi22$ (top) 3: $\phi22$ (bot)	3: $\phi16$ (top) 4: $\phi16$ (bot)	4: $\phi22$ (top) 3: $\phi22$ (bot)	16: $\phi25$	16: $\phi19$
Shear reinforcement	2: $\phi12$ @152 mm	2: $\phi12$ @76 mm	2: $\phi12$ @152 mm	2: $\phi12$ @76 mm	3: $\phi12$ @203 mm	3: $\phi12$ @152 mm

Table 2 Concrete material properties.

Sections	σ_c (MPa)	ε_c	σ_{cu} (MPa)	ε_{cu}	f_t (MPa)	E_{ts} (MPa)
Unconfined	27.6	0.002	0	0.004	$0.1\sigma_c$	$0.5E_c$
Confined						
Beams	33.1	0.004	6.6	0.033	$0.1\sigma_c$	E_c
Columns	30.4	0.003	6.1	0.013	$0.1\sigma_c$	E_c
Slabs	27.6	0.002	6.9	0.018	$0.1\sigma_c$	E_c

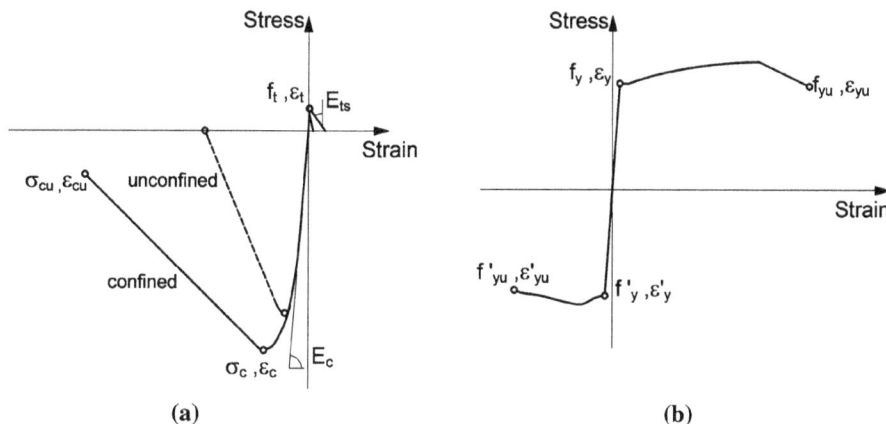

Fig. 2 Material constitutive models: **a** concrete and **b** reinforcing steel.

2.1 Localization of Inelastic Behavior

One of the considerations in specifying constitutive model properties, particularly for the Liu–Hughes beam–column elements used in the current simulation, is the likelihood of localization of inelastic behavior. Such localization can lead to a non-objective response during softening (Bao et al. 2014). Progressive collapse simulations involve large strains with failure occurring during the post-peak phase; hence care must be taken to adjust post-peak parameters to obtain the desired stress–strain response. Shown in Fig. 3 is the localized response if the target stress–strain curve is input without accounting for mesh size (in the case of the reinforcing steel bars, mesh size refers to the length of the discretized element). After adjusting the input stress–strain curve based on the chosen mesh size, the desired stress–strain curve is achieved as shown in the figure labeled "Non-local response". This is not truly a non-local response (resulting from a non-local model) but an indirect approach that modifies the input constitutive model to avoid localization. Strain estimates in the localized zones from such an approach will be inaccurate; however, the overall displacement estimates (the primary parameter of interest in the present collapse simulations) are reliable.

3. Sensitivity and Convergence Studies

Prior to finalizing the simulation model, it is essential to gain confidence in the level of discretization of elements as well as the choice of integration points in the cross-section. This is achieved by carrying out sensitivity studies with varying mesh sizes and integration points and examining the distribution of deformations and forces in critical elements. The objective of the simulations is to achieve a reasonable balance between accuracy and computational efficiency. In structural modeling for progressive collapse analysis, the model accuracy has to be verified for both linear and nonlinear responses, when large deformations involving material and geometrical nonlinearities are expected. As stated

earlier, the simulations in this study involve column-removal scenarios, consistent with APM, to examine the collapse resistance of concrete frame structures following an extreme event. Consequently, the focus of the modeling is related to the nonlinear response of a frame model to sequential column removal.

Since an explicit time integration method is used in all the simulations, damping across a frequency range needs to be defined. For the presented case studies the damping ratio is set equal to 0.05 and the frequency range is set to cover all relevant frequencies of the structure. The studies in this section were carried out on a single-story building because the validity of the sensitivity study is not affected by the height of the building: such as the localization of inelastic behavior, the element discretization and the cross section integration schemes. As for the modeling of gravity loads and the analysis of alternative grid beam models presented later in the paper, the one-story building case actually represents an upper bound condition—since additional stories would only increase the gravity loads supported by the first-story columns.

3.1 Element Discretization

In the first set of simulations, beams were discretized into 8, 10 and 12 elements of equal length. Since beams are deemed to be critical for the evaluation of the structural behavior in response to the sudden loss of a vertical load-carrying element, element discretization in beams were distributed into two different ways, resulting in six different cases to compare, as displayed in Fig. 4. Another consideration in determining an appropriate discretization was motivated by the fact that plastic hinge zones are crucial for the activation of catenary effect (Yi et al. 2008, 2014; Li et al. 2014). Two initial options were considered: in the first case 8, 10 and 12 elements were uniformly subdivided into equal lengths along the beam span—Option A, in the next case elements were concentrated in the hinge zones (25 % of beam span) leaving just one node at the mid-span—Option A*. For each case a nonlinear dynamic analysis is performed

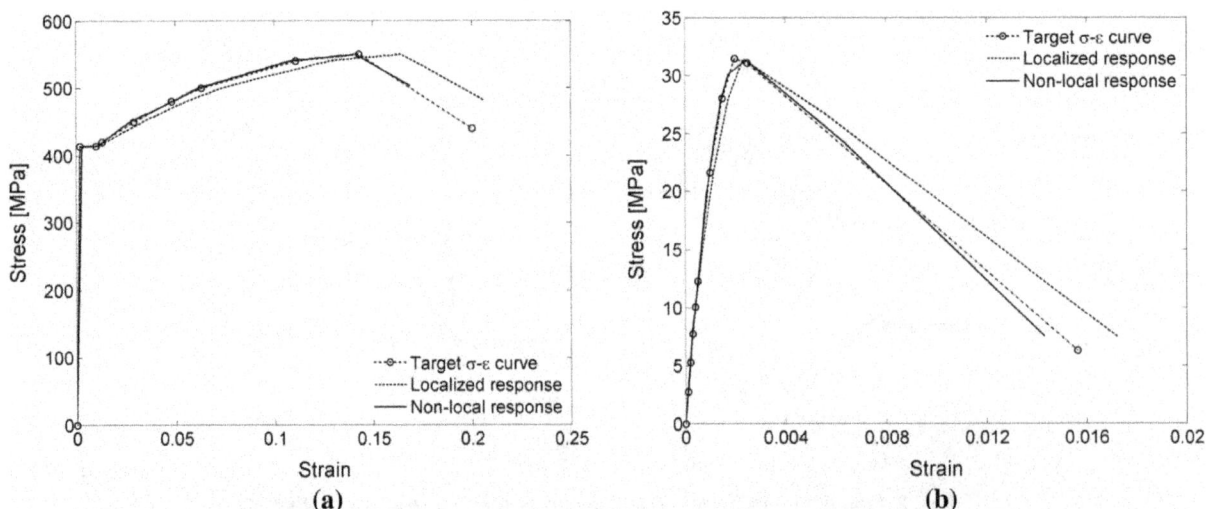

Fig. 3 Modified stress–strain input to avoid localization: **a** reinforcing steel model, and **b** concrete model.

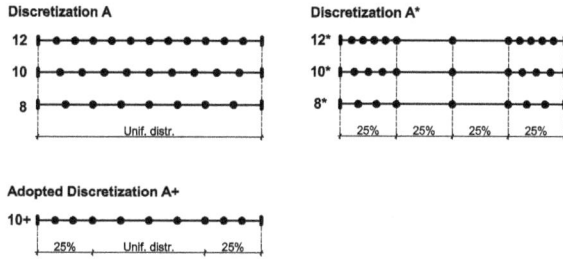

Fig. 4 Beam discretizations considered.

first under gravity loads (self-weight) followed by the removal of two columns.

Figure 5 shows the deflection of the continuous beam segments A1–3 under gravity loads for both discretization options. Option A resulted in generally similar shapes while Option A* expectedly led to different profiles at the mid-span of each beam. A closer view of the deflected shape in the support regions (potential plastic hinge locations) revealed that Option A* provides more consistent shape for all discretizations. However, the need to better represent the mid-span profile led to a third option (A+): starting with Option A comprising 8 elements two additional nodes were added at the mid-span to generate 10 elements evenly distributed in each zone (see Fig. 4).

The deflection profile resulting from this option is shown in all the plots of Fig. 5 and labeled as 10+. The accuracy of the adopted solution, when compared with Options A and A*, was also verified when considering larger deformations. Figure 6 shows the deflection profile of beams A1–3 after removing columns A1 and A2. It is seen that Option A+ (with only 10 elements) provides as accurate a response as attained with 12 elements. Next, considering a deformed profile in the short direction of the building, Fig. 7 shows the deflection profile of beams D2–C2–A2 following removal of columns A1 and A2. This confirms the previous observation with respect to beam deflection in the long direction that Option A+ provides the optimal discretization to achieve an acceptable response. The adopted discretization can capture

nearly the same deflection profile obtained with 12 elements in addition to saving about 30 % in computational time.

3.2 Cross-Section Integration

Once the optimal number and discretization of the beams were defined, three different options to model the cross-section of the beams were compared: the cases of 23, 42 and 52 integration points to model the reinforcing bars and cover and core concrete were considered. The accuracy of each option in predicting the distribution of stress/strain in the cross section was evaluated by comparing the variation of the axial force in a critical beam following the removal of two columns. Note that the axial force is computed by considering the average stress on the cross-section. The simulation consists of first applying the gravity load followed by removing column A1 at time $t = 2$ s and column A2 at $t = 8$ s. Figure 8 shows the time history of the axial force in beams A2–A3. Interesting fluctuations in the axial force in the beam are observed—the beam transitions from compression to tension after the removal of column A1 and then reverses back to compression after the removal of the second column. Although all three options give comparable responses after the first column removal, the section with 23 integration points overestimates the axial force following the second column removal. Since the results obtained with 42 and 52 integration points are nearly identical, the option with 42 integration points was adopted, representing an acceptable balance between solution accuracy and computational efficiency.

4. Modeling of Floor Slabs

A separate study was then conducted on slabs for defining the optimal mesh refinement for shells, modeled as four-node layered shell elements in LS-DYNA. Each of the 10 layers in the cross section—203 mm thick—is assigned a specific mechanical property, representing concrete and steel: concrete properties are those reported in Table 2, while

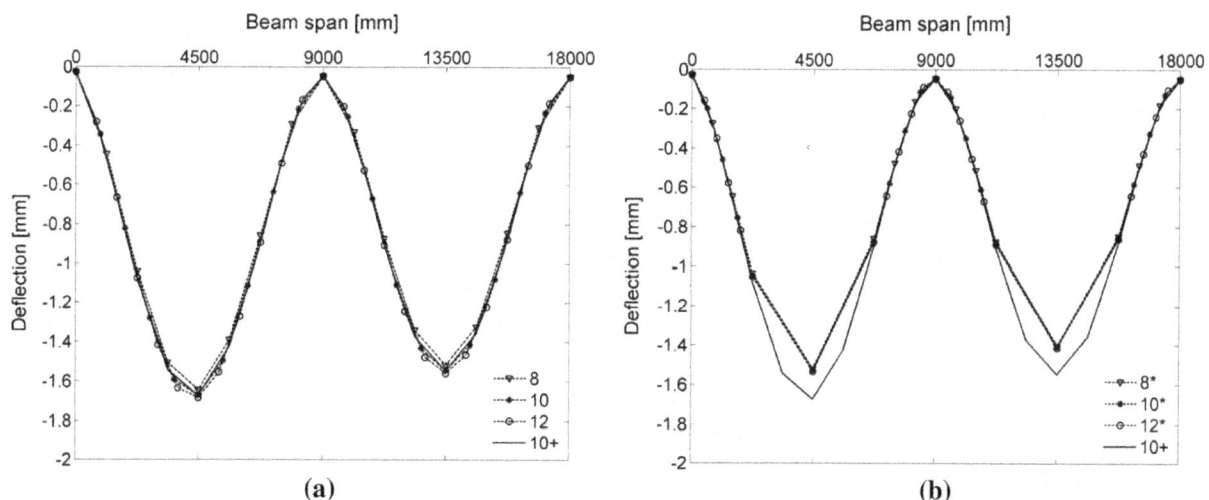

Fig. 5 Deflected shape of beams A1–3 for different discretizations: **a** Option A versus A+, and **b** Option A* versus A+.

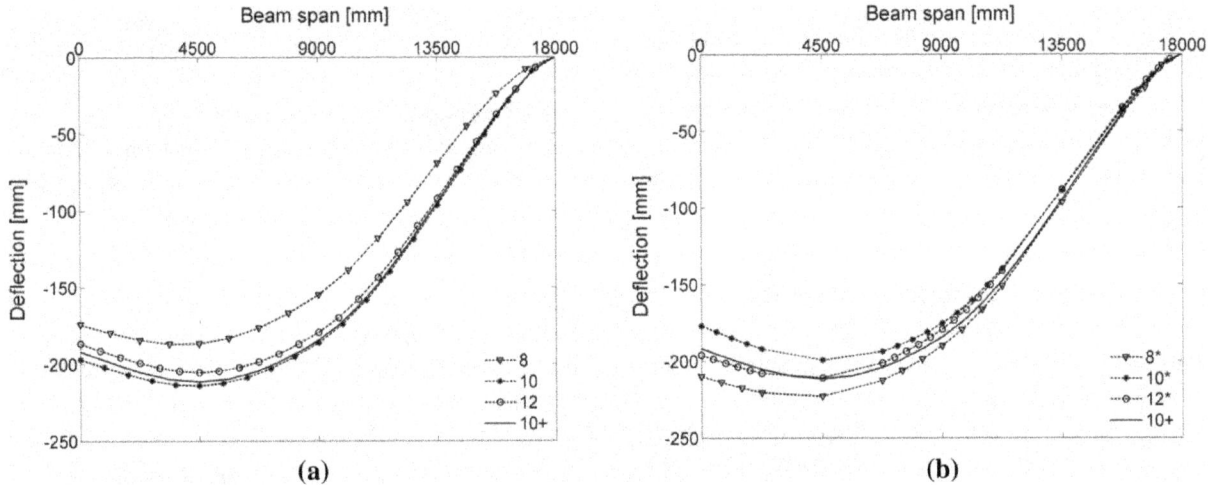

Fig. 6 Deflection profile of beams A1–3 following removal of two columns for different discretizations: **a** Option A versus A+, and **b** Option A* versus A+.

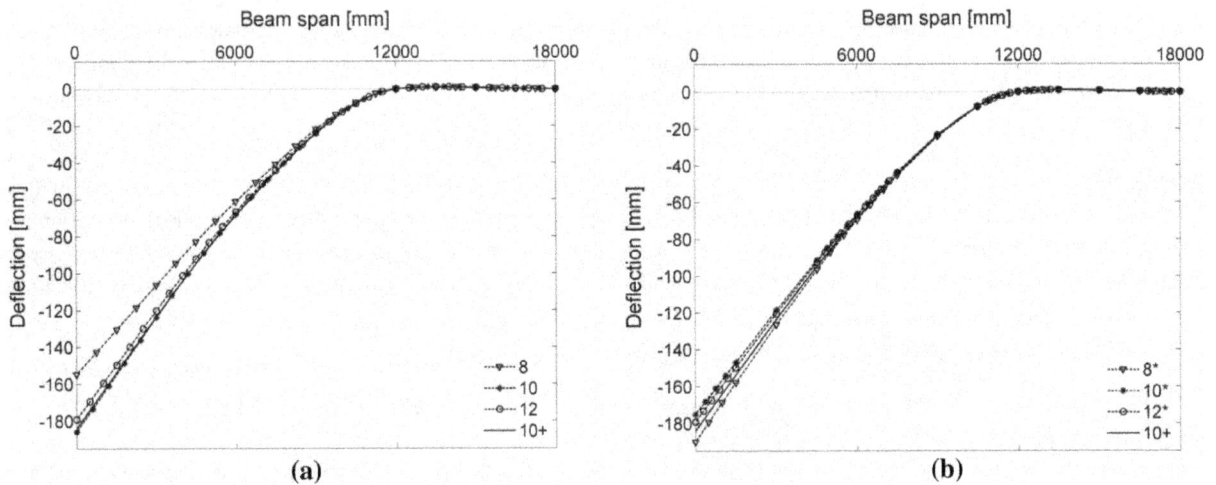

Fig. 7 Deflection profile of beams C2–B2–A2 following removal of two columns for different discretizations: **a** Option A versus A+, and **b** Option A* versus A+.

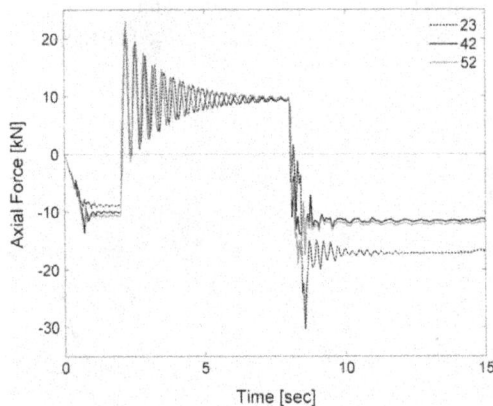

Fig. 8 Axial force–time history in beam A2–3.

for steel are the same adopted for beams and columns. Reinforcement is modeled through a smeared area corresponding to 13 mm dia @203 mm reinforcing bars at the top and bottom (in both directions). For the sensitivity analysis, mesh sizes of 300×300, 500×500 and 700×700 (mm²)

were investigated. Deflected shapes and axial force history along the mid-line of the slabs under gravity loads and after two column removals—columns A1 and A2—were assumed as the basic loading scenarios to compare the accuracy/acceptability of the three different mesh sizes.

Figure 9a shows the deflected shape along the mid-line of slab panel A1–B1–B3–A3 in the long direction under gravity loads. The comparison at points of maximum deflection suggests that 500×500 mesh size gives a solution fairly close to 300×300, in addition to savings in computational time. This observation is confirmed after examining the deflected shape following the removal of columns A1 and A2, as shown in Fig. 9b. Figure 10 presents the results of the same analysis for slab panel A1–C1–C2–A2.

The three shell refinements were also compared in terms of axial force variation in the slab. Figures 11a and 11b show axial force–time history at the mid-line of slab panel A1–A2–B1–B2 in long and short direction, respectively, under gravity load between 0 and 2 s and then after removing

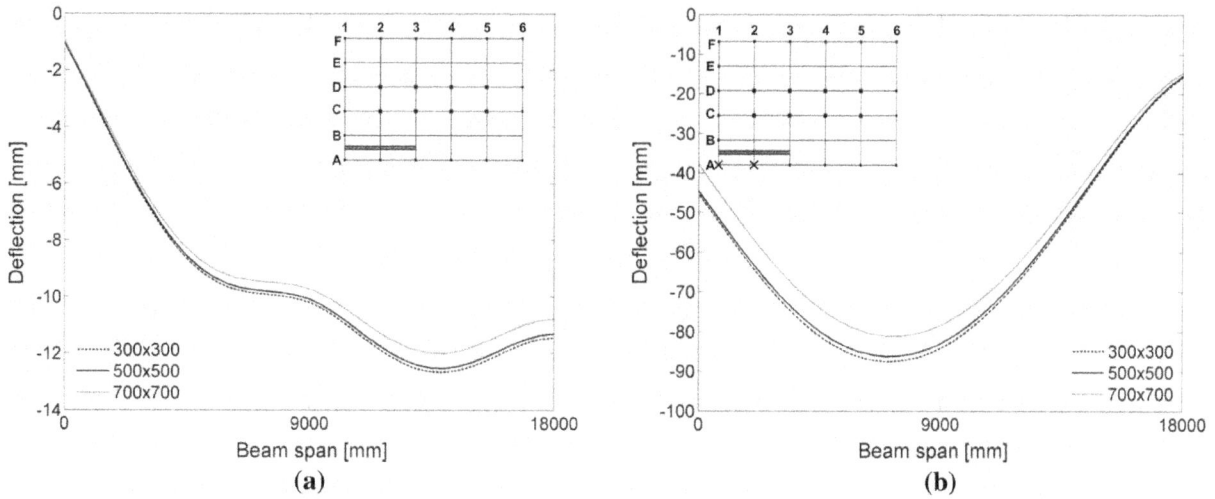

Fig. 9 Deflected shape along *mid-line* of *panel A1–B1–B3–A3* in long direction: **a** under gravity loads, and **b** after removing *columns A1* and *A2*.

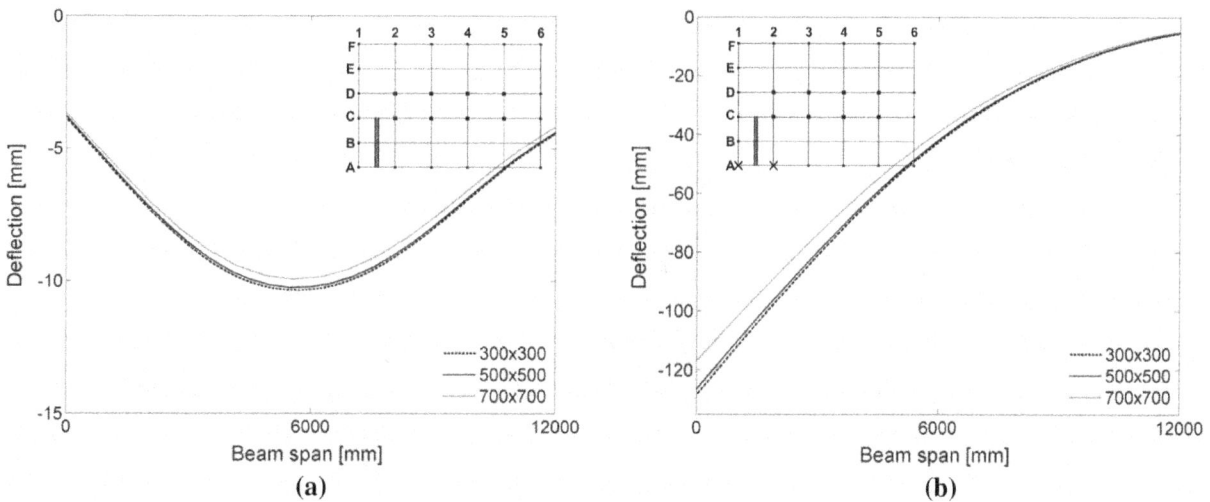

Fig. 10 Deflected shape along *mid-line* of *panel A1–C1–C2–A2* in short direction: **a** under gravity loads, and **b** after removing *columns A1* and *A2*.

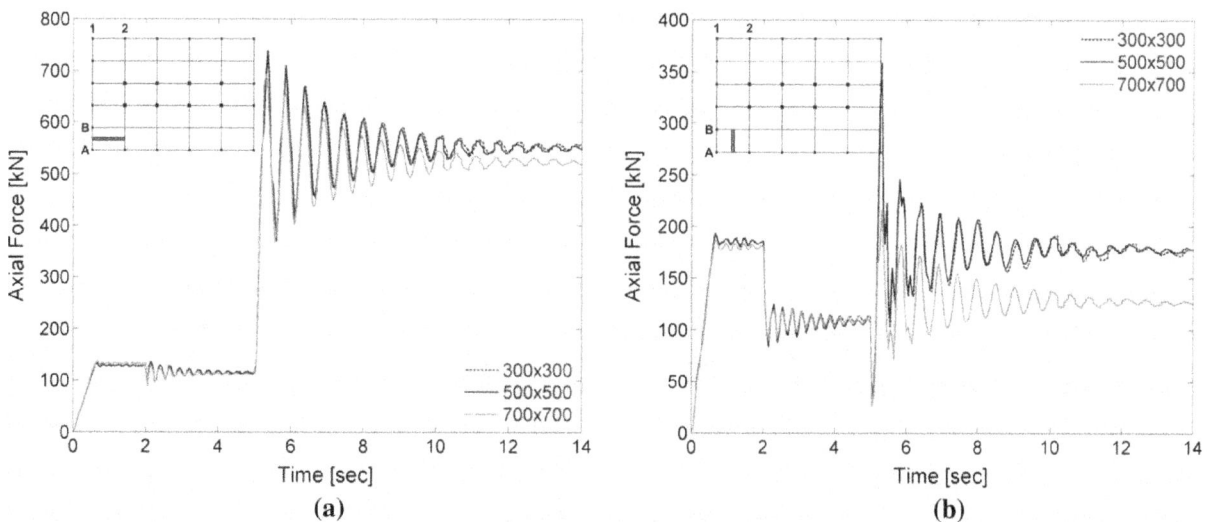

Fig. 11 Axial force–time history at the *mid-line* of slab *panel A1–A2–B2–B1*: **a** long direction, and **b** short direction.

column A1 at 2.0 s and after removing column A2 at 5.0 s. The difference in the three discretizations becomes significant after the second column removal, especially in short direction, where the 700×700 mesh size underestimates the force while the other two solutions give nearly the same values. This difference is confirmed in plots (a) and (b) of Fig. 11, showing the variation of the peak axial force in the exterior slab panels in the long and short direction: A1–B1–B6–A6 and A1–F1–F2–A2, respectively—after columns A1 and A2 are removed. The resultant axial force, calculated along the mid-line of the slab, follows the expected trend when moving from the panel next to the removed column to those further away and confirms that the mesh refinement of 700×700 underestimates the magnitude of the axial force by about 12 % on average. Based on these findings and those reported previously on slab deflections, the 500×500 mesh was adopted, having the added merit of saving approximately 40 % in total computational time.

5. Modeling of Gravity Loads

In some progressive collapse simulations reported in the literature, floors slabs are not explicitly modeled to save computational time and modeling effort. Loads from the slabs are therefore applied on the beams. However, when simulating column removal scenarios the load re-distribution that occurs after each removal alters the way loads are transferred to the beams. A study was hence undertaken to examine load distributions that simulate load transfer from slabs to beams for different scenarios.

Structure self-weight can typically be specified in most structural analysis software through body force loads after both the mass density of the adopted materials and the geometry of the structure are defined. This feature allows the automatic evaluation of the actual self-weight when slabs are included in the model. When slabs are not modeled, the self-weight of slabs (a significant source of gravity loading on buildings) and additional live loads should be distributed on beams. The nature of this distribution is not straight-forward and depends on the configuration of the building in plan, the stiffness of structural elements and the sequence of column removals.

In this study three different models were compared. In all cases, beams and columns are modeled with Hughes–Liu elements, having characteristics derived from the previously described sensitivity study: 10 elements for each beams based on discretization A+, 5 evenly distributed elements for columns and fiber sections with 42 integration points. The first model represents the "exact" solution incorporating two-way 203 mm thick slab, modeled with 500 mm × 500 mm shells connected to beams with shell edge-to-surface contacts. The second model is a frame model, where the self-weight of the slabs is evenly distributed on beams, representing the solution often adopted in structural design. The third model is also a frame without slabs, but a *modified* load distribution is used, derived to obtain the same beam deflection and the same magnitude of

shear forces calculated in the so-called "exact" model, resulting in the definition of a specific load distribution for each beam. The contribution of the slab in terms of stiffness is not considered in the frame models. Figure 12a and 12b show the deflection of beams A1–3 under gravity loads and after the loss of column A1, respectively, for the three models mentioned above.

It is observed that under gravity loads the frame model with uniform loads applied on the beams considerably overestimates the deflection of beams, whereas that with the modified distribution, as expected, gives the same deflected profile as the baseline "exact" model. Figure 12b shows the deformed shape of the same beams after removing column A1. It can be seen that the profile representing the frame model with uniform loads is missing because, right after column A1 is removed, beams A1–2 collapses under the imposed gravity load. Next, comparing the vertical displacement of node A1 using the frame model with modified load distribution it is seen that the predicted displacement is highly inaccurate. This is attributed to two factors: first, the distribution of loads based on a target deflection (shape and magnitude) and shear distribution that result from gravity loads leads to a concentration of loads in the plastic hinge zones and consequently to very large displacements after column removal, secondly, unlike models which incorporate slabs, frame models are unable to adequately account for alternate load paths and force redistribution after sudden column removals.

5.1 Grid Beam Models

Efficient slab models are not commonly available in structural analysis software thereby prompting engineers and researchers to seek alternative schemes using available beam models. Previous studies (Nurhuda and Lie 2004; Tian et al. 2012) have identified some configurations of beam grids suitable for the analysis of RC flat-plate structures. Sasani et al. (2011) used a two-directional grid of beams (similar to Fig. 13a) with modified torsional properties and an effective width to capture nonlinear effects in the slab region and reported satisfactory results.

To verify the efficiency and accuracy of alternative modeling methods for simulating the contribution of slabs in progressive collapse analysis, the nonlinear response of some simplified grid beam models under gravity loads and column removal scenarios was investigated. The validity of four different grid beam configurations, as depicted in Fig. 13, was verified by comparing their responses with the so-called "exact" solution (the response obtained with a model that simulates the slab using layered shell elements with nonlinear material properties). Additionally, for each configuration a separate parametric study was also conducted, wherein finer discretizations and modified stiffness properties were considered. The validation included comparisons of deflection (shape and magnitude), shear and axial forces in the beams and axial forces in the columns, under both gravity loads and column removal scenarios, when large deformations and pronounced nonlinear behavior are expected.

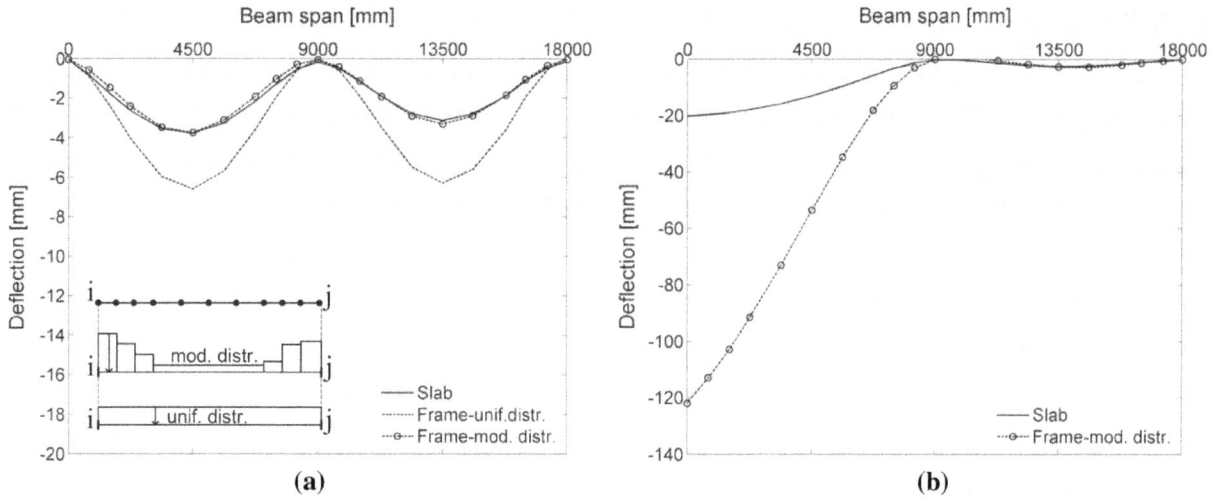

Fig. 12 Deflected profile of beams A1–3 for the three considered models: **a** under gravity loads and **b** after removal of column A1.

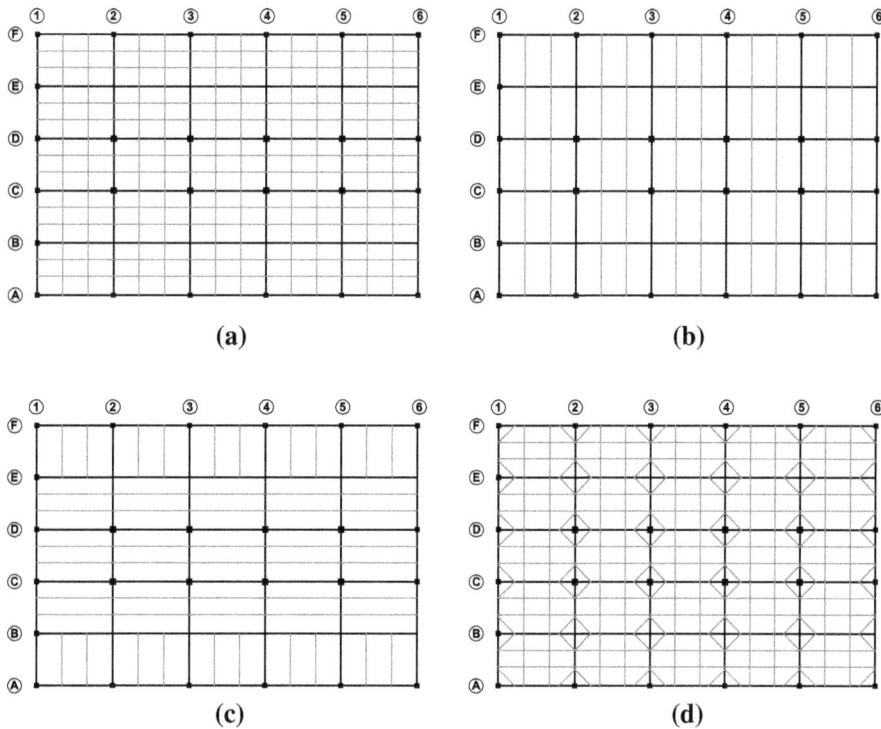

Fig. 13 Beam grids to replace floor slabs: **a** beams in both directions, **b** beams in short direction, **c** mixed configuration, and **d** Case 'a' with added diagonals.

The first analyzed configuration has two-way grid beams, as shown in Fig. 13a: the cases of 1, 3, 5, and 7 evenly distributed girders for each panel were analyzed. In all cases grid beams were assigned the same depth—203 mm—and the same reinforcement ratio as the slab. When compared to the "exact" model that incorporates nonlinear slabs, the two-way grid beam model produces acceptable results when comparing axial forces under gravity loads, however the errors in peak displacements range from 10 to 25 % (in the region of the missing column lines) and even larger errors are noted in predicting shear forces in the beams. Increased refinement of the grids and modifying the stiffness properties of the equivalent beams do result in improved predictions

but the overall effort in identifying a suitable grid and stiffness modifiers was deemed too cumbersome to merit further investigation.

Both the second and the third configurations, Figs. 13b and 13c, represent an attempt in redistributing loads on the main beams based on the column configuration and expected load transfer and both these configurations were evaluated for 3, 5 and 7 grid beams. The second configuration is that of one-way grid beams, running in the short direction. When tested under gravity loads, this configuration with 7 grid beams in each panel gave acceptable approximations of both resultant forces and displacements. However, when evaluated for the scenario involving a column removal, the errors were significant (with displacements being overestimated by a factor of 3 or more). Likewise, the third configuration, wherein one-way grid beams run in the short direction in the exterior panels and in the long direction in the interior panels, was also unsatisfactory in predicting either displacements or internal forces to an acceptable degree.

The fourth configuration, depicted in Fig. 13d, basically represents a modification of the first configuration (a), where elastic diagonals were added at the corners of the panels. The overall arrangement of the grid beams was conceived with the idea of reproducing in each panel the distribution of bending moments and the formation of yield lines observed in slab panels. A separate parametric study was conducted to calibrate the number of grid beams and the stiffness of diagonals to produce acceptable agreement with the "exact" model under large deformations. Though this configuration provided the best estimates of displacement, the model was incapable of reproducing correct internal forces.

This part of the study demonstrated the difficulty in simulating true slab effects using a grid of beams. While calibrating the stiffness properties of the beams to produce acceptable results under gravity loads appeared feasible, the task of recalibrating the models for each successive column removal makes this approach unreliable and unfeasible.

6. Energy-Based Approach to Assess Proximity to Collapse

This final section presents an energy-based approach to assess the proximity of the building to progressive collapse. Starting from the method proposed by Dusenberry and Hamburger (2006), in which the potential of a structure to survive disproportionate collapse is determined by an indicator proportional to the kinetic energy (KE), a more comprehensive approach is developed in this study based on a careful consideration of the energy contributions of the different structural components to the redistribution of forces following each column removal. A "collapse index" I_c is derived that tracks the variation in the system energy following each sudden removal of a critical element up to the collapse of the structure. The collapse index is used in parallel with another damage measure, the displacement ratio of the model that is defined as the ratio of the peak downward displacement of any removed column to the floor height. It is postulated that a collapse condition is imminent when the displacement ratio exceeds 0.6. Though the numerical solution may yet converge to an equilibrium condition, it is evident that the physical damage from a displacement that exceeds half the story height is likely to be severe and irreparable.

The proposed approach recognizes the fact that the system KE and strain energy (SE) are reliable indicators of the motion of a system caused by the sudden loss of bearing elements and also of the energy absorbed by the system as new load paths are established and equilibrium is being restored, if possible. It is also well-recognized that the variation of these two quantities is not uniform across the

entire structure, but mostly involves the local area where the structural elements are removed. For this reason a separate evaluation and a subsequent comparison of global and local energy change is assumed as an indicator of how much the local failure of the structure affects the stability of the overall system. Consequently, the collapse index I_c is defined as follows:

$$I_c = \frac{\kappa_{global}}{\kappa_{local}} \leq 1.0 \quad \text{where: } \kappa = \frac{KE}{SE} \qquad (1)$$

In the above expression, κ_{global} and κ_{local} represent the ratio of the KE to the SE calculated over the entire system and over the local portion involved in the collapse, respectively. In the present study, these energy quantities are directly computed in LS-DYNA. However, the computation of SE and KE can be accomplished using well-known relationships as follows:

$$SE = U_{beam} + U_{slab}, \qquad (2a)$$

$$U_{beam} = \frac{1}{2} \left(\int_0^L \frac{N^2}{EA} dx + \int_0^L \frac{V^2}{GA^*} dx + \int_0^L \frac{M^2}{EI} dx + \int_0^L \frac{T^2}{GJ} dx \right), \qquad (2b)$$

$$U_{slab} = \int_V \{\sigma\}^T \{\varepsilon\} dV, \qquad (2c)$$

$$KE = \frac{1}{2} \left(\sum_{j=1}^{nt} m_j \dot{u}_j^2 + \sum_{j=1}^{nr} I_\theta \dot{\theta}_j^2 \right) \qquad (2d)$$

The following notations are used: L = length, N, V, M, T = axial force, shear, moment and torsion, respectively, E, G, A, A^*, I, J = Young's modulus, shear modulus, cross-sectional area, shear area, moment of inertia and polar moment of inertia, respectively, $\{\sigma\}$, $\{\varepsilon\}$ = vectors of stresses and corresponding strains, nt, nr = number of translational and rotational degrees-of-freedom, m_j, I_θ = translational mass and rotational inertia, $\dot{u}$, $\dot{\theta}$ = translational and rotational velocities, respectively. When following the APM, I_c indicates the proximity to collapse of the structural system during a sequence of column removals. Ideally, at the ultimate condition $I_c = 1$ though it was necessary to introduce the displacement ratio as an additional criterion to assess the severity of the damage state.

The proposed method was applied to three buildings (1-, 3- and 6-story), and both the models with and without floor slabs were analyzed. A strain-based failure criterion was also assigned in which the rupture strain of reinforcing steel was specified as 0.15. However, as noted previously, this failure strain is not a reliable measure since a modified post-peak slope was specified for the reinforcing steel material to avoid localization. Yet, it does provide an indication that the ultimate stress in the bar has been exceeded. Figures 14a and 14b show the variation of I_c and the corresponding displacement ratios for three different column removal sequences (paths A–C)

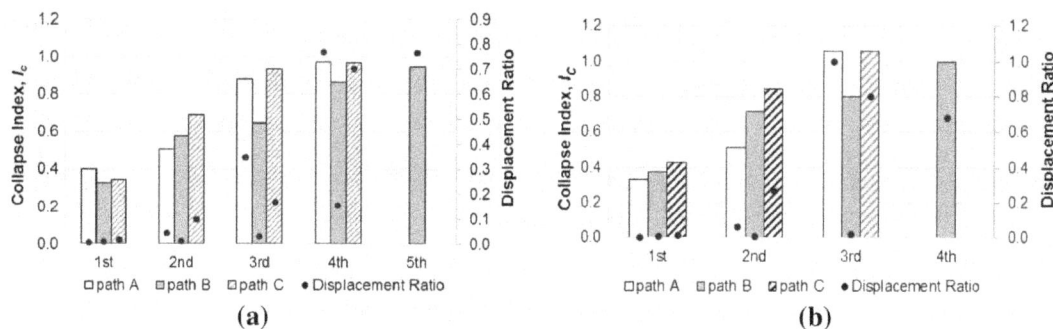

Fig. 14 Comparison of different column removal scenarios for 1-story building: **a** model with slab, and **b** bare frame model.

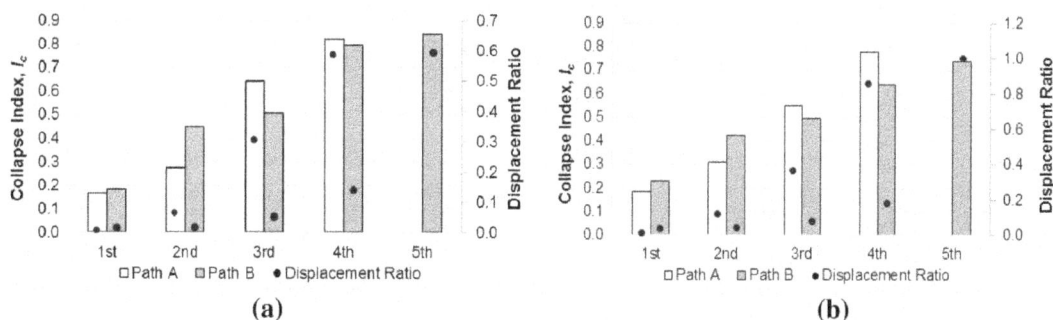

Fig. 15 Energy-based collapse index for multi-story buildings: **a** 3-story building, and **b** 6-story building.

for the single-story building for both the bare frame model and the full model with slabs. Likewise, Figs. 15a and 15b present the change in the collapse index with successive column removals for the 3- and 6-story models. Based on the results shown in these plots, the following facts emerge:

(1) The collapse index, I_c, shows an increasing trend with each subsequent column removal hence validating the normalized energy-based approach for defining the onset of progressive collapse.

(2) Incipient "collapse" is initiated when the displacement ratio exceeds 0.6. This corresponds to an I_c value close to unity for the 1-story building and approaches a ratio of approximately 0.8 for the 3- and 6-story buildings. This damage state was also accompanied by simultaneous rupture of reinforcing steel bars in critical locations where the largest deformations are localized.

(3) In all the analyzed cases, the bare frame model (without slabs) approaches the collapse state with fewer column removals than the model incorporating floor slabs—highlighting the role of slabs in finding alternate load paths after column removal and contributing to progressive collapse resistance of the system.

(4) An assessment of the plots shown in Figs. 14 and 15 indicate that progressive collapse is "sequence-dependent" and that there exists a critical sequence of column removals that results in the minimum number of removed columns before collapse is initiated.

In the case of the model incorporating floor slabs, path A corresponds to the sequential removal of the following columns: A1–A2–B1–A3, path B corresponds to removal of C2–C1–B1–

A1–A2 and path C refers to A3–A2–A1–B1. For the bar frame models, the removal sequences were as follows: path A: A1–A2–B1, path B: C2–C1–B1–A1, path C: A3–A2–A1. For the 1-story frame model, paths A and C require only three column removals before collapse is initiated whereas path B involves the removal of four columns. Similar observations are evident from the response of the full model incorporating slab elements (Fig. 14a) though an additional column (compared to the frame model) needs to be removed to initiate failure. The results shown in Fig. 15a and b are for the models in which the floor slabs are included. Results for the case of the bare frame are not shown here and trends similar to that for the single-story building were observed. The proposed method, therefore, in addition to providing a measure of the reserve capacity of the structure, can also be used to identify critical load-bearing elements in the structure that provide the greatest resistance to progressive collapse.

7. Conclusions

The results presented in this study offer useful guidelines on modeling and simulation of progressive collapse of RC frame structures within the context of the APM. Using a simple case study of a RC frame structure, various issues in modeling of materials and elements are presented with a view to providing practical insights into progressive collapse simulations. The importance of adequately modeling the floor slab is highlighted and the often employed approach of using a grid of beams to model the slab system is shown to be impracticable for simulating collapse within the APM framework that involves the removal of column elements.

Finally, a new collapse index is proposed that can be used to assess both the damage state and the reserve capacity of the system and can also serve as a means to identify critical load-bearing elements in the structure that provide the greatest resistance to progressive collapse.

One of the objectives of the study was to consider simplified approaches for progressive collapse analysis. Consequently, though an advanced finite element software was used in the simulations presented in this study, beams and columns were modeled as line elements with cross-section integration (available in both open-source and commercial software used in engineering practice) and slabs were modeled using layered shells with smeared reinforcement. Some structural analysis software may not provide convenient options for sudden column removal. If dynamic analysis is feasible, the application of short-duration pulse loads corresponding to the axial force in the removed columns is a viable option. Another alternative is to carry out a static analysis in the absence of the removed columns but to magnify the applied loads to consider dynamic effects. But both these approaches become more involved and complex for multiple column removals.

Another critical aspect in the development of a simulation model is its validation using experimental data. The availability of appropriate experimental data to validate the response to sudden column removals in a building is extremely limited. A number of recent tests carried out by researchers in China (Yi et al. 2008,2014; Li et al. 2014; Xiao et al. 2015) provide an initial starting point for such validations but additional experimental data on large-scale experiments is still needed to calibrate and assess the adequacy of the proposed modeling schemes.

Acknowledgments

Funding for this research was sponsored by the National Science Foundation through Grant CMMI-0928953. Any opinions, findings, conclusions, and recommendations expressed in this paper are those of the authors and do not necessarily reflect the views of the sponsors.

References

ACI. (2014). *Building code requirements for structural concrete*. ACI 318-14. Farmington Hills, MI: American Concrete Society.

Alashker, Y., Li, H., & El-Tawil, S. (2011). Approximations in progressive collapse modeling. *Journal of Structural Engineering, 137*(Special Issue: Commemorating 10 Years of Research Since 9/11), 914–924.

Arora, J. S., Haskell, D. F., & Govil, A. K. (1980). Optimal design of large structures for damage tolerance. *AIAA Journal, 18*(5), 563–570.

ASCE. (2010). *Minimum design loads for buildings and other structures*. ASCE-7, 2010. Reston, VA: American Society of Civil Engineers.

Bao, Y., & Kunnath, S. K. (2010). Simplified progressive collapse simulation of RC frame-wall structures. *Engineering Structures, 32*(10), 3153–3162.

Bao, Y., Kunnath, S., El-Tawil, S., & Lew, H. (2008). Macro-model-based simulation of progressive collapse: RC frame structures. *Journal of Structural Engineering, 134*(7), 1079–1091.

Bao, Y., Lew, H. S., & Kunnath, S. K. (2014). Modeling of reinforced concrete assemblies under column-removal scenario. *Journal of Structural Engineering, 140*(1), 04013027.

Dusenberry, D., & Hamburger, R. (2006). Practical means for energy-based analyses of disproportionate collapse potential. *Journal of Performance of Constructed Facilities, 20*, 336–348.

El-Tawil, S., Li, H., & Kunnath, S. (2014). Computational simulation of gravity-induced progressive collapse of steel-frame buildings: Current trends and future research needs. *Journal of Structural Engineering, 140*(8), A2513001.

Gomes, A., & Appleton, J. (1997). Nonlinear cyclic stress–strain relationship of reinforcing bars including buckling. *Engineering Structures, 19*(10), 822–826.

Gross, J. L., & McGuire, W. (1983). Progressive collapse resistant design. *Journal of Structural Engineering, 109*, 1–15.

Hallquist, J. (2007). *LS-DYNA keyword user's manual*. Livermore, CA: Livermore Software Technology Corporation.

Jeong, J.-P., & Kim, W. (2014). Shear resistant mechanism into base components: Beam action and arch action in shear-critical RC members. *International Journal of Concrete Structures and Materials, 8*(1), 1–14.

Kaewkulchai, G., & Williamson, E. B. (2004). Beam element formulation and solution procedure for dynamic progressive collapse analysis. *Computers and Structures, 82*(7–8), 639–651.

Khandelwal, K., & El-Tawil, S. (2007). Collapse behavior of steel special moment resisting frame connections. *Journal of Structural Engineering, 133*(5), 646–655.

Kim, H. S., Kim, J., & An, D. W. (2009). Development of integrated system for progressive collapse analysis of building structures considering dynamic effects. *Advances in Engineering Software, 40*(1), 1–8.

Krauthammer, T., Hall, R. L., Woodson, S. C., Baylot, J. T., Hayes, J. R., & Sohn, Y. (2003). Development of progressive collapse analysis procedure and condition assessment for structures. In *National workshop on prevention of progressive collapse*. Washington, DC: Multihazard

Mitigation Council of the National Institute of Building Sciences.

Kwasniewski, L. (2010). Nonlinear dynamic simulations of progressive collapse for a multistory building. *Engineering Structures, 32*(5), 1223–1235.

Lewicki, B., & Olesen, S. O. (1974). Limiting the possibility of progressive collapse. *Building Research and Practice, 2*(1), 10–13.

Li, H., & El-Tawil, S. (2014). Three-dimensional effects and collapse resistance mechanisms in steel frame buildings. *Journal of Structural Engineering, 140*(8), A4014017.

Li, Y., Lu, Z., Guan, H., & Ye, L. (2014). Progressive collapse resistance demand of reinforced concrete frames under catenary mechanism. *ACI Structural Journal, 111*, 433–439.

Mander, J. B., Priestley, M. J. N., & Park, R. (1988). Theoretical stress–strain model for confined concrete. *Journal of Structural Engineering, 114*(8), 1804–1826.

Marjanishvili, S., & Agnew, E. (2006). Comparison of various procedures for progressive collapse analysis. *Journal of Performance of Constructed Facilities, 20*, 365–374.

McConnel, R. E., & Kelly, S. J. (1983). Structural aspects of progressive collapse of warehouse racking. *Engineering Structures, 61A*(11), 343–347.

Nurhuda, I., & Lie, H. A. (2004). Three dimensionally analysis of flat plate structures by equivalent grid method. In *29th Conference on our world in concrete and structures*, 25–26 August 2004, Singapore.

OpenSees. (2015). *Open system for earthquake engineering simulation*. Berkeley, CA: University of California. Retrieved March 4, 2015, from http://opensees.berkeley.edu/.

Pearson, C., & Delatte, N. (2005). Ronan point apartment tower collapse and its effect on building codes. *Journal of Performance of Constructed Facilities, 19*(5), 172–177.

Ruth, P., Marchand, K., & Williamson, E. (2006). Static equivalency in progressive collapse alternate path analysis:

Reducing conservatism while retaining structural integrity. *Journal of Performance of Constructed Facilities, 20*, 349–364.

Sasani, M., Kazemi, A., Sagiroglu, S., & Forest, S. (2011). Progressive collapse resistance of an actual 11-story structure subjected to severe initial damage. *Journal of Structural Engineering, ASCE, 137*(9), 893–902.

Sasani, M., & Kropelnicki, J. (2008). Progressive collapse analysis of an RC structure. *The Structural Design of Tall and Special Buildings, 17*, 757–771.

Scott, B. D., Park, R., & Priestley, M. J. N. (1982). Stress–strain behavior of concrete confined by overlapping hoops at low and high strain rates. *Journal of the American Concrete Institute, 79*, 13–27.

Tian, Y., Chen, J., Said, A., & Zhao, J. (2012). Nonlinear modeling of flat-plate structures using grid beam elements. *Computers and Concrete, 10*(5), 489–505.

Xiao, Y., Kunnath, S. K., Li, F. W., Zhao, Y. B., Lew, H. S., & Bao, Y. (2015). Collapse test of a 3-story 3-span half-scale RC frame building. *ACI Structural Journal, 112*(4), 429–438.

Yi, W.-J., He, Q.-F., Xiao, Y., & Kunnath, S. K. (2008). Experimental study on progressive collapse-resistant behavior of reinforced concrete frame structures. *ACI Structural Journal, 105*(4), 433–439.

Yi, W.-J., Zhang, F. Z., & Kunnath, S. K. (2014). Progressive collapse performance of RC flat plate frame structures. *Journal of Structural Engineering*. doi:10.1061/(ASCE)ST.1943-541X.0000963.

Yu, M., Zha, X., & Ye, J. (2010). The influence of joints and composite floor slabs on effective tying of steel structures in preventing progressive collapse. *Journal of Constructional Steel Research, 66*(3), 442–451.

2

Seismic Response Analysis of Reinforced Concrete Wall Structure Using Macro Model

Dong-Kwan Kim*

Abstract: During earthquake, reinforced concrete walls show complicated post-yield behavior varying with shear span-to-depth ratio, re-bar detail, and loading condition. In the present study, a macro-model for the nonlinear analysis of multi-story wall structures was developed. To conveniently describe the coupled flexure-compression and shear responses, a reinforced concrete wall was idealized with longitudinal and diagonal uniaxial elements. Simplified cyclic material models were used to describe the cyclic behavior of concrete and re-bars. For verification, the proposed method was applied to various existing test specimens of isolated and coupled walls. The results showed that the predictions agreed well with the test results including the load-carrying capacity, deformation capacity, and failure mode. Further the proposed model was applied to an existing wall structure tested on a shaking table. Three-dimensional nonlinear time history analyses using the proposed model were performed for the test specimen. The time history responses of the proposed method agreed with the test results including the lateral displacements and base shear.

Keywords: nonlinear analysis, macro model, cyclic loads, earthquake loads, structural wall, reinforced concrete.

1. Introduction

Recently, nonlinear analysis has became popular in the earthquake design and evaluation of structures due to the advances in earthquake engineering and numerical analysis [ATC 40 (1996), FEMA 356 (ASCE 2000); FEMA 440 (ASCE 2005)]. In particular, reinforced concrete walls are used in many high-rise buildings as the primary lateral load-resistant system. Thus, an effective analytical method for walls is required to evaluate the overall inelastic response of buildings with walls. Both microscopic finite element models and macroscopic models can be used for the nonlinear analysis of wall systems. The microscopic finite element models can provide detailed local responses of walls with accuracy (Park and Klingner 1997; Okamura and Maekawa 1991; Stevens et al. 1991; Feenstra and de Borst 1993; Mansour and Hsu 2005; Wong and Vecchio 2002; Palermo and Vecchio 2007; Petrangeli et al. 1999; D'Ambrisi and Filippou 1999). However, it requires tremendous efforts and time for modeling and numerical computations. The macroscopic models, on the other hand, are simple and practical though their application is limited depending on the assumptions that each model is based on (Kabeyasawa et al. 1982; Vulcano and Bertero 1987; Orakcal 2004; Orakcal et al. 2006; Park and Eom 2007; Monti and Spacone 2000;

Wallace 2012). Currently, because of the efficiency of the macroscopic models, existing structural analysis platforms such as Perform 3D (Computer and Structures Inc. 2006), OpenSEES (PEER 2001), and DRAIN-2DX (Prakash et al. 1993) use macroscopic models for the nonlinear analysis of wall systems.

Figure 1 shows various macroscopic models for the nonlinear analysis of walls subjected to cyclic loading. In Fig. 1, the existing macroscopic models use multiple vertical uniaxial elements of concrete and re-bar in order to describe the flexure-compression responses of walls such as the relocation of neutral axis in wall cross-sections, tension stiffening behavior of concrete, flexural crack opening and closing, and the confinement effect of the concrete in boundary elements. Previous studies (Vulcano and Bertero 1987; Orakcal 2004; Orakcal et al. 2006) show that the flexure-compression response of walls can be accurately predicted by using the existing macroscopic models. On the other hand, the existing macroscopic models use a horizontal uniaxial element (i.e. shear spring element) in order to describe the shear response of walls. However, the horizontal uniaxial element does not accurately describe the shear action associated with inclined cracking and diagonal strut action of the web concrete. Thus, it is difficult to accurately estimate the shear response of short and medium-rise wall, which is significantly affected by diagonal strut action.

In the present study, a macro finite element model was developed to predict the coupled flexure-compression and shear responses of the reinforced concrete walls subjected to cyclic loading. The macro-model was idealized with multiple longitudinal uniaxial elements of concrete and re-bar to describe the flexure- compression responses of walls. To describe the shear

SEN Structural Engineers Co., R&D Team, Seoul 07229, Korea.
*Corresponding Author; E-mail: dkkim@senkuzo.com

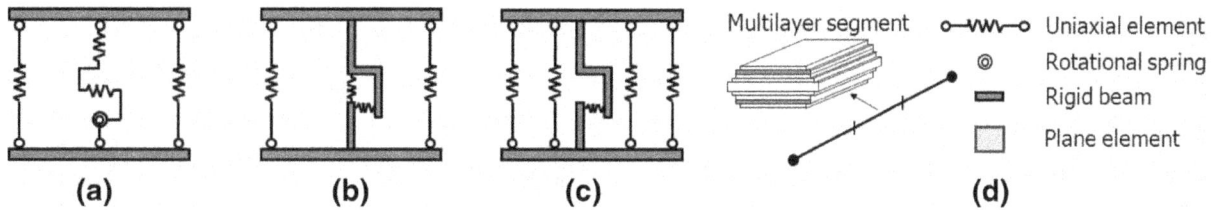

Fig. 1 Existing macro-models for structural walls. **a** Kabeyasawa et al. (1982), **b** Linda and Bachmann (1994), **c** Vulcano and Bertero (1987), **d** Monti and Spacone (2000).

response significantly affected by cyclic loading, diagonal strut elements were used. In the diagonal strut elements, for the web concrete, uniaxial cyclic models of concrete and steel re-bar were used. In the cyclic models, the confinement effect of concrete in wall boundary elements and the compressive softening of web concrete were considered. For verification, the proposed macro-model was applied to existing slender, low-rise, and coupled wall specimens subjected to cyclic loading. Further, the proposed macro-model was applied to three-dimensional nonlinear time history analysis for a 1:5 scale 10-story R.C. wall specimen of residential building specimen, which was tested on a shaking table.

2. Proposed Macro-Model

Figure 2 shows the proposed macro-model for walls. The proposed model consists of dimensionless lateral rigid beam elements, longitudinal uniaxial elements, diagonal uniaxial elements of concrete and re-bar. The longitudinal uniaxial elements of concrete and re-bar connecting the top and bottom rigid beams are used to describe the flexure-compression action of the wall. The plane section assumption was used by the top and bottom rigid beams. The diagonal elements of concrete are used to describe the shear response of the wall associated with the inclined cracking and strut action of the web concrete.

Each longitudinal element consists of concrete and re-bar, as shown in Fig. 2a. Perfect bond is assumed between the concrete and re-bar. Therefore, the concrete and re-bar have identical axial elongation or shortening due to the flexure-compression response of walls. The section areas A_{lc} and A_{lr} of the concrete and re-bar in a longitudinal element are defined with the tributary areas of the concrete and re-bars assigned to the longitudinal element at the wall cross-section (see Fig. 2b).

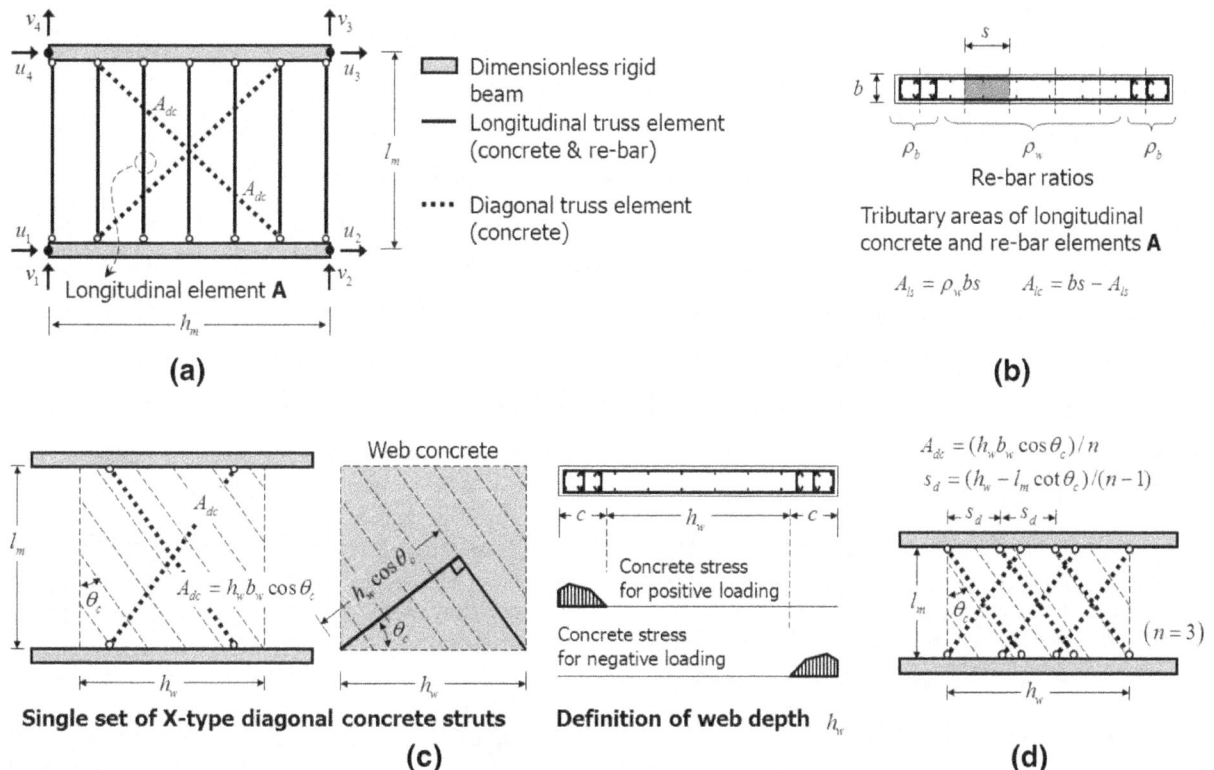

Fig. 2 Proposed macro-element for reinforced concrete wall. **a** Longitudinal-and-Diagonal-Line-Element-Model (LDLEM) element, **b** tributary areas of concrete and re-bar of each longitudinal uniaxial element, **c** diagonal concrete strut of wall web concrete, **d** multiple sets of X-type diagonal concrete struts.

$$A_{lc} = (1 - \rho)bs \tag{1}$$

$$A_{lr} = \rho bs \tag{2}$$

where b and s are the width and depth of the wall cross-section segment assigned to a longitudinal element and ρ is the longitudinal reinforcement ratio of the cross-section segment bs. The number of the longitudinal uniaxial elements can be increased to accurately describe the axial stress and strain distribution in the wall cross-section.

The diagonal element is defined with the concrete strut with an inclination angle of θ_c, which is determined as the angle of the inclined cracking in the web concrete, as shown in Fig. 2c. Basically, the diagonal element is symmetrically arranged at the center of the web concrete. The section area A_{dc} of the diagonal element is defined as the area of the web concrete transverse to the inclined cracking (θ_c) as follows.

$$A_{dc} = b_w h_w \cos \theta_c \tag{3}$$

where b_w and h_w are the width and depth of the web concrete. The depth h_w of the web concrete is defined as the net tension zone depth of the wall cross section, which is calculated by subtracting the compression zone depths subjected to the positive and negative moments from the overall wall depth (see Fig. 2c). If a very high axial load is applied to a wall, the depth h_w of the web concrete can be decreased to zero by definition, as shown in Fig. 2c. In such case, the proposed macro-model should not be used. If diagonal re-bars with area A_{dr} and inclination angle θ_r are used in the wall web, the diagonal uniaxial element of re-bar (A_{dr}) can be added to the macro-element.

The height l_m of a macro-element may significantly affect the overall response of walls, because the strains of longitudinal and diagonal uniaxial elements due to the flexure-compression and shear response of walls are assumed to be uniform over the height l_m. Thus, the height l_m of the macro-element should be sufficiently small to accurately predict the flexure-compression response which is affected by the moment gradient along the wall height as well as the shear response influenced by the inclined cracking (θ_c) and strut action of the web concrete. Basically, l_m should be smaller than $h_w \cot \theta_c$. If a refined wall model is required to address detailed response of the flexure-compression and shear actions, a smaller l_m can be used.

For detailed shear response of short and medium-rise walls, multiple X-type diagonal uniaxial elements of concrete can be used, as shown in Fig. 2d. When n sets of X-type diagonal elements are used, the section area of each diagonal concrete element is determined as $b_w h_w \cos \theta_c / n$ and the spacing between the parallel diagonal concrete elements is $s_d = (h_w - l_m \tan \theta_c)/(n - 1)$. In this case, the height l_m of a macro-element should be smaller than $h_w \cot \theta_c$. $l_m = h_w \cot \theta_c / n$ is recommended.

The depth h_w of the web concrete can be determined by sectional analysis about the axial load applied to the wall, as shown in Fig. 2c. However, it is difficult to determine the depth h_w of the web concrete in actual building structures

without nonlinear analysis because the arrangement and cross-section shape of walls are complicate. In that case, h_w can be approximately determined as follows. First, a trial h_w which is less than the overall depth of the wall cross section is assumed to establish an initial macro-model. Nonlinear analysis is then performed for the wall model. If the compression zone of wall cross section resulting from the nonlinear analysis is located outside the assumed h_w, the assumed h_w is acceptable. Otherwise, a smaller h_w should be used, on the basis of the numerical analysis result.

The inelastic behavior of reinforced concrete walls are significantly affected by the inclined cracking angle θ_c of the web concrete. However, it is difficult to accurately estimate θ_c without sophisticated nonlinear numerical analysis because the web concrete is subjected to biaxial stresses. According to Oesterle et al. (1984), Vecchino and Collins (1986), and Bentz et al. (2006), the inclined cracking angle θ_c varies from 35 to 55° depending on vertical and horizontal reinforcement ratios of the web concrete. In the present study, for convenience, $\theta_c = 45°$ is used for diagonal uniaxial elements of concrete. For more accurate analysis, the angle θ_c can be determined by using the modified compression field theory (Vecchino and Collins 1986; Bentz et al. 2006) or the softened truss model (Hsu and Mo 1985).

If shear reinforcement (i.e. horizontal reinforcement) is not sufficient, premature yielding of shear reinforcement occurs before flexural yielding of walls, which can increase the shear deformation of the wall under cyclic loading. However, the proposed macro-element cannot address the premature yielding of shear reinforcement since the dimensionless rigid beam elements restrain the tensile strain of shear reinforcement at the top and bottom of the macro-element. Therefore, the proposed method should be applied to the walls that have sufficient horizontal shear reinforcement.

The proposed macro-element can be easily incorporated into existing numerical analysis programs. Though a macro-element have a number of uniaxial elements of concrete and re-bar, the number of degree-of-freedoms of a macro-element can be reduced to a rectangular element with only 6 degree-of-freedoms (u_1, u_4, v_1, v_2, v_3, and v_4, see Fig. 2a) by static condensation. Thus, the number of degree-of-freedoms can be significantly reduced, which save time and efforts in modeling and numerical computations.

In the present study, it was assumed that the shear force was carried by the non-flexural area (grey area in Fig. 4), which was determined by the longitudinal reinforcement ratio. Also the diagonal elements, which affect the lateral stiffness of the wall, were determined by the non-flexural area. This indicates that the lateral stiffness of the wall can be easily simulated by the geometry and the longitudinal reinforcement ratio of the wall.

3. Material Models of Concrete and Re-bar

The cyclic stress–strain relationship of concrete developed by Chang and Mander (1994) was used for uniaxial concrete elements of the proposed macro-element. In Fig. 3a, three

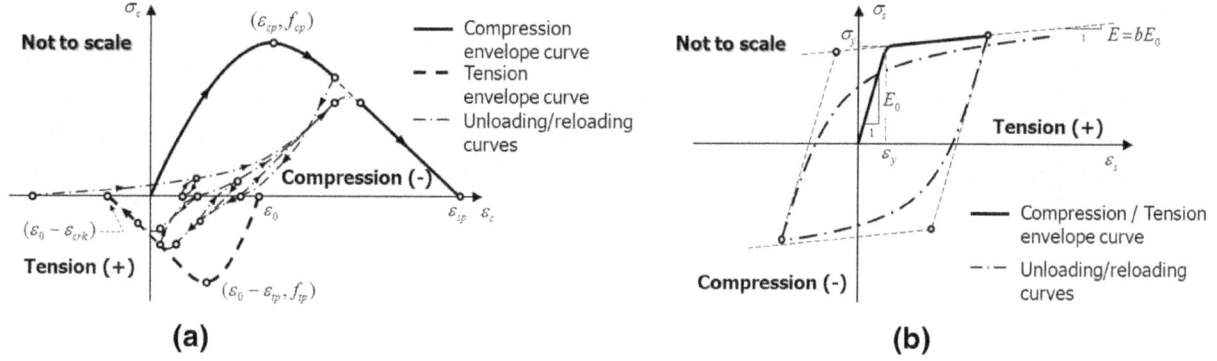

Fig. 3 Cyclic stress–strain relationships of concrete and re-bar uniaxial elements. **a** Cyclic model for concrete (Chang and Mander's model), **b** cyclic model for re-bar (Menegotto and Pinto's model).

types of curves are used to describe the cyclic stress–strain relationship: the compressive envelope curve, the tensile envelope curve, and the unloading/reloading curve connecting smoothly the two envelope curves. Basically, the Chang and Mander's concrete model is defined by the initial modulus of elasticity (E_c), stress and strain at the compression peak strength (f_{cp} and ε_{cp}), post-crushing strain at zero compression stress (ε_{sp}), stress and strain (f_{tp} and ε_{tp}) at the tension peak strength, and post-tensile cracking strain at zero tensile stress (ε_{crk}) (see Fig. 3a). Detailed equations defining the stress–strain relationship can be found in the relevant reference (Chang and Mander 1994).

For longitudinal concrete elements, the compressive stresses and strains are defined as $f_{cp} = f_c'$, $\varepsilon_{cp} = \varepsilon_{co}$, and $\varepsilon_{sp} = \varepsilon_{cu}$. f_c' is the compressive strength, ε_{co} is the strain corresponding to f_c', and ε_{cu} ($= 2\varepsilon_{co}$, Zhang and Hsu 1998) is the ultimate strain. When concrete is confined by lateral reinforcement, the compressive strength and deformation capacity are increased. In this case, the increased material properties f_{cc}, ε_{cco}, and ε_{ccu} of the confined concrete replace f_{cp}, ε_{cp}, and ε_{sp}, respectively (Mander et al. 1988).

$$f_{cc} = K f_c' \tag{4a}$$

$$\varepsilon_{cco} = -0.002(1 + 5(K - 1)) \tag{4b}$$

$$\varepsilon_{ccu} = -0.012 \tag{4c}$$

$$K = -1.254 + 2.254\sqrt{1 + 7.94 f_l'/f_c'} - 2 f_l'/f_c' \tag{4d}$$

where f_l' is the effective confining stress due to the lateral confinement reinforcement. f_l' can be determined based on the volumetric ratio and yield stress of the confinement reinforcement (Mander et al. 1988). The tensile stresses and strains in the Chang and Mander's model are defined as $f_{tp} = 0.31\sqrt{f_c'}$ (Zhang and Hsu 1998), $\varepsilon_{tp} = f_{tp}/E_c$, and $\varepsilon_{crk} = 2\varepsilon_{tp}$.

For diagonal concrete elements, the compressive strength of concrete was defined considering the effect of coexisting diagonal cracking: when inclined cracking occurs in the web concrete, the compressive strength of the web concrete decreases (Vecchino and Collins 1986; Zhang and Hsu

1998). According to Vecchino and Collins (1986), the effective compressive strength of a concrete strut is defined as a function of the transverse tensile strain ε_t orthogonal to the concrete strut.

$$f_{ce} = \frac{f_c'}{0.8 - 0.34 \varepsilon_t / \varepsilon_{co}} \leq f_c' \tag{5}$$

As shown in Fig. 4, the transverse tensile strain ε_t can be calculated by using horizontal and vertical displacements of a virtual four-node plane element enclosing the diagonal concrete element as follows.

$$\varepsilon_t = \varepsilon_y \sin^2 \theta_c + \gamma \cos \theta_c \sin \theta_c \tag{6}$$

where,

$$\varepsilon_y = \frac{(v_4 - v_1) + (v_3 - v_2)}{2 l_m} \text{ and } \gamma$$
$$= \frac{(u_4 - u_1)}{l_m} + \frac{(v_4 - v_1) + (v_3 - v_2)}{2 h_w} \tag{7}$$

In Eqs. (6) and (7), u_1, u_4, v_1, v_2, v_3, and v_4 are the horizontal and vertical displacements of the virtual four-node plane element; ε_y and γ are the mean vertical strain and shear strain of the virtual plane element (see Fig. 4). Note that the horizontal strain ε_x of the virtual plane element is zero because the dimensionless rigid beam elements at the top and bottom of the macro-element restrain the horizontal expansion due to shear.

In the proposed method, web concrete crushing is assumed to occur when the compressive stress of the diagonal concrete element is greater than the effective compressive strength f_{ce} calculated by Eq. (3).

For the longitudinal and diagonal uniaxial re-bar elements, the Menegotto and Pinto's model addressing tension stiffening effect and the Bauschinger effect was used (Menegotto and Pinto 1973). As shown in Fig. 3b, two slopes E_s and $b E_s$ are used to define the nonlinear stress–strain relationship of re-bars under cyclic loading (E_s = the modulus of elasticity of re-bar; and b = the strain hardening ratio). Further details of the Menegotto and Pinto's model can be found in the reference (Menegotto and Pinto 1973).

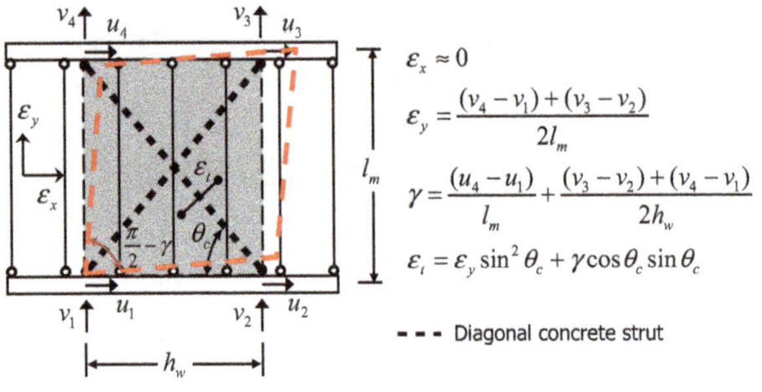

$$\varepsilon_x \approx 0$$

$$\varepsilon_y = \frac{(v_4 - v_1) + (v_3 - v_2)}{2l_m}$$

$$\gamma = \frac{(u_4 - u_1)}{l_m} + \frac{(v_3 - v_2) + (v_4 - v_1)}{2h_w}$$

$$\varepsilon_i = \varepsilon_y \sin^2 \theta_c + \gamma \cos \theta_c \sin \theta_c$$

- - - Diagonal concrete strut

Fig. 4 Transverse tensile strain orthogonal to diagonal concrete strut.

4. Verification of Proposed Model

For verification, the proposed macro-model was applied to existing isolated and coupled wall specimens subjected to cyclic loading. Table 1 shows the material and geometric properties of the wall specimens. For nonlinear analysis, a computer program for structural analysis, OpenSEES, was used OpenSEES (PEER 2001). The OpenSEES material models, Concrete07 and Steel02, are the same as the concrete and re-bar models of Sect. 3. Therefore, Concrete07 and Steel02 were used for the longitudinal and diagonal uniaxial elements of the macro-element (OpenSEES PEER 2001).

4.1 Slender Walls

Figure 5a shows the reinforcement details and geometric properties of slender wall specimens RW2 and TW2 (Thomsen and Wallace 2004). The shear span length and the overall depth of the cross section were $l = 3810$ mm and $h = 1219$ mm, respectively (shear span-to-depth ratio

$l/h = 3.13$). RW2 with rectangular cross section and TW2 with T-shaped cross section were subjected to axial compression loads $= 0.07A_g f_c'$ and $0.075A_g f_c'$, respectively, where A_g is the gross area of cross section. Due to the large shear span-to-depth ratio and the axial compression loads, the cyclic behavior of RW2 and TW2 was dominated by flexure-compression (Thomsen and Wallace 2004). The compressive strength of concrete was $f_c' = 34.0$ MPa for RW2 and 41.7 MPa for TW2. The yield stress of re-bars was $f_y = 414$ MPa. Reinforcement details of RW2 and TW2 are presented in detail in Thomsen and Wallace (2004) and Massone and Wallace (2009).

Figure 5b shows the macro-models of RW2 and TW2 for nonlinear analysis. RW2 and TW2 were idealized with five macro-elements ($l_m = 762$ mm $\leq h_w \cot \theta_c$). Each macro-element consisted of eight longitudinal uniaxial elements of concrete and re-bar (**L1** and **L2**). A set of X-type diagonal uniaxial concrete elements (**D**, $\theta_c = 45°$) was located at the center of the wall web. The section areas of the concrete and re-bar elements are presented in Fig. 5b. The concrete confined with hoops and cross ties were represented as the shaded areas at the cross sections, as shown in Fig. 5a. Since the end zone of the cross sections of RW2 and TW2 included confined and unconfined concretes, the confined and unconfined concretes were separately considered as the section areas A_{lcc} and A_{lc} (see **L1** element of Fig. 5b). The section area of the diagonal concrete element **D** was calculated using $h_w = 1019$ mm for RW2 and 889 mm for TW2. The depths h_w of web concrete in RW2 and TW2 were estimated from sectional analysis using the actual material strengths [see Eq. (3)].

Figure 5c compares the lateral load–drift ratio relationships of RW2 and TW2 resulting from the macro-model analysis and the test. In the case of RW2 with rectangular

Table 1 Dimensions and properties of existing wall specimens.

Specimens	Cross section	f_c' (MPa)	f_y (MPa)	A_s/A_s' (mm²)	A_w (mm²)	d (mm)	$P/(A_g f_c')$ (%)
Thomsen 2004							
RW2	Rectangular	34.0	448	570/570	253	1200	7.0
TW2	T-Shaped	41.7	448	1965/713	253	1200	7.5
Salonikios 1999							
MSW2	Rectangular	26.2	585	301/301	193	1100	0.0
MSW3		24.1	585	301/301	193	1100	7.0
Sittipunt 2001							
W1	Barbell-type	36.6	390	1432/1432	392	1300	0.0
W2		35.8	390	1432/1432	549	1300	0.0
Massone 2009							
Test1	Rectangular	25.5	424	804/804	796	1456	0.0
Test6		31.4	424	258/258	398	1456	10.0
Lee 2010							
RCSW	T-shaped	23.3	552	942/942	2984	1555	0.0
RCSW-B		23.5	552	942/942	2984	1555	0.0

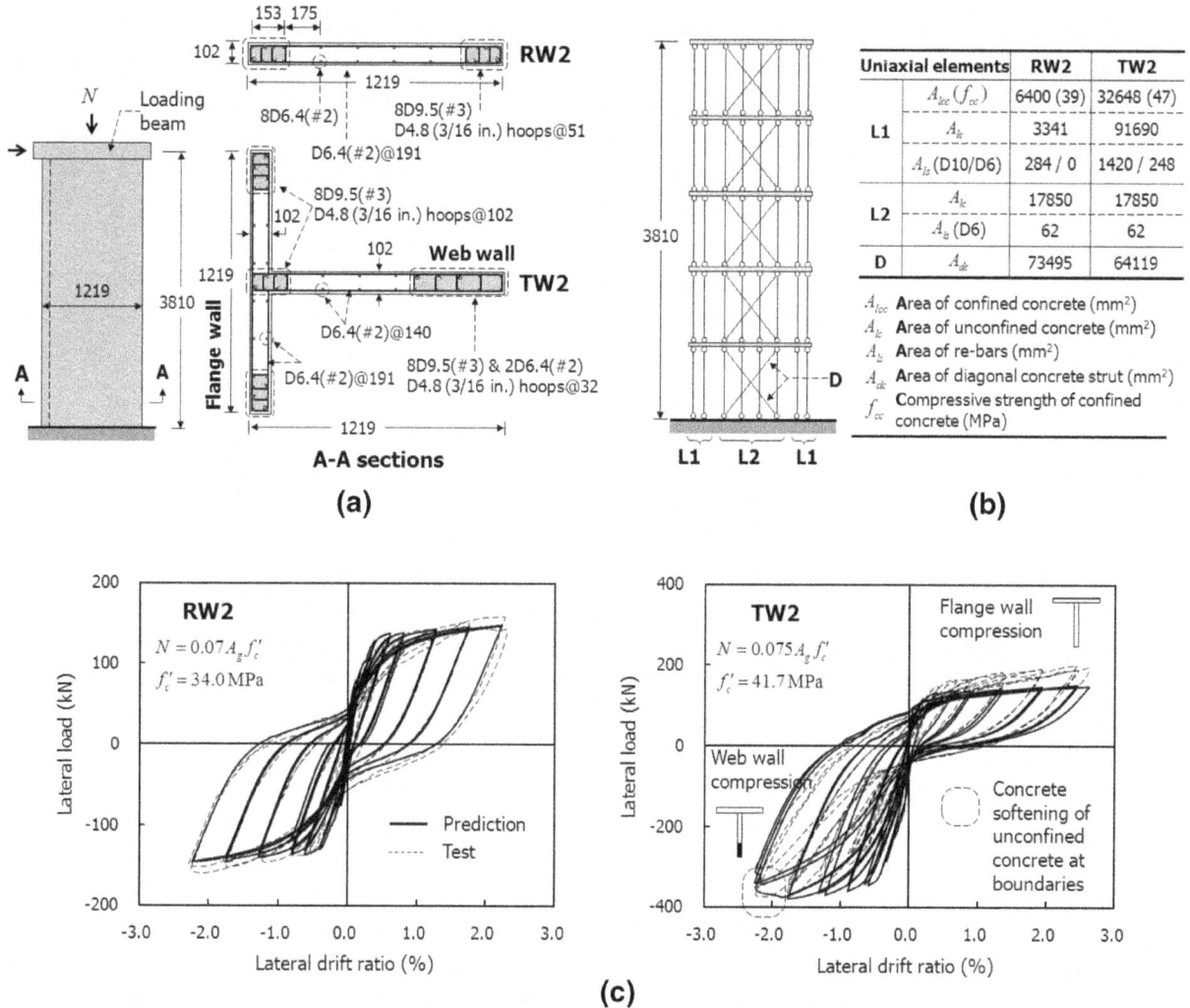

Fig. 5 Slender wall specimens RW2 and TW2 (Thomsen and Wallace 2004). **a** Dimensions and re-bar details (mm), **b** LDLEM modelling of wall specimens, **c** comparison of LDLEM analysis and test results.

cross section, the proposed method accurately predicted well the initial stiffness, load-carrying capacity, and unloading/reloading behaviors during cyclic loading. The flexural pinching caused by axial compression load was also predicted well. On the other hand, for TW2 with T-shaped cross section, the load-carrying capacity and initial stiffness were slightly overestimated under the negative loading in which the flange wall was in compression (see Fig. 5c). This is because TW2 with T-shaped cross section was idealized with the two-dimensional macro-elements; therefore, the shear-lag behavior of the flange wall in the vertical direction could not be addressed (see Fig. 5a).

In the test, TW2 failed due to the compressive softening of the concrete and the subsequent spalling of the cover concrete (see dotted lines in Fig. 5c). On the other hand, web concrete crushing did not occur in RW2 and TW2 because the axial compression load restrained the transverse tensile strain ε_t of the web concrete. The failure mode of the numerical analysis correlated with the test results.

4.2 Low-Rise Walls

The proposed method was applied to low-rise wall specimens, MSW2 and MSW3 (Salonikios et al. 1999, 2000). In Fig. 6a, the shear span length and the overall depth of cross section were $l = 1800$ mm and $h = 1200$ mm, respectively ($l/h = 1.50$). Both MSW2 and MSW3 had rectangular cross sections with confined end zones. No axial compression load was applied to MSW2. Small axial force, $N = 0.07A_gf_c'$ ($= 210$ kN) was applied to MSW3. The concrete strength was $f_c' = 26.2$ MPa for MSW2 and 24.1 MPa for MSW3. The yield stress was $f_y = 585$ MPa for D8 bars and 610 MPa for D4 bars. The reinforcement details, geometric properties, and material properties are presented in Salonikios et al. (1999, 2000).

Figure 6b shows the macro-models. Four macro-elements were used for each wall specimen ($l_m = 450$ mm $\approx (h_w \cot \theta_c)/n$). Seventeen longitudinal uniaxial elements of concrete and re-bar were used for each macro-element. For the longitudinal element **L1** in the end zones of the cross sections,

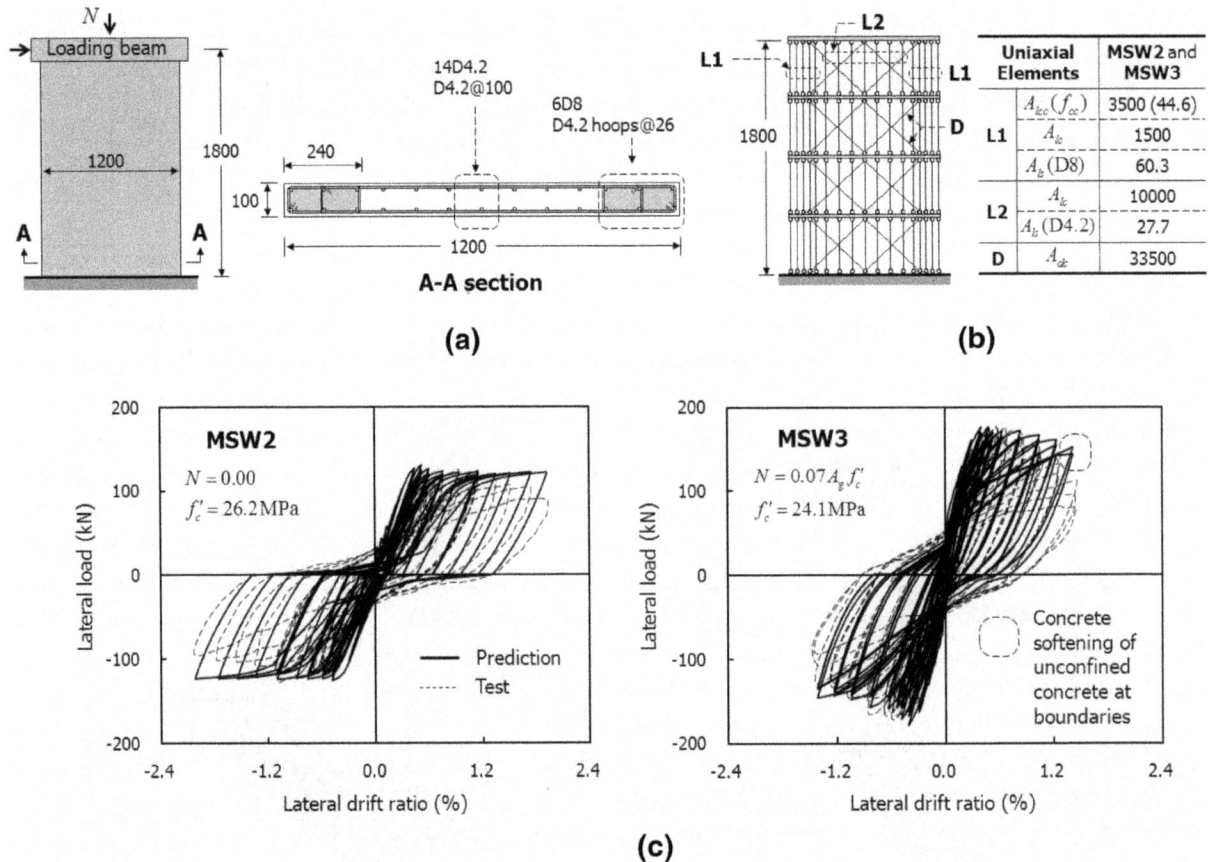

Fig. 6 Low-rise wall specimens MSW2 and MSW3 (Salonikios et al. 1999). **a** Dimensions and re-bar details (mm), **b** analysis model and element areas (mm², MPa), **c** comparison of LDLEM analysis and test results.

the confined and unconfined concretes were considered (see Fig. 6b). To accurately evaluate the shear response of the wall specimens, two sets of X-type diagonal concrete elements D were used for each macro-element ($\theta_c = 45°$; $h_w = 948$ mm for MSW2 and MSW3; and $n = 2$). The section areas of concrete and re-bar of the longitudinal and diagonal elements are presented in Fig. 6b.

Figure 6c compares the lateral load–drift ratio relationships predicted by the proposed method with the test results. In MSW2 without no axial compression load, pinching and shear-slip deformation were significant in the cyclic response. On the other hand, in MSW3 subjected to axial compression load $N = 0.07A_gf'_c$, the pinching and shear-slip deformation were decreased as the axial compression load restrained inclined diagonal cracking in the web. As shown in Fig. 6c, the proposed method predicted the shear responses of MSW2 and MSW3 with reasonable precision. Ultimately, strength degradation due to softening of the unconfined cover concrete occurred in MSW2 (see dashed line in Fig. 6c). However, the proposed method did not properly capture such strength degradation (see solid line in Fig. 6c). In MSW3 subjected to $N = 0.07A_gf'_c$, the load-carrying capacity was decreased due to the second-order effect and softening of the unconfined cover concrete. The analysis results correlated well with the test results. Web concrete crushing didn't occur in both specimens. Neither

the test nor the proposed method predicted the web concrete crushing failure of MSW2 and MSW3.

The proposed method was applied to wall specimens W1 and W2 with barbell-shaped cross sections (Sittipunt et al. 2001). As shown in Fig. 7a, the shear span length and the overall depth of cross section were $l = 2150$ mm (= 1900 + 500/2 mm) and $h = 1500$ mm, respectively ($l/h = 1.43$). Figure 7b shows the macro-models. Three macro-elements were used for each wall specimen. Thirteen longitudinal elements of concrete and re-bar and two sets of X-type diagonal elements of concrete were used for each macro-element ($\theta_c = 45°$; $h_w = 1252$ mm for W1 and 1366 mm for W2; and $n = 2$). The concrete strength was $f'_c = 36.6$ MPa for W1 and 35.8 MPa for W2. The yield stress of re-bar was $f_y = 473$ MPa for D16 bars, 425 MPa for D12 bars, and 450 MPa for D10 bars. The vertical and horizontal reinforcement ratios were $\rho_v = 0.0052$ and $\rho_h = 0.0039$ for W1 and $\rho_v = 0.0079$ and $\rho_h = 0.0052$ for W2.

Figure 7c compares the lateral load–drift ratio relationships of W1 and W2 predicted by the proposed method with the test results. The solid and dashed lines indicate the predictions and the test results, respectively. When compared to MSW2 and MSW3, W1 and W2 showed significant shear-slip deformation and pinching in the cyclic responses. As shown in Fig. 7c, the proposed method predicted the

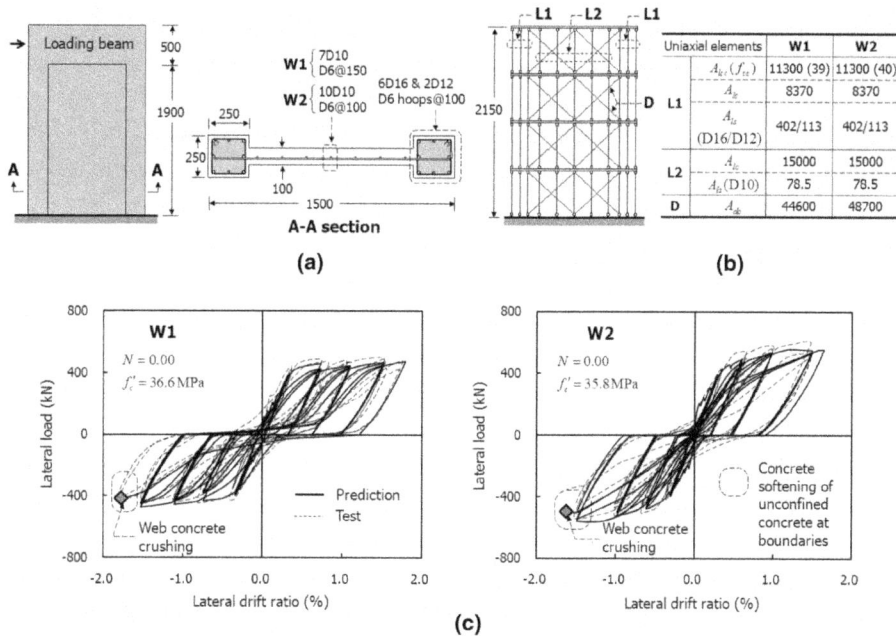

Fig. 7 Low-rise wall specimens W1 and W2 with barbell-shaped cross section (Sittipunt et al. 2001). **a** Dimensions and re-bar details (mm), **b** analysis model and element areas (mm², MPa), **c** comparison of LDLEM analysis and test results.

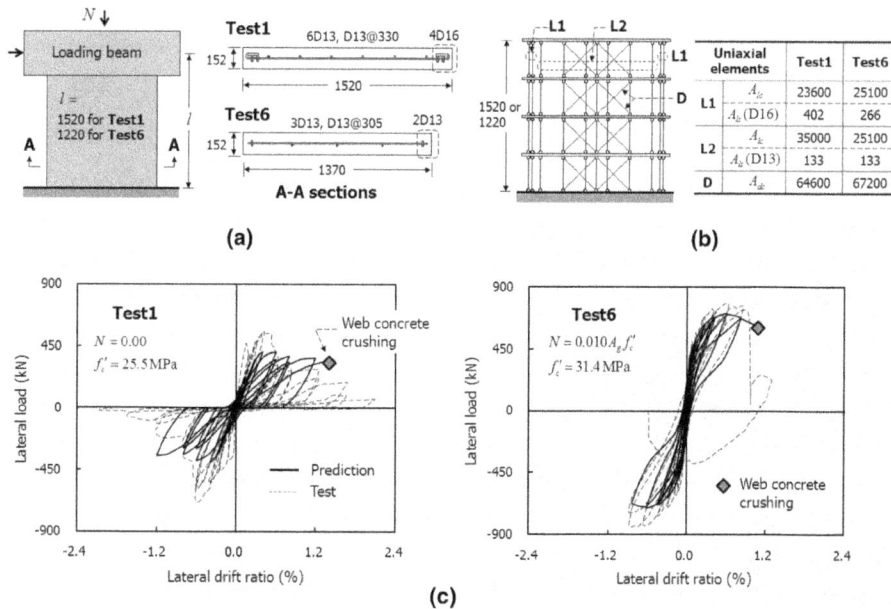

Fig. 8 Squat wall specimens Test1 and Test6 (Massone et al. 2009). **a** Dimensions and re-bar details (mm), **b** analysis model and element areas, **c** comparison of LDLEM analysis and test results.

shear responses of W1 and W2 with reasonable precision. The numerical analysis predicted web crushing of diagonal element **D** at lateral drift ratios −1.80 % (W1) and −1.70 % (W2) (see grey diamond remarks in Fig. 7). The predicted failure modes agree with the test results.

4.3 Squat Walls

The proposed method was applied to squat wall specimens Test1 and Test6 (Massone et al. 2009). In Fig. 8a, the shear span length and the overall depth of the cross section were

$l = 1520$ mm and $h = 1520$ mm for Test1 ($l/h = 1.00$) and $l = 1220$ mm and $h = 1370$ mm for Test6 ($l/h = 0.89$). Test6 was subjected to a moderate axial compression load $N = 0.10\ A_g f_c'$, while Test1 was not subjected to axial compression load ($N = 0$). The concrete strength was $f_c' = 25.5$ MPa for Test1 and 31.4 MPa for Test6. The yield stress was $f_y = 448$ MPa for D16 bars and 427 MPa for D13 bars.

Figure 8b shows the macro-models of Test1 and Test6. Four macro-elements were used for each wall specimen.

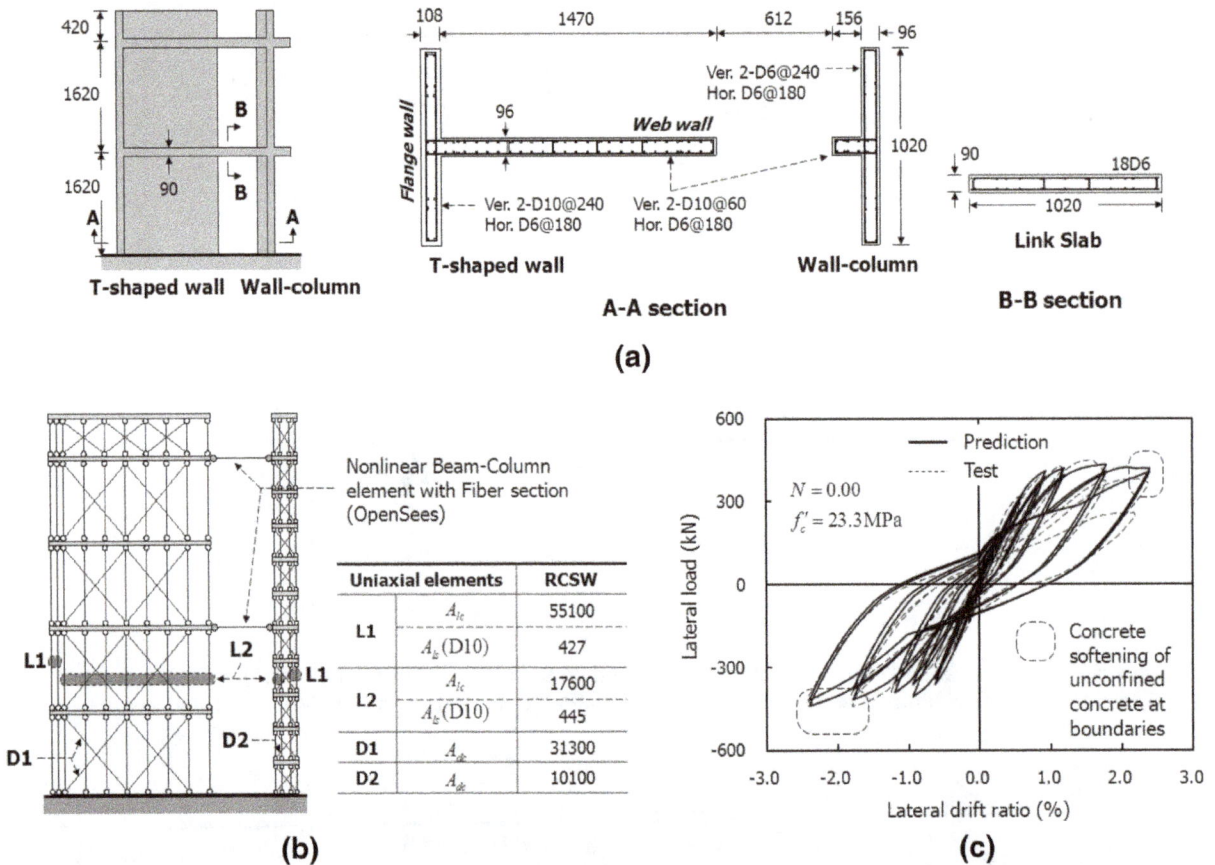

Fig. 9 Coupled wall specimens RCSW (Lee et al. 2010). **a** Dimensions and re-bar details (mm), **b** analysis model and element areas, **c** comparison of LDLEM analysis and test results.

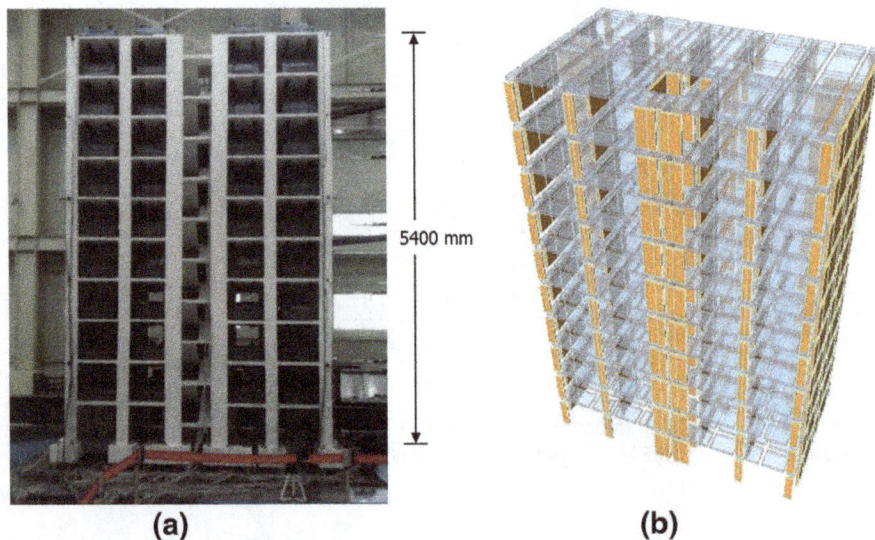

Fig. 10 3-Dimesional 1/5 scale RC wall building model and analysis model. **a** 1/5 scale test model, **b** scheme of analysis model.

Eleven longitudinal elements of concrete and re-bar were used for each macro-element. For the shear response of the squat wall specimens, two sets of X-type diagonal elements **D** were used for each macro-element ($\theta_c = 45°$; $h_w = 1202$ mm for Test1 and $h_w = 1251$ mm for Test6; and $n = 2$). The section areas of concrete and re-bar of the longitudinal and diagonal elements are presented in Fig. 8b.

Figure 8c compares the lateral load–drift ratio relationships predicted by the proposed method with the test results. Although the proposed method predicted well the pinching and web concrete crushing in the cyclic responses varying with the axial compression load, the load-carrying capacities of Test1 and Test6 were underestimated.

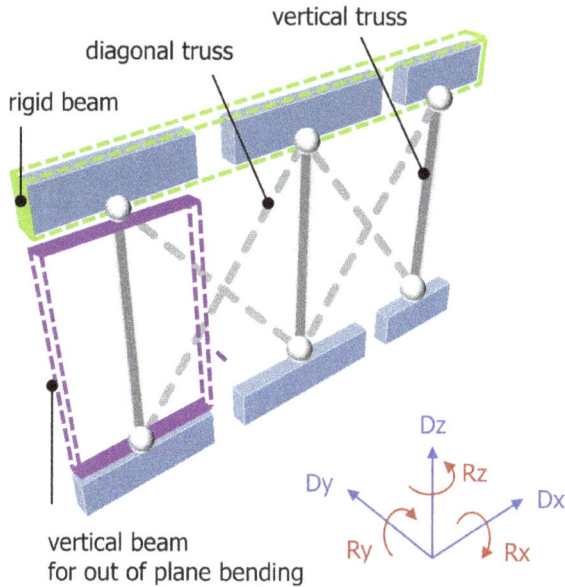

rigid beam

diagonal truss

vertical truss

Dz

Dy   Rz   Dx

Ry   Rx

vertical beam
for out of plane bending

Fig. 11 Unit model for 3-D wall building.

4.4 Coupled Walls

The proposed method was applied to a coupled wall specimen RCSW (Lee et al. 2010). Figure 9a shows the configurations, cross sections, and reinforcement details of RCSW. RCSW consisted of a T-shaped wall, a wall-column, and slabs (see Fig. 9a). The wall and wall-column were connected by the link slabs (thickness = 90 mm). The shear span of RCSW was $l = 3910$ mm. The total depth of the cross section including the T-shaped wall, wall-column, and opening was $h = 2442$ mm. The concrete strength was $f_c' = 23.3$ MPa. The yield stress of re-bars was $f_y = 553$ MPa for D10 bars and 340 MPa for D6 bars. The hoop re-bars for concrete confinement were not used in the T-shaped wall and wall-column (see Fig. 9a).

Figure 9b shows the macro-model. The T-shaped wall and wall-column were modeled with five and eleven macro-elements, respectively. In the T-shaped wall, ten longitudinal elements (**L1** and **L2**) of concrete and re-bar were used for each macro-element. Since the cyclic response of the T-shaped wall was expected to be dominated by shear, two sets of X-type diagonal elements (**D1**, $h_w = 922$ mm) of concrete were used. On the other hand, for the wall-column which was dominated by flexure-compression, four longitudinal elements (**L1** and **L2**) and a set of X-type diagonal elements (**D2**, $h_w = 149$ mm) were used. The link slabs (thickness = 90 mm) which are dominated by flexural action were modeled with equivalent beam elements (nonlinear Beam-Column Element of OpenSEES (PEER 2001) with fiber section).

Figure 9c compares the lateral load–drift ratio relationship. The solid and dashed lines indicate the prediction and the test result, respectively. The proposed method accurately predicted the overall cyclic responses. As shown in Fig. 9c, at a lateral drift ratio of 1.7 %, compression softening and subsequent spalling of the unconfined cover concrete occurred at the boundary of the web wall in the T-shaped wall. The failure mode agreed with the test result.

4.5 Shaking Table Test for Wall-Slab Structure

Lee et al. (2011) reported results of a shaking table test on a 1:5 scale 10-story R.C. wall specimen of a residential building. Using the proposed macro-model, three-dimensional nonlinear time history analyses were performed. The predicted results were compared with the test result.

Figure 10a shows the test specimen. The external dimensions of the specimen were 3560 mm × 2220 mm × 5400 mm (x-length × y-length × height). The thicknesses of the exterior wall, interior wall, and slab were 36, 32, and 40 mm, respectively. In the wall, 3 mm diameter bars were

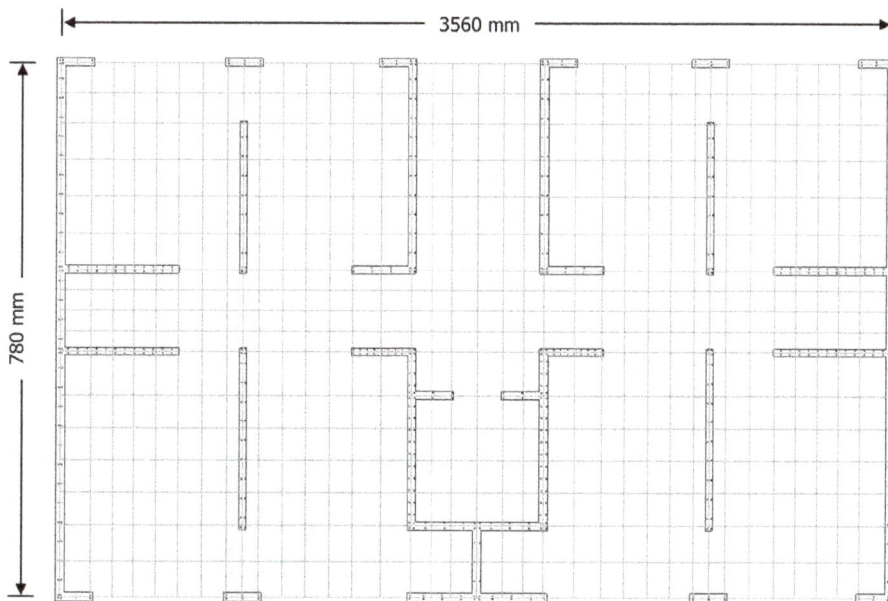

Fig. 12 Shell element grid for slabs in 3-D wall building.

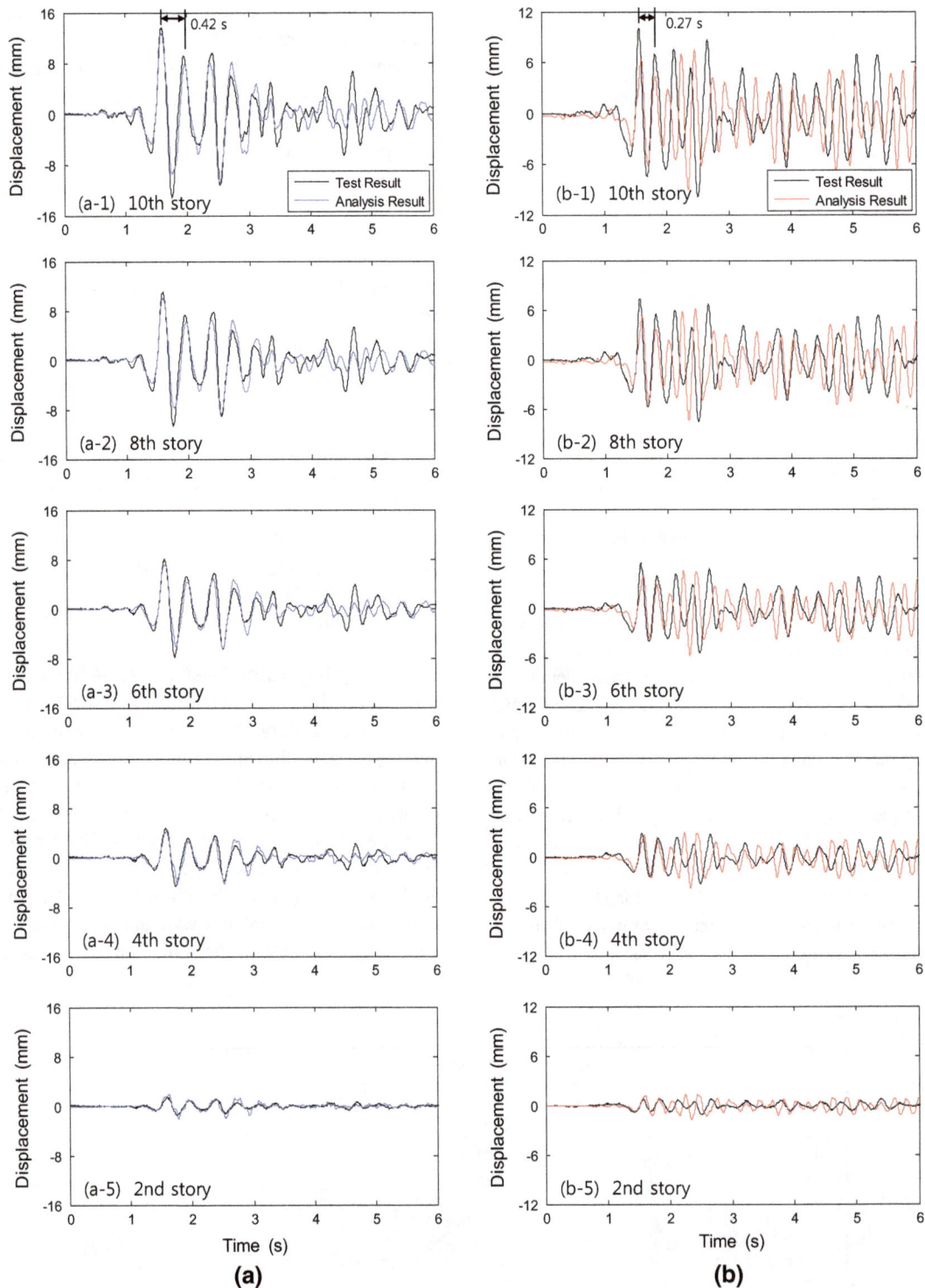

Fig. 13 Displacement of each story in 3-D wall building. **a** X-direction, **b** Y-direction.

used for the vertical bars, and 2 mm diameter bars were used for the horizontal bars. For slabs, 2 mm diameter bars were used.

Taft earthquake acceleration (N69E, S21E) was applied to the shaking table in the x and y-directions, simultaneously. The peak input acceleration levels of the earthquake events were gradually increased from 0.07 to 0.525 g. In the present study, test results of the peak ground acceleration (PGA) of

0.374 g, which was close to the design acceleration for site class S_C in Korea Building Code, were used to compare the shaking table test and the numerical analysis result.

The dynamic periods of the 1:5 scale specimen were evaluated by the white noise input and the FFT analysis. Before the earthquake test, the dynamic periods in the x and y-directions were 0.24 and 0.18 s, respectively. During the earthquake of 0.374 g, the dynamic periods in the x and

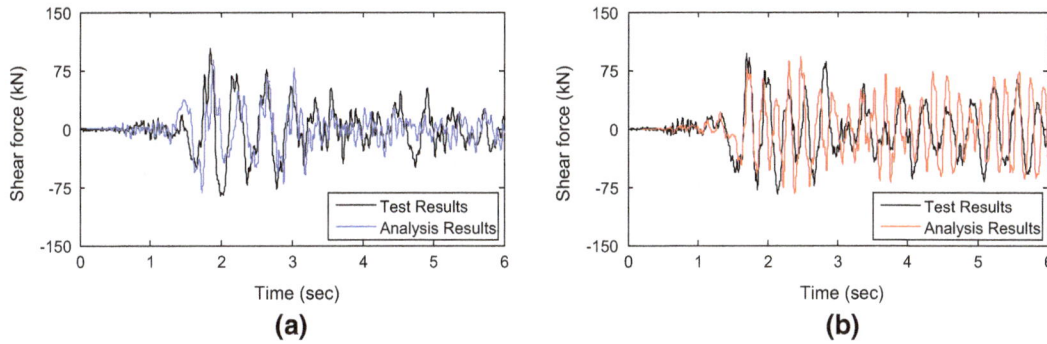

Fig. 14 Base shear of 3-D wall building. **a** X-direction, **b** Y-direction.

y-directions were increased to 0.40 and 0.30 s due to structural damages of the specimen.

Figure 10b shows scheme of the nonlinear numerical analysis model for the R.C. wall specimen. To describe the proposed macro-model, out-of-plane bending of the wall, and effect of the slab, the truss element, the elastic beam-column element, and the shell element of the OpenSEES (PEER 2001) were used, respectively. Nonlinear materials of the OpenSEES, Concrete07 and Steel02, were used for the longitudinal and diagonal uniaxial elements of the macro-element.

Figure 11 presents a unit model for the three-dimensional behavior of a wall, and the contribution of each component to the six-degrees of freedom. For in-plan-action of the wall, the proposed macro-model using vertical and diagonal trusses was used, horizontal rigid beams were used. For out-of-plane bending of the wall, vertical flexural element was used. As shown in Fig. 12, the slab was divided into 37×18 grid. Shell elements were used to describe the six-degrees of freedom of the slab.

The out-of-plane flexural stiffness of the wall and the flexural stiffness of the slab should be considered as cracked sections. Considering translation of the neutral axis, 15 % of the gross flexural stiffness (= $0.5I_{gx}$) was used for the out-of-plane flexural stiffness of the wall, and 20 % of the gross flexural stiffness was used for the slab (shell element). From the eigenvalues of the analysis model, the Rayleigh damping of 5 % was applied to the analysis model.

Figure 13 shows the net displacements of each story excluding the foundation displacement. In the x-direction (Fig. 13a), the results of the analysis model were similar to those of the shaking table test. The dynamic period of the analysis model was 0.42 s which was close to that of the test specimen (= 0.40 s).

In the y-direction (Fig. 13b), the period of the analysis model was 0.26 s which was smaller than that of the test specimen (= 0.30 s). For this reason the displacements of the numerical analysis were smaller than those of the test specimen. This result indicates that the damages of the walls in the y-direction were not well predicted by the numerical analysis.

In Fig. 14, the base shear forces of the shaking table test and the numerical analysis were compared. The base shear force in the x-direction was well predicted by the analysis model. For the y-direction, the base shear force predicted by the numerical analysis model was similar to that of the shaking table test.

5. Summary and Conclusions

In the present study, a macro-model for the nonlinear analysis of wall structures was developed. For convenience in modeling and numerical computation, the macro-model is idealized with longitudinal and diagonal uniaxial elements of concrete and re-bar. The proposed model is similar to the truss model that is popular in the design of reinforced concrete members. The proposed model was intended to describe both the flexure-compression and the shear responses, with reasonable precision. Particularly, the proposed model focused on accurate prediction of the shear response of walls associated with inclined cracking and diagonal strut action of the web concrete. The longitudinal and diagonal uniaxial elements consist of concrete and re-bar. Simplified cyclic models for the concrete and re-bar were used.

For verification, the proposed macro-model was applied to isolated wall specimens and a coupled wall specimen subjected to cyclic loading. The results showed that the proposed macro-model predicted well the flexure-compression and shear responses of the slender, short, and coupled walls, addressing the effects of various design parameters (e.g. shear span-to-depth ratio, axial compression force, reinforcement ratio, and the shape of cross sections). The characteristics of cyclic responses including strength- and stiffness-degradations, pinching behavior, and the overall shape of cyclic curves were reasonably captured.

Further, the proposed macro-model was applied to three-dimensional nonlinear time history analyses for a 1:5 scale 10-story R.C. wall-type residential building specimen, which was tested on shaking table. The predictions were compared with the shaking table test results. The results showed that the displacements, the base shear forces, and the global deformation correlated well with the shaking test results.

However, the following limitations should be considered when the proposed macro-model is used: 1) the macro-model should not be used for walls subjected to very high axial compression load; 2) The walls should have sufficient shear reinforcement to resist in-plane shear.

Acknowledgments

This work was supported by the National Research Foundation of Korea (NRF) grant (2010-0027593) funded by the Korean government's Ministry of Education, Science and Technology.

References

American Society of Civil Engineers (ASCE). (2000). *Prestandard and commentary for the seismic rehabilitation of buildings*. Reston, VA: FEMA-356.

American Society of Civil Engineers (ASCE). (2005). *Improvement of nonlinear static seismic analysis Procedures*. Reston, VA: FEMA-440.

Applied Technology Council (ATC). (1996). Seismic evaluation and retrofit of concrete buildings. Rep. No. ATC-40, Redwood City, CA.

Bentz, E. C., Vecchio, F. J., & Collins, M. P. (2006). Simplified modified compression field theory for calculating shear strength of reinforced concrete elements. *ACI Structural Journal, 103*(4), 614–624.

Chang, G. A., & Mander, J. B. (1994). Seismic Energy Based Fatigue Damage Analysis of Bridge Columns: Part I-Evaluation of Seismic Capacity. Rep. No. NCEER-94-0006, State University of New York at Buffalo, New York, NY.

Computer and Structures Inc. (2006). *Nonlinear analysis and performance assessment for 3D structures*, Berkeley, CA: Computer and Structures Inc.

D'Ambrisi, A., & Filippou, F. C. (1999). Modeling of cyclic shear behavior in RC members. *ASCE Journal of Structural Engineering, 125*(10), 1143–1150.

Feenstra, P. H., & de Borst, R. (1993). Aspects of robust computational modeling for plain and reinforced concrete. *Heron, 38*(4), 5–26.

Hsu, T. T. C., & Mo, Y. L. (1985). Softening of concrete in low-rise shear walls. *ACI Structural Journal, 82*(6), 883–889.

Kabeyasawa, T., Shiohara, T., Otani, S., & Aoyama, H. (1982). Analysis of the full-scale seven story reinforced concrete test structure: Test PSD3. In *3rd JTCC, US.-Japan Cooperative Earthquake Research Program*, BRI, Tsukuba, Japan.

Lee, S. H., Hwang, S. J., Lee, K. B., Kang, C. B., Lee, S. H., & Oh, S. H. (2011). Earthquake simulation tests on a 1:5 scale 10-story R.C. residential building model. *Earthquake Engineering Society of Korea, 15*(6), 67–80.

Lee, S. H., Oh, S. H., Hwang, W. T., Lee, K. B., & Lee, H. S. (2010). Static experiment for the seismic performance of a 2 story RC shear wall system. *Earthquake Engineering Society of Korea, 14*(6), 55–65.

Linda, P., & Bachmann, H. (1994). Dynamic modeling and design of earthquake-resistant walls. *Earthquake Engineering and Structural Dynamics, 23*, 1331–1350.

Mander, J. B., Priestley, M. J. N., & Park, R. (1988). Theoretical stress–strain model for confined concrete. *ASCE Journal of Structural Engineering, 114*(8), 1804–1826.

Mansour, M. Y., & Hsu, T. T. C. (2005). Behavior of reinforced concrete elements under cyclic shear. II: Theoretical model. *ASCE. Journal of Structural Engineering, 131*(1), 54–65.

Massone, L. M., Orakcal, K., & Wallace, J. W. (2009). Modeling of squat structural walls controlled by shear. *ACI Structural Journal, 106*(5), 646–655.

Massone, L. M., & Wallace, J. W. (2009). *RC wall shear-flexure interaction: Analytical and experimental responses*. Los Angeles, CA: College of Engineering, University of California (UCLA-SGEL-2009/2).

Menegotto, M., & Pinto, P. E. (1973). Method of analysis for cyclically loaded reinforced concrete plane frames including changes in geometry and non-elastic behavior of elements under combined normal force and bending. In *IABSE symposium on the resistance and ultimate deformability of structures acted on by well-defined repeated loads*, Lisbon.

Monti, G., & Spacone, E. (2000). Reinforced concrete fiber beam column element with bond slip. *ASCE Journal of Structural Engineering, 126*(6), 654–661.

Oesterle, R. G., Aristijabal-Ochoa, J. D., Shiu, K. N., & Corley, W. G. (1984). Web crushing of reinforced concrete structural walls. *ACI Structural Journal, 81*(3), 231–241.

Okamura, H., & Maekawa, K. (1991). *Nonlinear analysis and constitutive models of reinforced concrete*. Tokyo, Japan: Gihodo-Shuppan.

Orakcal, K. (2004). *Nonlinear modeling and analysis of slender reinforced concrete walls*. Dissertation, University of California, Los Angeles, CA.

Orakcal, K., Massone, L. M., & Wallace, J. W. (2006). Analytical modeling of reinforced concrete walls for predicting flexural and coupled–shear-flexural responses. PEER Report 2006/07. Pacific Earthquake Engineering Research Center, University of California, Berkeley, CA.

Pacific Earthquake Engineering Research Center (PEER). (2001). *Open system for earthquake engineering simulation*. Berkeley, CA: University of California at Berkeley.

Palermo, D., & Vecchio, F. J. (2007). Simulation of cyclically loaded concrete structures based on the finite-element method. *ASCE Journal of Structural Engineering, 133*(5), 728–738.

Park, H., & Eom, T. (2007). Truss model for nonlinear analysis of RC members subject to cyclic loading. *ASCE Journal of Structural Engineering, 133*(10), 1351–1363.

Park, H., & Klingner, R. E. (1997). Nonlinear analysis of RC members using plasticity with multiple failure criteria. *ASEC Journal of Structural Engineering, 123*(5), 643–651.

Petrangeli, M., Pinto, P. E., & Ciampi, V. (1999). Fiber element for cyclic bending and shear of RC structures—I: Theory.

ASCE Journal of Engineering Mechanics, 125(9), 994–1001.

Prakash, V., Powell, G. H., & Campbell, S. (1993). DRAIN-2DX Base Program Description and User Guide-Version 1.10. Rep. No. UCB/SEMM-93/17, Proceedings, Structural Engineering Mechanics and Materials, University of California, Berkeley, CA.

Salonikios, T. N., Kappos, A. J., Tegos, I. A., & Penelis, G. G. (1999). Cyclic load behavior of low-slenderness reinforced concrete walls: Design basis and test results. *ACI Structural Journal, 96*(4), 649–660.

Salonikios, T. N., Kappos, A. J., Tegos, I. A., & Penelis, G. G. (2000). Cyclic load behavior of low-slenderness reinforced concrete walls: Failure modes, strength and deformation analysis, and design implications. *ACI Structural Journal, 97*(1), 132–142.

Sittipunt, C., Wood, L. S., Lukkunaprasit, P., & Pattararat-tanakul, P. (2001). Cyclic behavior of reinforced concrete structural walls with diagonal web reinforcement. *ACI Structural Journal, 98*(4), 554–562.

Stevens, N. J., Uzumeri, S. M., Collins, M. P., & Will, G. T. (1991). Reinforced concrete subjected to reversed cyclic shear-experiments and constitutive model. *ACI Structural Journal, 88*(2), 135–146.

Thomsen, J. H., & Wallace, J. W. (2004). Displacement-based design of slender reinforced concrete structural walls—Experimental verification. *ASCE Journal of Structural Engineering, 130*(4), 618–630.

Vecchino, F., & Collins, M. P. (1986). The modified compression field theory for reinforced concrete elements subject to shear. *ACI Structural Journal, 83*(2), 219–231.

Vulcano, A., & Bertero, V. (1987). Analytical model for predicating the lateral response of RC shear walls: Evaluation of their reliability. Report No. UCB/EERC-87/19, USA.

Wallace, J. W. (2012). Behavior, design, and modeling of structural walls and coupling beams—Lessons from recent laboratory tests and earthquakes. *International Journal of Concrete Structures and Materials, 6*(1), 3–18.

Wong, P. S., & Vecchio, F. J. (2002). *VecTor2 & Formworks User's Manuals*. Toronto, Canada: Department of Civil Engineering, University of Toronto, Canada.

Zhang, L. X., & Hsu, T. T. C. (1998). Behavior and analysis of 100 MPa concrete membrane elements. *ASCE Journal of Structural Engineering, 124*(1), 24–34.

3

Flexural Strength of RC Beam Strengthened by Partially De-bonded Near Surface-Mounted FRP Strip

Soo-yeon Seo[1),*], **Ki-bong Choi**[2)], **Young-sun Kwon**[3)], **and Kang-seok Lee**[4)]

Abstract: This paper presents an experimental work to study the flexural strength of reinforced concrete (RC) beams strengthened by partially de-bonded near surface-mounted (NSM) fiber reinforced polymer (FRP) strip with various de-bonded length. Especially, considering high anchorage capacity at end of a FRP strip, the effect of de-bonded region at a central part was investigated. In order to check the improvement of strength or deformation capacity when the bonded surface area only increased without changing the FRP area, single and triple lines of FRP were planned. In addition, the flexural strength of the RC member strengthened by a partially de-bonded NSM FRP strip was evaluated by using the existing researchers' strength equation to predict the flexural strength after retrofit. From the study, it was found that where de-bonded region exists in the central part of a flexural member, the deformation capacity of the member is expected to be improved, because FRP strain is not to be concentrated on the center but to be extended uniformly in the de-bonded region. Where NSM FRP strips are distributed in triple lines, a relatively high strength can be exerted due to the increase of bond strength in the anchorage.

Keywords: flexural strength, partially de-bonded NSM FRP strip, de-bonded length, anchorage capacity, deformation capacity, single and triple lines of FRP.

1. Introduction

The retrofit of concrete structures using fiber reinforced polymer (FRP) is rooted in the retrofit using reinforcements in Europe in the 1950s. Afterward, externally bonded retrofit (EBR) which is a method of bonding a sheet or plate shaped FRP to the surface of concrete, has been widely used with development of FRP materials. Near surface-mounted retrofit (NSMR) method is currently being applied to sites.

Accordingly, ACI 440 (2002) and fib code (2001) have reflected the contents of NSMR for their utilization in design. However, the existing studies on NSMR of FRP members have produced only very limited outcome. That is why it is necessary to conduct a study on the systemization of construction method and the completion of design method.

NSMR that strip-typed FRP is vertically mounted can increase retrofit efficiency, because it can relatively facilitate grooving and increase bond area. However, as a groove needs more than a certain depth, there are, comparatively, limits to applying this method to members with thin cover concrete. In contrast, the bond area in embedding rectangular or round shaped FRP is comparatively smaller compared with plate typed FRP. However, rectangular or round shaped FRP needs to make a relatively wide groove instead of making a deep groove. Thus, that is easy to apply to members with thin cover concrete.

According to the result of the existing studies, the method of making a narrow and deep groove in a concrete member and embedding FRP strip in the groove vertically has higher retrofit efficiency compared with a method using NSMR using a round or rectangular bar, because the bond area of FRP plate inserted in a groove is more. Besides, the former's grooving work is easier than the latter's.

This study intends to conduct an experimental study on the retrofit effect of NSMR using a FRP strip and the anchorage effect of the FRP strip in reinforced concrete (RC) member. Especially, considering high anchorage capacity at end of a FRP strip, this study intends to set up a de-bonded region at a central part and to study a behavior according to a change in a de-bonded length. In order to check the improvement of strength or deformation capacity when the bonded surface area only increased without changing the FRP area, two cases of FRP line were planned: an installation in a line for beam section; a distributed installation in triple line. The sectional areas of FRP in these two cases are almost the same.

Besides, this study intends to evaluate the bond strength of a partially de-bonded flexural member, based on the existing

---

[1)]Department of Architectural Engineering, Korea National University of Transportation, Chungju, Korea.
*Corresponding Author; E-mail: syseo@ut.ac.kr
[2)]Department of Architectural Engineering, Gachon University, Seongnam, Korea.

[3)]Jaeshin CTNG, Co., Ltd., Seoul, Korea.

[4)]Department of Architectural Engineering, Cheonnam National University, Gwangju, Korea.

researchers' strength equation for calculation of bond strength.

2. Previous Researches

The superiority of the NSMR that forms a groove on concrete within the cover concrete and then filling an FRP for reinforcement is known from the flexural experiment for a highway bridge carried out by Nanni (2000). In the study, for bonding a carbon FRP (CFRP) plate on the external surface, the CFRP plate peeled off at the final step, but for embedding a bar shaped CFRP in cover concrete, the CFRP got a tensile fracture, and the contributions to overall resistance were 17 and 29 %, respectively, so it turned out that the NSMR was more effective.

After that, several researchers performed bond test about NSM FRP (Jose and Barros 2007; Ceroni et al. 2012; Al-Mahmoud et al. 2012; Lee et al. 2013; Ali et al. 2008; Seracino et al. 2007a, b; Seo et al. 2013) and found that it is effective in acquiring higher bond capacity. Among them, a few studies (Ali et al. 2008; Seracino et al. 2007a, b; Seo et al. 2013) about both bond theory and bond strength equation of NSM FRP strip have been performed based on the previous bond test results. As a recent research, Seo et al. (2013) performed a bond test on NSM FRP plate with various bonded lengths as well as number of shear keys. From the test result, they found that the member strengthened by NSMR has almost 1.5 times higher bond strength than that by EBR.

From a flexural test of RC beam strengthened by NSM FRP retrofit (El-Hacha and Rizkalla 2004; Hassan and Rizkalla 2003; Sharaky et al. 2014), the NSMR has a much higher retrofit capacity than EBR, and presented that especially the case in which the FRP plate was vertically inserted in the member for reinforcement exhibits the most excellent performance. As a recent research, Rezazadeh and Barros (2015) performed an experimental work for the beams retrofitted by pre-stressed CFRP whose shape is very thin. In the manner of construction labor, the groove for vertical embedment of a thin FRP strip is little bit narrower and deeper than that for square or round one. The work for making the narrow groove is relatively easy and the FRP can be easily fixed into the groove during the curing of epoxy. And the rigidity of the FRP for vertical embedment is relatively high for vertical direction so that it is possible to reduce the vertical deflection of the FRP by self-weight during the construction. In case of the strengthening of a long member, an excessive vertical deflection of FRP material may occur during the construction and it can give a negative effect in inserting the FRP into the groove and fixing it. Considering this merit, several researchers (Ali et al. 2008; Seracino et al. 2007a, b; Seo et al. 2013; Rezazadeh and Barros 2015; Yost et al. 2007; Seo 2012) have performed various experimental and analytical works about the NSM FRP retrofit to find both the bond capacity and a proper design process.

In spite of the excellence of NSM FRP retrofit, there is a possibility of a reduction of energy and displacement ductility or a bond failure of NSM FRP (Yost et al. 2007; Lorenzis et al. 2000).

Chahrour and Soudki (2005) investigated the flexural behavior of RC beams strengthened with end-anchored partially bonded CFRP strips through an experimental and analytical study. The experimental results revealed that end-anchored partially bonded CFRP strips significantly enhanced the ultimate capacity of the control beam and performed better than the fully bonded strip with no end-anchorage.

To find a method for improving ductility, Choi et al. (2008) conducted an investigation on the flexural behavior of partially bonded FRP strengthened concrete beams. Also, to predict the behavior of the beams, an analytical model was developed based on the curvature approach. The result of the analysis showed that ductility of the partially bonded system was improved while sustaining high load carrying capacity in comparison to the fully bonded system.

Choi et al. (2011) further investigated a partially bonded strengthening approach for RC beams utilizing NSM CFRP bars with the specific objective of improving deformability. Test results of beams with NSM CFRP bars of various unbounded lengths showed a decrease of the stiffness at the post-yield stage of the load–deflection response in the partially bonded beams. This is caused by the delayed increase of the FRP strain within the de-bonded length. As a result, the beam deformability was increased as the unbounded length increased at the same applied load. Internal slip of the FRP bar and gradual concrete failure were observed near the ultimate state, which caused a complicate nonlinear behavior of the beams. An analytical model was proposed to address the complete beam behavior including the effect of slip of FRP reinforcement and gradual concrete crushing.

From the research results of Choi et al. (2008, 2011), the deformability of RC beam retrofitted by NSM FRP strip can be improved by putting de-bonded region in the middle area of member. However, the experimental result is not enough to explain the overall behavior of RC beam strengthened by NSM FRP strip.

In this manner, this study carried out a flexural experiment of RC beams strengthened by EBR and NSMR with CFRP strip based on the existing results of bond tests by Seo et al. (2013). Especially, for the case of that strip is partially de-bonded in a region of central part to improve the deformability of FRP, the effect of de-bonded length is studied. Also, in order to improve the bond strength in the bonded region with keeping same sectional area to avoid the bond failure that may occur at where de-bonded length increases, the retrofit by using triple lines strip (the thickness of each strip is 1/3 of the single strip) is planned and the retrofit effect is compared to that of single strip.

3. Experiment

The parameters in this experiment are a retrofit method using FRP strip and de-bonded length. The retrofit methods are NSMR and EBR, as shown in Fig. 1. NSMR is targeted at the method of making a deep and narrow groove in the surface of a member and inserting a FRP plate in it with epoxy. In order to check the improvement of strength or

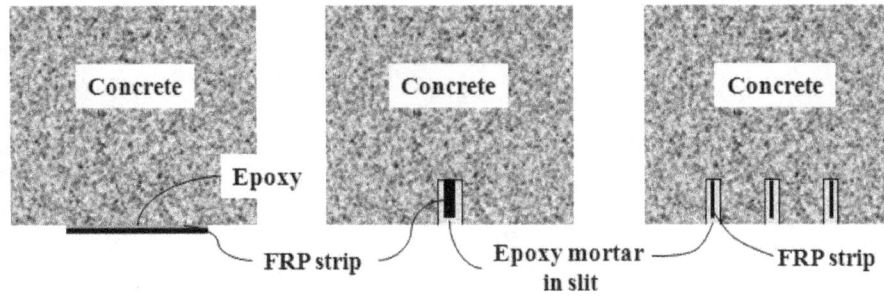

Fig. 1 Retrofit method by using FRP strip.

Table 1 Specimen list.

Specimen names	Retrofit methods	Bonded length (mm)	De-bonded length (mm)	Strip dimension (width × height) (mm × mm)
BC2000	None	–	–	
BP1600	EBR	1600	–	50 × 1.2
CP1600-1	NSMR	1600	–	3.6 × 16
CP1600-3				1.2 × 16
CP500-1		500 at each ends	600	3.6 × 16
CP500-3				1.2 × 16
CP400-1		400 at each ends	800	3.6 × 16
CP400-3				1.2 × 16
CP300-1		300 at each ends	1000	3.6 × 16
CP300-3				1.2 × 16

deformation capacity when the bonded surface area only increased without changing the FRP area, two cases of FRP line were planned: an installation in a line for beam section; a distributed installation in triple line. The sectional areas of FRP in these two cases are almost the same.

3.1 Specimen Design and Manufacture

Table 1 shows the list of specimens. Figure 2 shows specimen shape and detail. After manufacturing a RC beam with the dimensions of 200 mm × 400 mm × 1800 mm, flexural retrofit was performed at the bottom of the beam using a FRP strip, according to the purpose of experiment.

FRP used in experiment is composed of carbon fiber, with the thickness 1.2 mm and width 50 mm. One unit of FRP was bonded for EB retrofitting. In NSMR using triple line, a FRP strip with the width 50 mm was divided by the width 16 mm into three FRP strips with the dimensions of 1.2 mm × 16 mm to be used. In NSMR using one line, three 1.2 mm × 16 mm strips were overlapped and bonded with epoxy into a 3.6 mm-thick FRP bar to be used. Accordingly, the area of FRP in a specimen is almost similar to each other in all the specimens.

The 28-day compressive strength of concrete used for the specimens is 21 MPa, and FRP strip is 1.2 mm-thick

Carbodur-Plate S512/80 made by Sika. The FRP strip has 2800 MPa tensile strength and 160,000 MPa elastic module. According to the data presented by Sika, adhesive resin is Sikadur®-30 two-component resin, which has 70 MPa compressive strength, 28 MPa tensile strength, 18 MPa shear strength, and 128,000 MPa elastic module. The yield and ultimate strength of D10 steel bar used as main bar are 486.7 and 833.2 MPa, respectively.

3.2 Experimental Method

As shown in Fig. 3, specimens were installed to keep both ends in simple support condition and gradually increasing load was applied to the central part. Their failures were observed until the specimen reaches final failure. In order to measure the deformation of specimens, deflection was measured at the center of the bottom and at the 1/4 point of the specimen length. In order to measure the deformations of the steel bars at the bottom and FRP, strain gauges were bonded as shown in Fig. 4. In the specimens with de-bonded region, a gauge was bonded at the center since strain is uniform in the de-bonded region. In the bonded region, also, a gauge was bonded at the point of 50 mm to the middle from a support and at the point of any multiple of 100 mm away from that point.

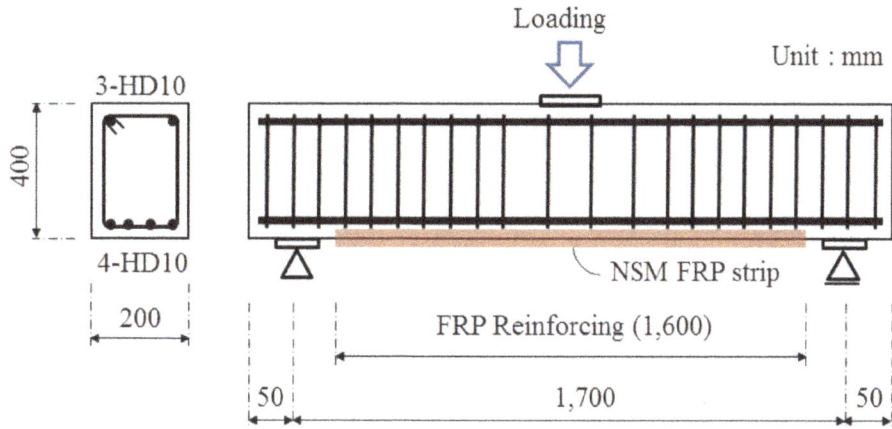

Fig. 2 Dimension and reinforcements detail.

Fig. 3 Test set up.

3.3 Test Result and Analysis
3.3.1 Failure Shape
The failure conditions of each specimen are shown in Figs. 5 and 6. As BC2000 specimen, which was not retrofitted, developed typical flexural cracks with occurrence of initial flexural crack, the central part was plasticized and the specimen reached final failure.

In the case of BP1600 specimen retrofitted using the external bonding for the retrofit region of 1600 mm, with incremental load, flexural cracks began at a central part and relatively concentrated on the central part, compared with non-retrofitted BC2000 specimen. Finally, a failure that FRP peeled off happened.

In contrast, CP1600-1 and CP1600-3 specimens that had the whole region retrofitted by NSMR showed slightly distributed cracks without a failure that cracks appeared concentrated on a central part like BP1600 specimen. Around ultimate load, mounted FRPs fell off the central part of a beam, together with a part of cover concrete at the bottom. Especially, according to figure of the bottom of tension side in Figs. 6c and 6e, the falling-off area of CP1600-3 specimen retrofitted using the embedding in triple line was larger, compared with CP1600-1 specimen retrofitted using the embedding in a line, where a central part was de-bonded, cracks developed at regular intervals, showed an aspect that they were distributed in the whole de-bonded region, and concrete did not fall off. FRP ruptured in all the de-bonded regions. Afterward, it showed a failure almost similar to that of non-retrofit specimen.

Further, it can be found that both CP1600-1 specimen using NSMR and BP1600 specimen retrofitted using EBR for the whole bottom had cracks concentrated on a central part and that, accordingly, concrete fell off. In contrast, a non-retrofit specimen and a specimen using partial de-bonding for a central part had flexural cracks distributed wholly, and accordingly, there was no seriously concentrated falling off area. Thus, a failure can be distributed without being concentrated on a central part by de-bonding for the central part.

3.3.2 Load–Deflection Curve
Figure 7 shows each comparative load–deflection curve based on the bonded length for a specimen using NSMR and

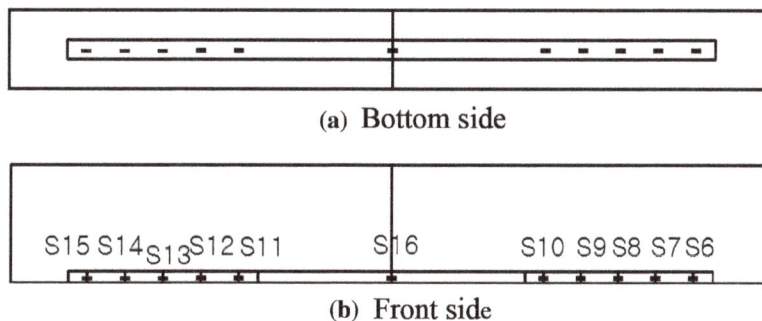

(a) Bottom side

S15 S14 S13 S12 S11   S16   S10 S9 S8 S7 S6

(b) Front side

Fig. 4 Locations of strain gauges on FRP.

Table 2 Test and calculated results.

Specimens	P_{cr} (kN)	Yield		Ultimate		μ (δ_u/δ_y)	ΔP (kN)	P_n (kN)	$\frac{P_u}{P_n}$	Failure patterns
		P_y (kN)	δ_y (mm)	P_u (kN)	δ_u (mm)					
BC2000	42.65	143.18	6.03	190.24	79.61	13.20	–	109.65	1.73	Ductile failure
BP1600	56.93	167.20	5.97	190.25	10.12	1.70	45.6	186.97	1.02	FRP peel off
CP1600-1	66.20	200.55	6.72	225.55	15.98	2.38	72.08	206.32	1.09	FRP fracture
CP1600-3	46.58	173.58	6.63	233.40	26.13	3.94	72.08	206.32	1.13	
CP500-1	42.17	161.32	6.35	208.39	19.06	3.00	54.92	193.10	1.08	
CP500-3	53.45	162.79	6.04	233.40	26.13	4.33	72.08	206.32	1.13	
CP400-1[a]	52.50	163.50	–	225.00	–	–	–	185.81	1.21	
CP400-3	41.19	151.02	6.60	224.08	29.67	4.50	59.82	206.32	1.09	
CP300-1	40.21	154.95	6.63	194.17	19.16	2.89	40.7	177.23	1.10	
CP300-3	53.94	145.14	6.70	212.31	24.46	3.65	52.95	203.93	1.04	

P_{cr} is cracking load, P_y and δ_y are load and displacement at yield state, P_u and δ_u are load and displacement at ultimate state, ΔP is load increase by FRP retrofit and P_n is calculated load through the calculation process of flexural strength by Eq. (1).
[a] Displacement data of this specimen were not recorded.

Table 3 Internal tensile forces of NSM FRP strip.

Resisting force	single line			Total of triple lines		
	300	400	500	300	400	500
T_{f1} (kN)	161.28	161.28	161.28	161.28	161.28	**161.28**
T_{f2} (kN)	172.80	230.40	288.00	518.40	691.20	864.00
T_{f3} (kN)	**122.05**	**134.01**	**144.09**	**157.60**	**160.69**	164.05
T_f (kN)	122.05	134.01	144.09	157.60	160.69	161.28
T_f/T_{f1}	0.76	0.83	0.89	0.98	1.00	1.00

Bold one represents the minimum value governing the failure

a specimen using EBR. CP400-1 specimen was omitted from the graph as its data was not stored due to errors in the experiment. According to Fig. 7, in spite of the same retrofit amount, the specimen strengthened by NSMR has greater strength, compared with the specimen by EBR. Besides, the arrival time of ultimate load of a member is more delayed for the former than the latter. This is because of that FRP is mounted in a slit and its bond strength is increased by the confining effect of concrete around the slit.

In NSMR, as the de-bonded length of a specimen is longer, the ultimate strength gets lower and the stiffness becomes lower up to the ultimate load after yield. The reason is that, as de-bonded region is longer, the deformation of a FRP increases uniformly in de-bonded region, and accordingly deformation gets greater.

When bonded length is the same-proportioned, NSMR using triple line showed greater ultimate strength and deflection compared with NSMR using a line. The reason is that where FRP strip is distributed in triple line despite the same bonded length, the anchorage capacity of FRP in

bonded region is improved as a bonded area gets comparatively larger.

Although the de-bonded length in a central part of CP500-3 specimen is 600 mm, the load–deflection curve up to ultimate load of the specimen is nearly identical to that of CP1600-3 specimen using complete bonding for the total length. This means that if bonded length secures 500 mm at both ends, a specimen using partial de-bonding for a central part has the same capacity as a specimen retrofitted using complete bonding. In the comparison of the fully bonded NSMR specimens, CP1600-3 with triple lines has higher ultimate strength than CP1600-1 with single line. But on the other hand CP1600-1 showed linear stiffness until 200 kN which is higher than that of CP1600-3(170 kN). It is thought that the reason is that the three lines of strips reached their elastic limit at different time, respectively so that the elastic limit of the RC beam strengthened by triple lines of thin FRP strips reduced while the inelastic deformation capacity increased. From this, a special consideration is necessary when a beam is strengthened with full bonding NSMR by using too thin FRP strip.

The initial crack of all specimens occurred at the bottom of the beams below loading point. It is known that the cracking load closely depends on the concrete strength and will be increased when its initial stiffness increases. All specimens have same concrete strength and their initial stiffness is almost same regardless of FRP strengthening (from Fig. 7). This means that the FRP strengthening the bottom of RC beam does not contribute the increase of initial stiffness.

3.3.3 Load–Strain Curve

As the result of an observation of a change in the value of strain gauges bonded at regular intervals, as shown in Fig. 4, aimed to find the strain distribution of FRP as load acts, strain was high in a central part that a lot of cracks happen on

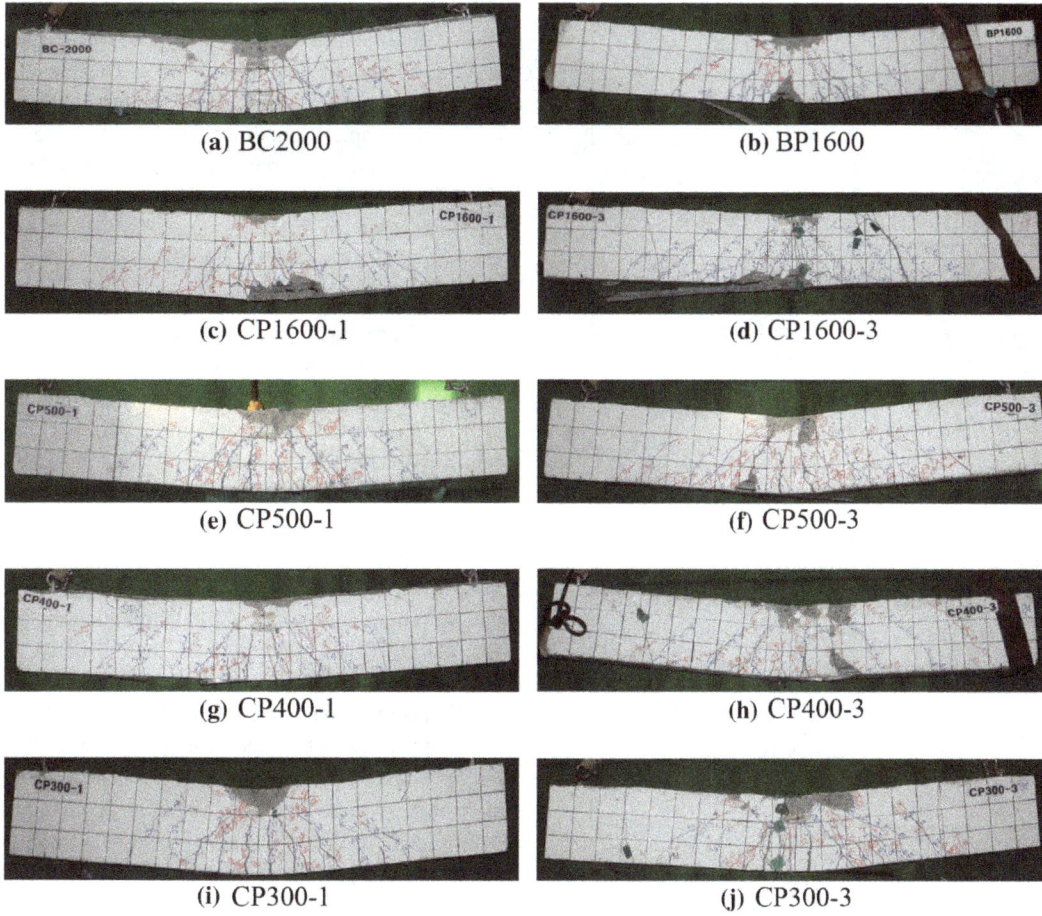

(a) BC2000

(b) BP1600

(c) CP1600-1

(d) CP1600-3

(e) CP500-1

(f) CP500-3

(g) CP400-1

(h) CP400-3

(i) CP300-1

(j) CP300-3

Fig. 5 Failure shape of specimens.

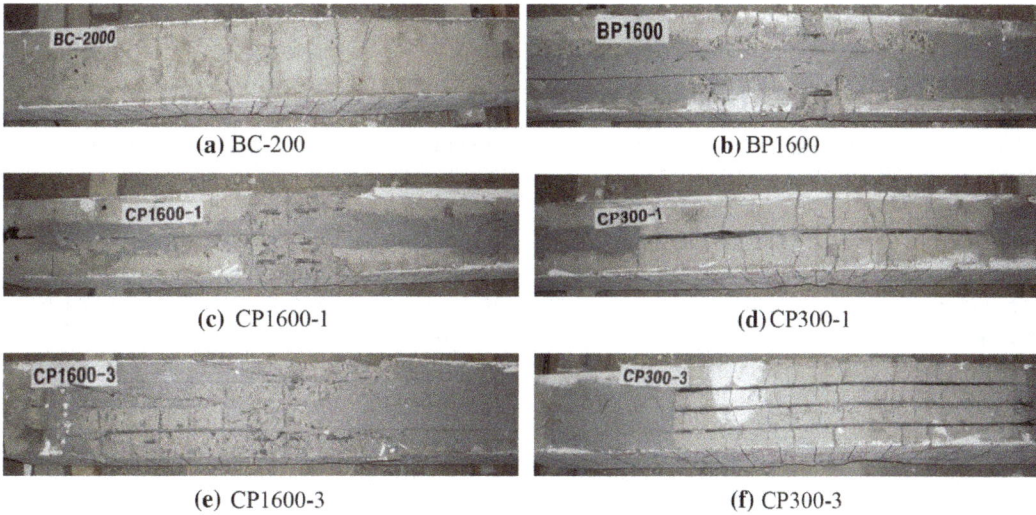

(a) BC-200

(b) BP1600

(c) CP1600-1

(d) CP300-1

(e) CP1600-3

(f) CP300-3

Fig. 6 Crack pattern of bottom surface failed in tension.

the whole. In order to compare strain distributions by experimental variables, Fig. 8 shows the comparison of strains at the time that the deflection of a member is $L/250$.

All FRP strains are deemed the same in de-bonded region, therefore the strain values in the de-bonded region are shown to be uniform.

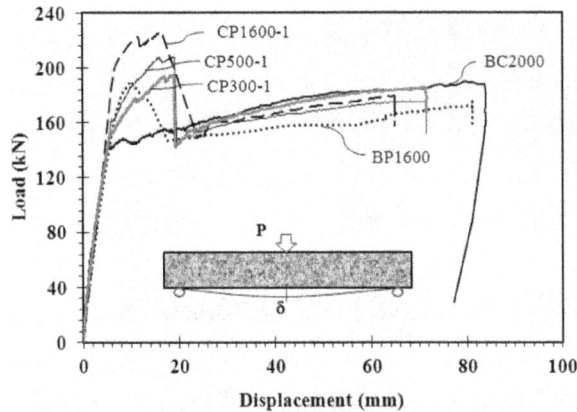

(a) Specimens with single line strip

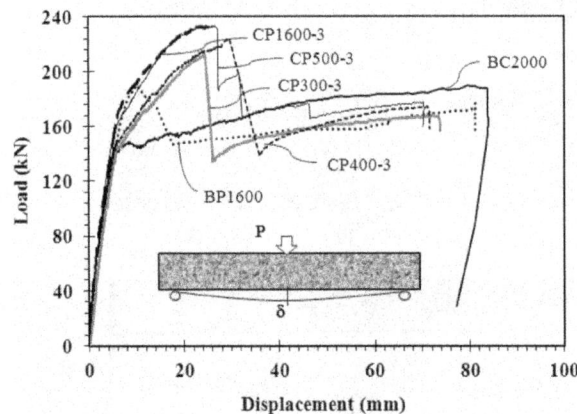

(b) Specimens with triple lines strip

Fig. 7 Load–deflection curves of specimens.

The strain of FRP strip used for NSMR was higher than that of FRP strip used for EBR, and the retrofit using triple line showed a higher strain distribution than that using a line. In the case of NSMR using a line, CP1600-1 specimen without de-bonding showed the strain values which is very similar to CP300-1 whose de-bonded length in a central part is 1000 mm. Although BP1600 and CP1600-1 specimens were expected to show high strain in a central part, their strain distributions are uniformly formed in a certain region of a central part. The data of a strain gauge bonded to a central part was not measured for CP500-1 specimen due to the gage problem.

As shown in Fig. 8b, a specimen using NSRM in triple line showed higher strain distribution under $L/212$ at the same deflection as described above, compared with a specimen using NSMR in a line. Especially, CP1600-3 specimen using triple line embedding for the total length showed remarkably high FRP strain in a central part, unlike CP1600-1 specimen using a line embedding. This means that deformation is concentrated on a central part of FRP. The ultimate strain value of CP500-3 specimen, whose de-bonded central part is 600 mm in length, is similar to that of

CP1600-3 specimen. Actual load–deflection curve in Fig. 7b shows that the behaviors of two specimens up to ultimate load are nearly identical to each other. This means that where anchorage capacity at end of FRP is sufficiently secured, the effect of complete bonding for the whole region can be gained even though some of a central part is de-bonded. As the de-bonded length of a central part is longer, the value of ultimate strain of FRP decreases, which means that the strength exerted by tensile side decreases.

3.3.4 Comparison of Test Results

Figure 9 indicates loads of all specimens at initial cracking, yield and ultimate. Also, the displacements at each critical load are summarized in Table 2. The specimens retrofitted with EB FRP or NSM FRP bonded along the entire length of the members showed similar cracking loads with the specimens retrofitted with partially de-bonded FRP for a central part. As mentioned previously, all specimens have same concrete strength and their initial stiffness is almost same regardless of CFRP strengthening. This means that the CFRP strengthening the bottom of RC beam does not contribute the increase of initial stiffness.

All specimens showed higher yield load than non-retrofitted specimen. The specimen that showed highest yield strength was CP1600-1 specimen, followed by BP1600 specimen. A specimen using partially de-bonded FRP for a central part as NSMR, except CP300 series, showed an increase in displacement after yield with an increase in de-bonded length, as described in the load–displacement curve. In the case of completely bonded FRP for the total length, deformation was concentrated on FRP in a cracked area. In contrast, deformation happens uniformly to FRP in de-bonded region with an increase in the de-bonded length of a central part, which, supposedly, increased the amount of deformation on the whole.

The strength increase by retrofit was calculated to evaluate the retrofit effect by each specimen. As shown in Fig. 10, the strength increase was deemed as the strength difference between the values of non-retrofit specimen and retrofitted specimen for the same displacement at the occurrence of ultimate strength increased after yield.

Figure 11 shows the degree of strength increase by each specimen. In the case of a specimen using NSMR, as debonded length is shorter (bonded length is longer), strength increase by retrofit is larger. Compared with BP1600 specimen retrofitted using external bonding, the strengths of all specimens, except CP300-1 the bonded length of whose ends is 300 mm and which is retrofitted using the embedding in a line, is higher than that of a specimen retrofitted using external bonding.

In particular, CP300-3, whose bonded area was comparatively increased by embedding in triple line despite of a short bonded length of 300 mm, showed a bond failure, but its strength was higher than that of EBR. Although CP500-3

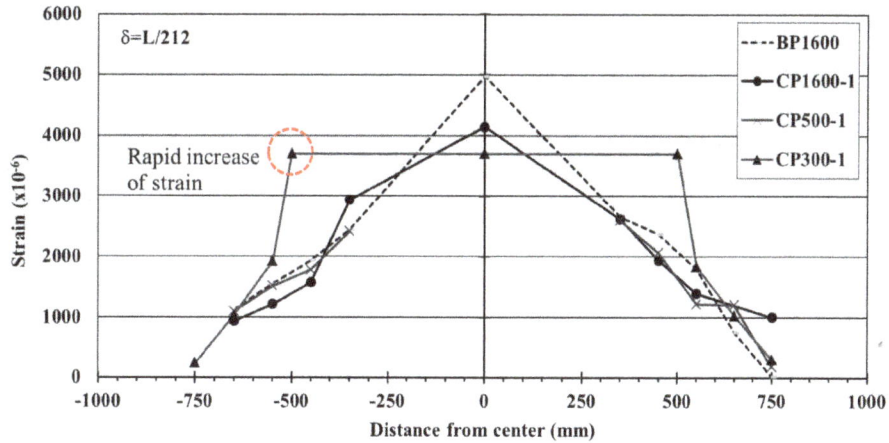

(a) Specimens with single line strip

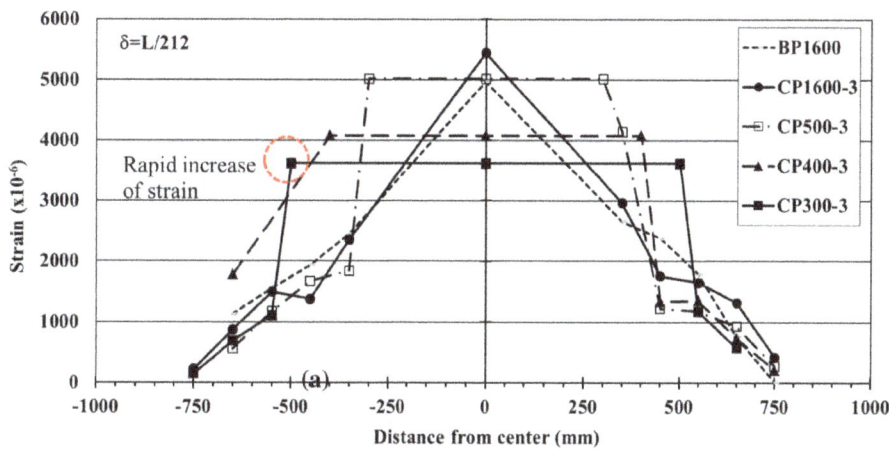

(b) Specimens with triple lines strip

Fig. 8 Comparison of FRP strains at $\delta = L/212$.

specimen had a de-bonded length of 600 mm, it showed the same strength as CP1600-3 specimen with fully bonded FRP for whole length. This means that, where sufficient length is secured at ends, strength decrease arising from de-bonding will never happen.

Figure 12 shows ductility ratio which is the ratio of displacement at ultimate strength and displacement at yield. According to this figure, in the case of all specimens except CP300 series specimens, ductility ratio increases with an increase in de-bonded length. CP300 specimen has very short bond length at both sides while it has relatively long de-bonded region in central part. In this case, the strain in the bonded region tends to rapidly increase from the support to the starting point of de-bonded region; there is a rapid increase of FRP strain along with the bonded length from the strain distribution of FRP in Fig. 8. If the bonded length is enough to resist the tension force acting to the FRP, the behavior of FRP in de-bonded region governs and shows relatively ductile behavior. However, if not, the bond failure in the bonded region governs overall behavior of the beam showing brittle behavior. The bond length of CP300 specimen is not enough to show ductile behavior. Where anchorage length is sufficiently secured, strength decrease

will not happen, and deformation capacity can be improved even though de-bonded length is increased.

4. Flexural Strength in Retrofit Using FRP Strip

Where FRP strip is bonded externally or mounted near surface for retrofit, the relation between strain and flexural stress according to ACI 440-2R (2002) is shown in Fig. 13 and flexural strength can be calculated as follows:

$$M_n = A'_s f_y (d - d_1) + (A_s - A'_s) f_s \left(d - \frac{\beta_1 c}{2} \right)$$
$$+ k_m A_f f_f \left(d_f - \frac{\beta_1 c}{2} \right) \qquad (1)$$

where A_s and A'_s are sectional areas of tensile and compression bar, respectively, β_1 is ratio of depth of rectangular stress block, c is distance from extreme-compression fiber to neutral axis, d and d_1 are effective depth and distance from extreme-compression fiber to centroid of compression reinforcement, respectively, A_f is sectional area of FRP

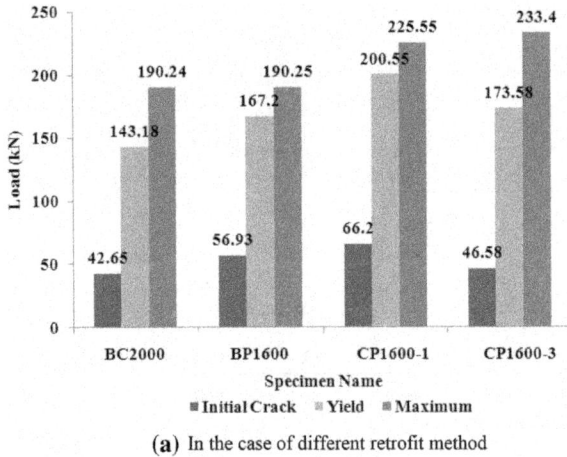

(a) In the case of different retrofit method

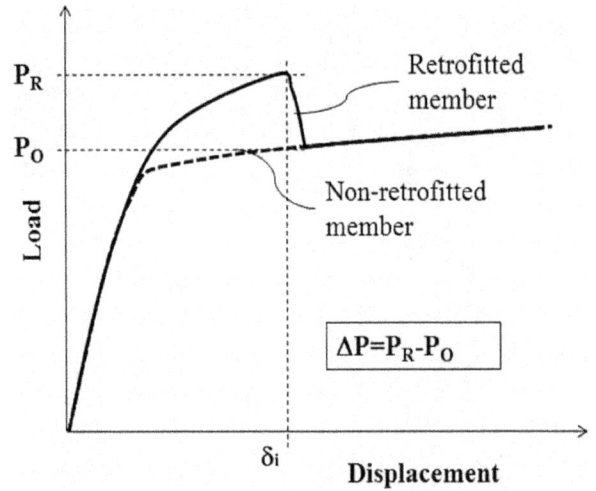

(b) In the case of different bond length in single line strip

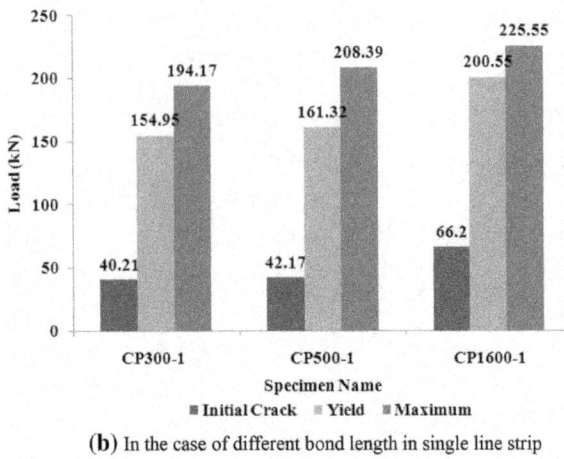

(c) In the case of different bond length in triple lines strip

Fig. 9 Strength comparison of specimens at initial crack, yield and ultimate.

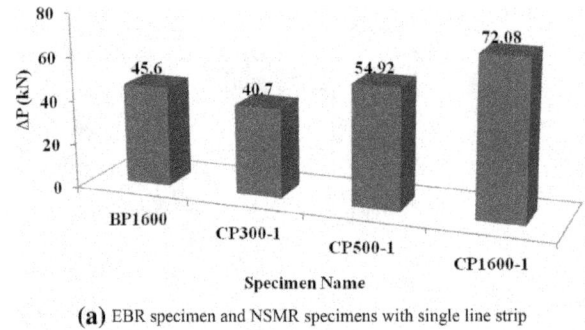

Fig. 10 Definition of strength increase by retrofit.

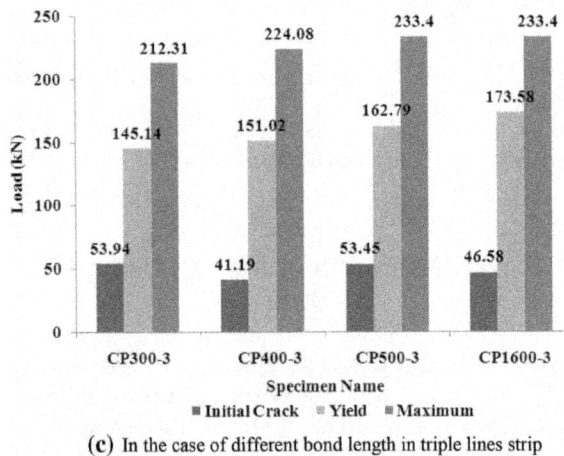

(a) EBR specimen and NSMR specimens with single line strip

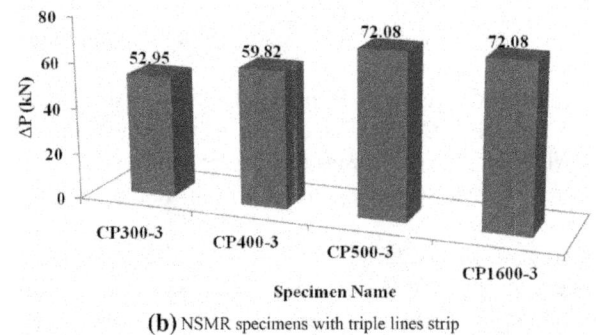

(b) NSMR specimens with triple lines strip

Fig. 11 Strength increases of specimens by retrofit.

reinforcement, f_y is yield strength of bar, f_s and f_f are stress in bar and FRP reinforcement, respectively, k_m is reduction factor considering the bond loss between FRP and concrete and d_f is distance from extreme-compression fiber to centroid of FRP reinforcement.

Strain of FRP and reinforcement can be calculated using the following formula (2002).

$$\varepsilon_{fe} = 0.003 \left(\frac{d_f - c}{c} \right) \leq k_m \varepsilon_{fu} \tag{2}$$

$$\varepsilon_s = \varepsilon_{fe} \left(\frac{d - c}{d_f - c} \right) \tag{3}$$

$$c = \frac{(A_s - A'_s)f_s + A_f f_{fe}}{\gamma f_{ck} \beta_1 b} \tag{4}$$

where ε_{fe} and ε_{fu} are effective strain and ultimate strain of FRP reinforcement, respectively, and ε_s is strain of tensile bar.

(a) NSMR specimens with single line strip

(b) NSMR specimens with triple lines strip

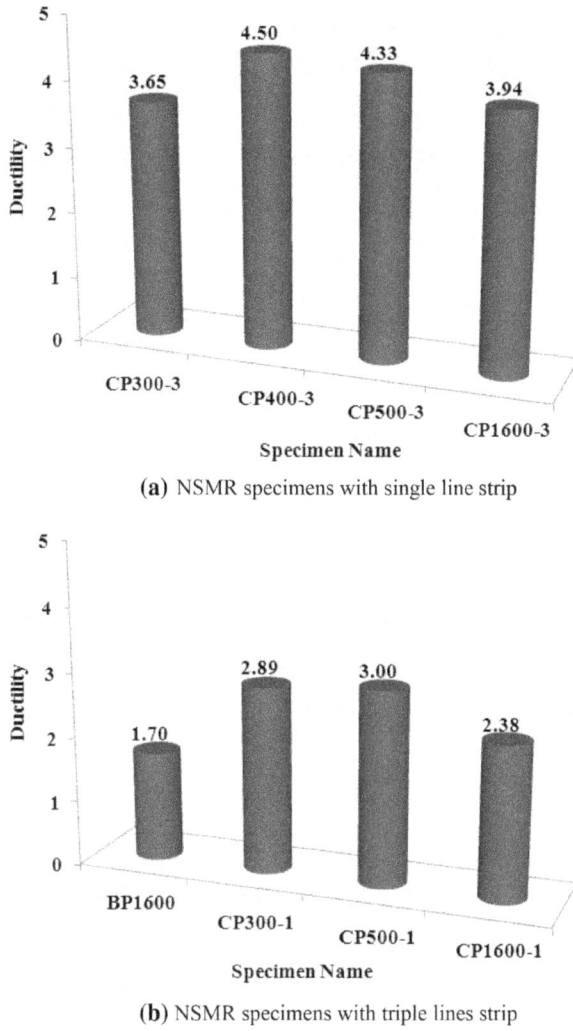

Fig. 12 Ductility of NSMR specimens.

In EBR, k_m is a reduction coefficient considering peeling off of FRP before reaching the ultimate stress and can be calculated using Eq. (5) (2002).

$$k_m = \begin{cases} \frac{1}{60\varepsilon_{fu}}\left(1 - \frac{nE_f t_f}{360,000}\right) \leq 0.9 & \text{for } nE_f t_f \leq 180,000 \\ \frac{1}{60\varepsilon_{fu}}\left(\frac{90,000}{nE_f t_f}\right) > 0.9 & \text{for } nE_f t_f > 180,000 \end{cases}$$

(5)

where n is layer number of FRP plate, E_f is elastic module of FRP, and t_f is thickness of FRP reinforcement.

As a result of calculation of k_m by Eq. (5), a bond reduction factor of a beam retrofitted using external bonding method was 0.5, and accordingly, the strength of a retrofitted member was calculated at 126.5 kN. This value is low, compared with 190.25 kN which is a result of actual experiment. Where 0.9, maximum value in Eq. (5), is used as bond reduction factor, the strength of a retrofitted member is calculated at 176.37 kN, which is still low compared with an experiment result. According to these findings, the

method for evaluating the strength of ACI 440-2R tends to underestimate an experimental result, in evaluating the strength of a beam retrofitted with EB FRP plate.

According to the result of experiment by Seo et al. (2013), in the case of NSMR, bond reduction factor varies according to embedment length. Especially, where a partial de-bonding method is used for a central part like this study, the extent that FRP is fastened to ends changes according to the bond length of ends. Consequently, it is desirable to calculate a reduction factor considering a mechanism of fracture of anchorage.

The mechanism of fracture in the areas of anchorage at ends of NSM FRP can be classified into tensile rupture of mounted FRP strip, shear failure of infilled epoxy, and failure by falling-off of concrete. As shown in Fig. 14, the resisting force is decided by the smallest value among FRP tensile strength T_{f1}, epoxy shear strength T_{f2} and force by bond strength of concrete T_{f3} (Seo et al. 2013).

$$T_f = \min \cdot \{T_{f1}, \ T_{f2}, \ T_{f3}\}$$

(6)

$$T_{f1} = \phi A_f f_f$$

(6 − 1)

$$T_{f2} = \lambda \tau_{ef}(2b_f h_e)$$

(6 − 2)

$$T_{f3} = \alpha_p 0.85 n_g \eta \phi_f^{0.25} f_{ck}^{0.33} \sqrt{(L_{per}\gamma)(EA)_f}$$

(6 − 3)

where λ is effective bond loss factor, τ_{ef} is shear strength of epoxy, b_f and h_e are width and bonded length of NSM FRP strip, respectively, $\gamma = (h_e/L_{per})^{0.65}$, $\eta = (L_{per}\gamma + B_e)/(nL_{per}\gamma) \leq 1.0$, $L_{per} = 2b_f + t_f$, $\phi_f = b_f/t_f$. And $(EA)_f$ is stiffness of FRP reinforcement, n_g is number of FRP strip in considering group effect, B_e is distance between centroids of FRP reinforcements located at extreme sides, and α_p is 1.0 as typical value and 0.85 as low bound.

Equation (6–3) suggested by Seo (2012) is a formula considering group effect arising out of a compact space between FRPs, based on empirical formula of Seracino et al. (2007). The resisting force at FRP member anchorage according to bond length, calculated based on Eq. (6), is shown in Table 3.

Where FRP strip is a line, the resisting force by bond failure, T_{f3} turns out lowest up to 500 mm in bond length. Where FRP strip is distributed in triple line, the tensile strength of FRP strip, T_{f1} turns out lowest from 500 mm in bond length. Compared with the resisting force at tensile rupture, the ratio from 0.76 to 0.89 is shown according to the bond length ranging from 300 to 500 mm in the case of embedding in a line. In the case of embedding in triple lined distribution, the ratio is calculated at 0.98 and 1.0 at 300 mm and more in bond length, respectively. Using these values as k_m which is bond reduction factor of Eq. (2), flexural strength that reflects the effect of bond failure can be calculated. The value is indicated by P_n shown in Table 2. Figure 15 shows the ratio of an experiment result to a calculation result, by bond lengths. From P_u/P_n values of the

(a) EBR

(b) NSMR

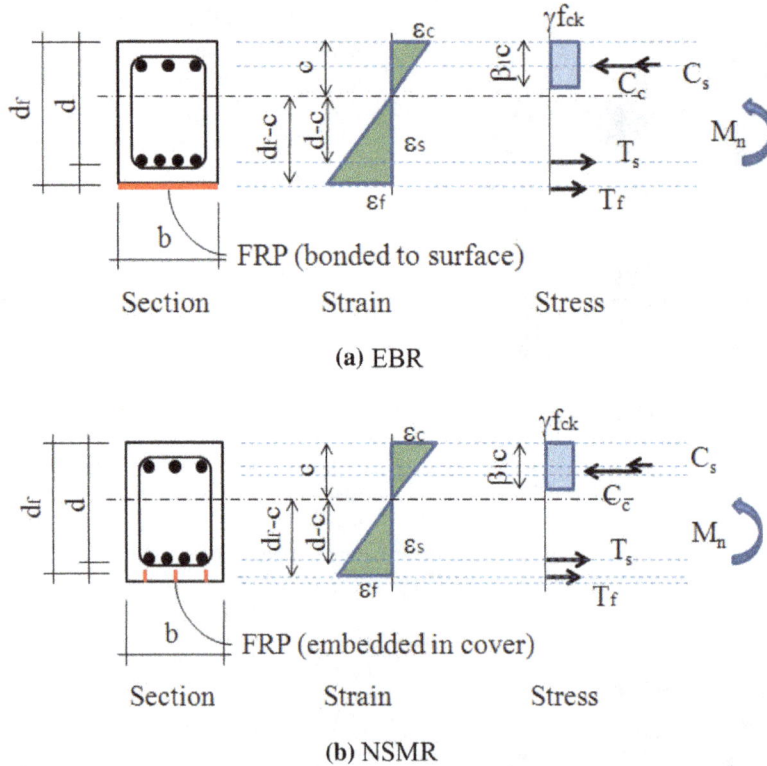

Fig. 13 Stress and strain relation of retrofitted beam according to retrofit method.

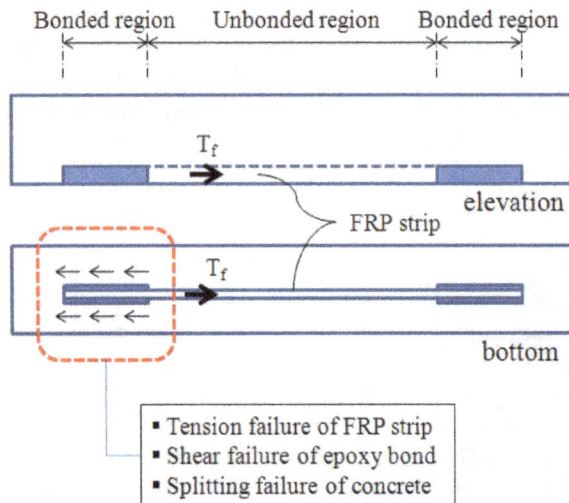

Fig. 14 Internal force of partially de-bonded FRP.

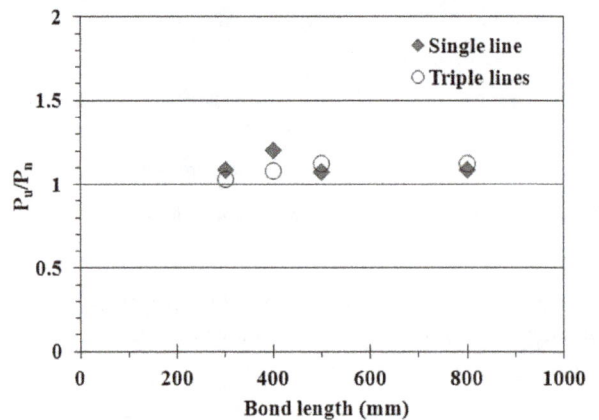

Fig. 15 Comparison of the calculated flexural strengths and test results.

specimens strengthened by FRP, the range of deviation is acceptable not too high (maximum is 21 % at CP400-1) and the flexural strength is safely predicted by using the process.

5. Conclusions

In this paper, an experimental work was presented to study the flexural strength of RC beams strengthened by partially de-bonded NSM FRP strip with various de-bonded length. Especially, considering high anchorage capacity at end of a NSM FRP strip, the effect of de-bonded region at a central part was investigated. In addition, the retrofit by using triple strips (the thickness of each strip is 1/3 of the single strip) was evaluated in order to improve the bond strength in the bonded region, The flexural strength of a RC member strengthened by a partially de-bonded NSM FRP strip was evaluated by using the existing researchers' strength equation for calculation of the strength after retrofit. From the study, the following conclusions were drawn.

The retrofit of mounting FRP strip in the cover concrete of RC beam, as the flexural retrofit method for RC members using FRP, exerts high retrofit capacity compared with the existing EBR method. The reason is that relatively high bond capacity of FRP can be secured by mounting FRP strip vertically.

Where NSM FRP strips are distributed in triple line, a relatively high strength can be exerted through an increase in the bond strength of anchorage. This means that where the retrofit method of mounting thin strip can fully exert tensile strength of FRP even though the bonded region is small. Accordingly, the de-bonded region of a central part can be increased. However, a special consideration is necessary when a beam is fully bonded by using NSMR with too thin FRP strip since the yield strength can be decreased.

The resisting force of a member retrofitted with NSM FRP turned out to be properly evaluated by following the method of calculating the resisting force in flexural retrofit of ACI 440-2R and considering three types of fracture of anchorage.

Where de-bonded region exists in the central part of a flexural member, the deformation capacity of a member is expected to be relatively improved, because FRP strain is not to be concentrated on the center but to be extended uniformly in the de-bonded region. Based on this, a design to induce ductile fracture of a member can be feasible. Consequently, it is necessary to conduct a study on the effect of the de-bonding of aramid and glass FRP other than CFRP targeted by this study. Moreover, it is also necessary to conduct an interpretative study on an evaluation of flexural displacement that reflects the deformation of a central part.

Acknowledgments

This research was supported by 2011 Basic Science Research Program through the National Research Foundation of Korea (NRF) funded by the Ministry of Education, Science and Technology (2011-0011350).

References

ACI Committee 440. (2002). *Guide for the design and construction of concrete reinforced with FRP bars (ACI 440.2R-02)*. American Concrete Institute.

Ali, M., Oehlers, D., Friffith, M., & Seracino, R. (2008). Interfacial stress transfer of near surface-mounted FRP-to-concrete joints. *Engineering Structures, 30*(7), 1861–1868.

Al-Mahmoud, F., Castel, A., & Francois, R. (2012). Failure modes and failure mechanisms of RC members strengthened by NSM CFRP composites—Analysis of pull-out

failure mode. *Composites Part B: Engineering, 43*(4), 1893–1901.

Ceroni, F., Pecce, M., Bilotta, A., & Nigro, E. (2012). Bond behavior of FRP NSM systems in concrete elements. *Composites Part B: Engineering, 43*(2), 99–109.

Chahrour, A., & Soudki, K. (2005). Flexural response of reinforced concrete beams strengthened with end-anchored partially bonded carbon fiber-reinforced polymer strips. *Journal of Composite Construction, 9*(2), 170–177.

Choi, H., West, J., & Soudki, K. (2008). Analysis of the flexural behavior of partially bonded FRP strengthened concrete beams. *Journal of Composites for Construction, 12*(4), 375–386.

Choi, H., West, J., & Soudki, K. (2011). Partially bonded near-surface-mounted CFRP bars for strengthened concrete T-beams. *Construction and Building Materials, 25*, 2441–2449.

EI-Hacha, R., & Rizkalla, S. (2004). Near surface-mounted fiber-reinforced polymer reinforcements for flexural strengthening of concrete structures. *ACI Structural Journal, 101*(5), 717–726.

Fib TG9.3. (2001). *Externally bonded FRP reinforcement for RC structures*. Technical report on the design and use of externally Bonded Fibre reinforced Polymer Reinforcement (FRP EBR) for reinforced Concrete Structures. International Federation for Structural Concrete.

Hassan, T., & Rizkalla, S. (2003). Investigation of bond in concrete structures strengthened with near surface mounted carbon fiber reinforced polymer strips. *Journal of Composite Construction, 7*(3), 248–257.

Jose, S., & Barros, J. (2007). A pullout test for the near surface mounted CFRP-concrete bond characterization. In *Conference of Sociedade Portuguesa de Materials*, 13, Porto, Portugal.

Lee, D., Cheng, L., & Hui, J. (2013). Bond characteristics of various NSM FRP reinforcements in concrete. *Journal of Composites for Construction, 17*(1), 117–129.

Lorenzis, D., Nanni, L., & Tegola, L. (2000). Flexural and shear strengthening of reinforced concrete structures with near surface mounted FRP rods. In *Proceeding of the 3rd international conference on advanced composite materials in bridges and structures (ACMBS III)* (pp. 521–528).

Nanni, A. (2000). FRP reinforcement for bridge structures. Proceedings of structural engineering conference beams. *Journal of Composites for Construction (ASCE), 5*(1), 12–17.

Rezazadeh, M., & Barros, J. (2015). Transfer zone of prestressed CFRP reinforcement applied according to NSM technique for strengthening of RC structures. *Composites Part B: Engineering, 79*(15), 581–594.

Seo, S. (2012). Bond strength of near surface-mounted FRP plate in RC member. *Journal of the Korea Concrete Institute, 24*(4), 415–422 (in Korean).

Seo, S., Feo, L., & Hui, D. (2013). Bond strength of near surface-mounted FRP plate for retrofit of concrete structures. *Composite Structures, 95*, 719–727.

Seracino, R., Jones, N., Ali, M., Page, M., & Oehlers, D. (2007a). Bond strength of near-surface mounted FRP strip-to-concrete joints. *Journal of Composite Construction, 11*(4), 401–409.

Seracino, R., Raizal, S., & Oehlers, D. (2007b). Generic debonding resistance of EB and NSM plate-to-concrete joints. *Journal of Composite Construction, 11*(1), 62–70.

Sharaky, I., Torres, L., Comas, J., & Barris, C. (2014). Flexural response of reinforced concrete (RC) beams strengthened with near surface mounted (NSM) fibre reinforced polymer (FRP) bars. *Composite Structures, 109*, 8–22.

Yost, J., Gross, S., Dinehart, D., & Mildenberg, J. (2007). Flexural behavior of concrete beams strengthened with near surface-mounted CFRP strips. *ACI Structural Journal, 104*(4), 430–437.

4

Investigation on the Effectiveness of Aqueous Carbonated Lime in Producing an Alternative Cementitious Material

Byung-Wan Jo*, Sumit Chakraborty, Ji Sun Choi, and Jun Ho Jo

Abstract: With the aim to reduce the atmospheric CO_2, utilization of the carbonated lime produced from the aqueous carbonation reaction for the synthesis of a cementitious material would be a promising approach. The present investigation deals with the aqueous carbonation of slaked lime, followed by hydrothermal synthesis of a cementitious material utilizing the carbonated lime, silica fume, and hydrated alumina. In this study, the aqueous carbonation reaction was performed under four different conditions. The TGA, FESEM, and XRD analysis of the carbonated product obtained from the four different reaction conditions was performed to evaluate the efficacy of the reaction conditions used for the production of the carbonated lime. Additionally, the performance of the cementitious material was verified analyzing the physical characteristics, mechanical property and setting time. Based on the results, it is demonstrated that the material produced by the hydrothermal method possesses the cementing ability. Additionally, it is revealed that the mortar prepared using the alternative cementitious material yields 33.8 ± 1.3 MPa compressive strength. Finally, a plausible reaction scheme has been proposed to explain the overall performances of the aqueous carbonation as well as the hydrothermal synthesis of the cementitious material.

Keywords: carbon dioxide emission, aqueous carbonation, CO_2 sequestration, hydrothermal synthesis, cementitious material.

1. Introduction

Nowadays, the world's increasing need is to control the global warming and environmental pollution minimizing the emission of the climate changing gasses (Jacobsen et al. 2013; Phair 2006). It is stated in the Kyoto protocol 6 that including the carbon dioxide, some other gasses such as methane, nitrous oxide, sulfur dioxide, sulfur hexafluoride, hydrocarbons, chlorofluorocarbons, and perfluorocarbons, etc., are responsible for the greenhouse effect and global warming. Usually, the fossil fuel is considered to be a major energy source worldwide, which emits a massive carbon dioxide during burning (Siegenthaler and Oeschger 1987; Keeling et al. 1995; Olajire 2013). Conversely, the cement production method emits an enormous amount of carbon dioxide. It is reported elsewhere that the production of the 1-ton cement emits ~ 830 kg of CO_2 (Amato 2013; Schrabback 2010). Additionally, it is also reported that the level of the CO_2 has increased ~ 30 % since the beginning of the industrial revolution (Siegenthaler and Oeschger 1987; Keeling et al. 1995). The current level of the atmospheric carbon dioxide (CO_2) is measured to be 400 ppm (Amato 2013; Schrabback 2010), which is projected to be increased up to ~ 800 ppm by

the end of this century without taking a courageous step (Feely et al. 2004). Hence, an immediate practical plan is required to reduce the CO_2 emission. Additionally, the awareness of the people needs to be increased in controlling the mentioned threats. Therefore, it is essential to develop an alternative technology for minimizing the CO_2 released by man-made and industrial activities.

With the aim to reduce the atmospheric CO_2, the attention of the scientists and the technologists was attracted towards the mineral carbonization. The capture or storage of CO_2 in the geological form is one of the most promising approaches to reduce the atmospheric CO_2 (Gerdemann et al. 2007; Chizmeshya et al. 2007; Chen et al. 2006). The numerous investigations were executed to reduce the atmospheric carbon dioxide by in-situ or the ex-situ mineral sequestration method. Mineral carbonation process can able to minimize the atmospheric CO_2 transforming into the stable carbonate minerals, such as calcite ($CaCO_3$), dolomite ($CaMg(CO_3)_2$), magnesite ($MgCO_3$) and siderite ($FeCO_3$) (Metz et al. 2005). The mineral carbonation is reported to be an exothermic process (Olajire 2013). It takes place at normal temperature. Equation 1 shows the chemistry of the mineral carbonation reaction.

$$MO + CO_2 \rightarrow MCO_3 + (Heat) \tag{1}$$

Although, an adequate research report is available related to the mineral carbonation (Huijgen and Comans 2005; Huijgen et al. 2007; Hanchen et al. 2008; Huijgen et al. 2006, 2004), however, a very few of those techniques are utilized practically. Recently, some other processes such as

Department of Civil and Environmental Engineering, Hanyang University, Seoul 133791, Korea.
*Corresponding Author; E-mail: joycon@hanmail.net

the precipitated calcium carbonate formation, carbonation using brines, accelerated weathering of limestone and straightforward carbonation are also reported to consume the atmospheric CO_2 (Chen et al. 2006; Huijgen and Comans 2005; Hanchen et al. 2008). Additionally, the carbonation of concrete, as well as the calcium hydroxide, is also reported to reduce the atmospheric CO_2 (Galan et al. 2010; Han et al. 2011). Currently, some investigations have been executed to reduce the CO_2 emission during the cement production by replacing the clinker with supplementary materials such as natural pozzolans, fly ash, and slag, etc (Schrabback 2010; Criado et al. 2010; Juenger et al. 2011; Amato 2013; Dinakar et al. 2013; Jeon et al. 2015; Kotwal et al. 2015; Roychand et al. 2016). Additionally, some alternative cementitious materials are produced by the alkali activation of the pozzolanic minerals such as ground granulated blast furnace slag, metakaolin and fly ash, etc (Dinakar et al. 2013; Kim et al. 2013; Kar et al. 2014; Jeon et al. 2015; Chindaprasirt and Cao 2015). Moreover, it is reported that an alternative cementitious material can be produced by the hydrothermal method utilizing geomaterials to reduce the CO_2 emission during the cement production (Jo et al. 2014a, b). The synthesized cementitious material reveals benefits in controlling the physical and mechanical performances of the mortar and concrete. In fact, the geopolymers are composed of aluminosilicate minerals, which has the ability to be hard and sturdy, eventually, lead to gain strength (Xu and Van Deventer 2000; Kar et al. 2014). Although an adequate attention has been paid to reduce the CO_2 emission. However, the utilization of the carbonated material in producing the cementitious material has yet to be investigated adequately. Very recently, we have reported the hydrothermal synthesis of a carbon dioxide-stored cementitious material in controlling the performances of the mortar (Jo et al. 2015).

Reviewing the literature, it is apparent that the utilization of the aqueous carbonated slaked lime for the production of a cementitious material has not studied yet. Hence, In this investigation, we have studied the aqueous carbonation of the slaked lime, followed by hydrothermal synthesis of a cementitious material utilizing the carbonated lime infused with hydrated silica and alumina. The overall process is expected to be beneficial in controlling the atmospheric CO_2. Although the aqueous carbonation of slaked lime can not reduce the ultimate CO_2 emission, nevertheless, it would transform the atmospheric CO_2 to a stable material.

2. Experimental Section

A systematic experimental program was arranged for the aqueous carbonation of slaked lime and the hydrothermal synthesis of an alternative cementitious material. The details of the experimental procedure and characterization techniques are described clearly in the succeeding sections.

2.1 Aqueous Carbonation of the Slaked Lime

The slaked lime ($Ca(OH)_2$) powder (98 % pure) was used as a primary reactant for the carbonation reaction. The used slaked lime contains some other oxide materials such as $MgO < 0.4 \%$, $SiO_2 < 0.8 \%$, $Al_2O_3 < 0.5 \%$ and $Fe_2O_3 < 0.3 \%$. An aqueous solution of the slaked lime was prepared using 0.865 g of dry slaked lime powder and 500 ml of distilled water in a beaker. It is reported elsewhere that the solubility of the hydrated lime in water is ~ 0.173 at 20 °C. Therefore, in this study, 0.173 % slaked lime solution was prepared. In this investigation, the carbonation reaction was performed in a closed reactor. Figure 1 represents a schematic diagram of the reactor used in this investigation. In the reactor, a mass flow meter (MFC-100) was installed to control the carbon dioxide injection speed as well as to measure the mass of the injected CO_2. Additionally, the instrument was associated with a pH measurement device (pHI-201), which was employed to record the change of pH during the carbonation reaction. After preparing the aqueous slaked lime solution, the beaker was placed into a Teflon made closed drum of the reactor to avoid the contamination of the other gasses and unwanted carbonation. Subsequently, the CO_2 gas was purged into the reacting vessel for 20 min maintaining the temperature of the reaction system 21 °C. In this study, the aqueous carbonation of the slaked lime was performed in four different conditions varying the agitation speed (viz., 200 and 450 RPM) and CO_2 injection speed (0.3 and 0.5 l/min). Table 1 presents the details of the reaction

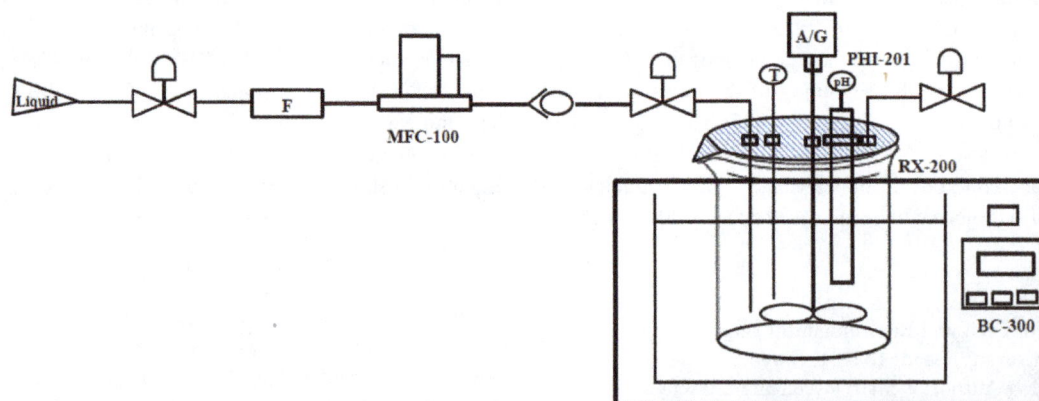

Fig. 1 Schematic presentation of the reactor for the aqueous carbonation reaction.

Table 1 Summary of the conditions used for the aqueous carbonization of the slaked lime.

Reaction condition code	Agitation speed (RPM)	Temperature (°C)	CO_2 injection speed (l/min)	Reaction time (min)
A	200	21	0.3	20
B	200	21	0.5	20
C	450	21	0.3	20
D	450	21	0.5	20

conditions used in this study. After completion of the reaction, the precipitates were filtered off using a Whatman 41 filter paper and an aspirator. Thereafter, the moisture was detached from the precipitates using a freeze dryer. Finally, the carbonated product was stored in a vacuum desiccator and allowed to characterize. In this investigation, the carbonation reaction was repeated three times to obtain an average result.

2.2 Hydrothermal Synthesis of the Cementitious Material

The carbonated lime produced from the aqueous carbonation method was used for the hydrothermal synthesis of an alternative cementitious material. Initially, 180 g of silica fume was mixed with the 700 ml of water in a beaker using a magnetic stirrer, followed by mixing with 120 g of carbonated lime. Simultaneously, 60 g of sodium hydroxide dissolved in 700 ml of water in an another beaker, followed by mixing with 100 g of sodium aluminate. Afterward, 140 g of triethanolamine (TEA) was added to the sodium aluminate solution for preventing the coagulation of aluminium hydroxide from the solution. Thereafter, the mixtures were agitated for 24 h. A thick gel was produced from the mixture of silica fume and carbonated lime. Conversely, a comparatively thin gel was produced from the hydrated alumina and TEA mixture. Subsequently, the thick gel

was added gently to the low-density gel with a constant stirring, followed by 3 h sonication. The mixed compound, thus, obtained was allowed to dry and crystallize in an oven at 105 °C. Finally, the solid crystalline material obtained after 7 days of controlled heating was treated as an alternative cementitious material (ACM). A pictographic model for the hydrothermal synthesis of the carbonated lime based ACM is presented in Fig. 2. The solid oven-dried material was, then, ground to make a powder for preparing the mortar samples.

2.3 Fabrication of the ACM Based Mortar Sample

In this study, the mortar samples were fabricated using the ACM, fine aggregate (sand), and the varying amounts of alkali activator (50 % NaOH solution) and water. For a particular batch mixing of the ACM based mortar, 100 g of ACM was mixed with the 300 g of fine aggregate, followed by mixing with the required amount of alkali activator and water. In this investigation, a control mortar sample was also prepared using 100 g ordinary Portland cement, 300 g of sand and 60 ml of water. Table 2 shows the mix proportion of the components used for the fabrication of the control and ACM based mortar. The mortar mixture obtained after mixing of the components was, then, cast immediately in the 50 mm side cubic molds and allowed to set for 24 h. After

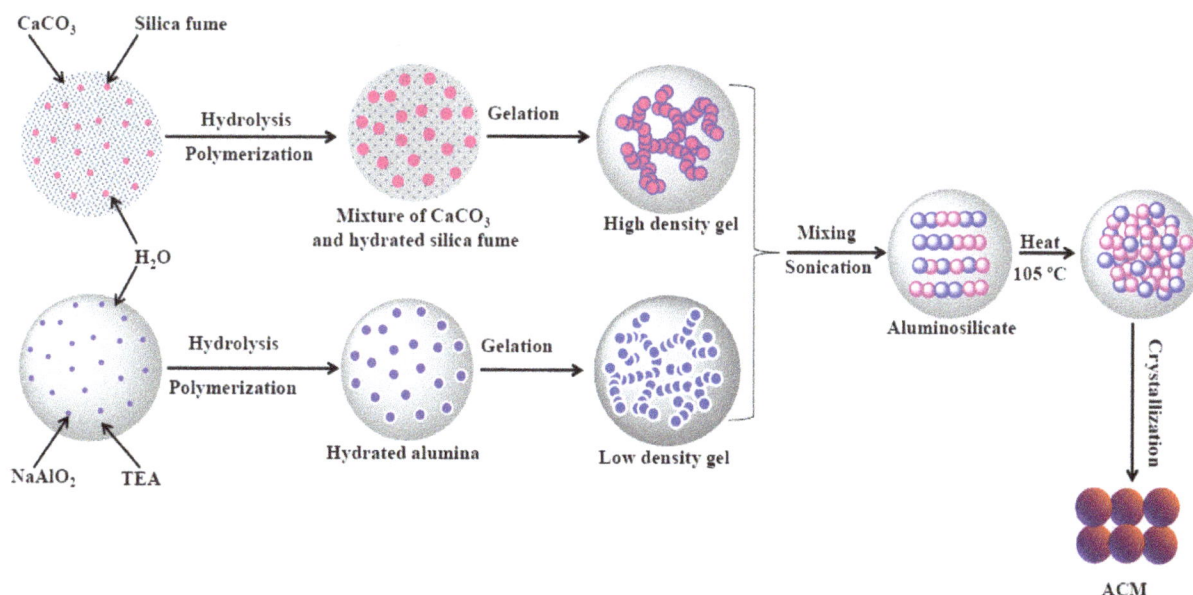

Fig. 2 The plausible model for the hydrothermal synthesis of the cementitious material utilizing carbonated material produced by the aqueous carbonation reaction.

Table 2 Formulation code and the mix proportions of the components for the fabrication of the control cement mortar and ACM based mortar.

Sample code	Mix proportioning of the components			
	Cement (g)	Fine aggregate (g)	Water (ml)	Alkali activator (ml)[c]
CCM	100[a]	300	60	–
ACM-M1	100[b]	300	60	–
ACM-M2	100[b]	300	30	30
ACM-M3	100[b]	300	–	60

[a] Ordinary portland cement.
[b] Alternative cementitious material.
[c] 50 % NaOH solution.

complete setting, the mortar samples were allowed to water cure for 28 days. The mortar samples were removed from the curing chamber and allowed to test after completion of the desired curing period.

2.4 Characterizations

The thermal analysis of the precipitates obtained from the aqueous carbonation reaction was performed using a thermogravimetric analyzer (TG-DTA 2020SA). In this experiment, exactly 10 mg of the powder sample was used for the analysis. The experiment was conducted under a dynamic N_2 atmosphere in the temperature range between 30 and 900 °C with a heating rate of 10 °C/min. In this investigation, each sample was analyzed three times to get an average result.

The X-ray diffraction pattern of the precipitates obtained from the aqueous carbonation reaction was recorded using an X-ray diffractometer (Ultima III). The instrument produced a monochromatic X-ray beam of the wavelength 1.54 Å using CuKα radiation (40 kV, 40 mA) and Ni filter. In this study, the X-ray diffraction pattern of the samples was recorded in the 2θ range between 10°–60°, maintaining a scan speed of 1 °/min with a step difference of 0.02″. Prior to analyzing the X-ray diffraction pattern, the sample was packed into a rectangular hollow area of the glass made sample holder, followed by placing in the instrument.

The microstructure of the precipitates obtained from the aqueous carbonation reaction, the product obtained from the hydrothermal synthesis and the 28 days cured mortar was recorded using a field scanning electron microscope (JEOL JSM-6700F). Prior to analyzing the microstructure, a very thin gold was sputter coated on the surface of the moisture-free dried samples to avoid charging. Thereafter, samples were placed on the SEM stub and allowed to analyze. The digital scanning electron micrographs were recorded in 10–20 kV accelerated voltage and 15 k× magnification.

The particle size analysis of the cementitious material obtained from the hydrothermal method was performed using LA-950 Laser particle size analyzer. For the particle size analysis, exactly 1 g of the oven-dried sample was fed into the PowderJet Dry Feeder of the LA-950 Laser particle size analyzer. Furthermore, the sample was analyzed based on the Mie scattering theory. In this instrument, two light sources viz., 5 mW, 650 nm red laser diode and 3 mW,

405 nm blue LED were used to analyze the particle size of the material. In the measurement array, the high-quality photodiodes were used to detect the scattered light over a wide range of angles.

The initial and final setting time of the alternative cementitious material obtained from the hydrothermal method was measured according to the KSL 5108 (2007). Usually, this standard method is used to measure the setting time of the hydraulic cement.

The compressive strength of the 50 mm side cubic alternative cementitious material (ACM) based mortar sample was measured using a universal testing machine (Shimadzu, CCM-200A) with a loading rate 0.06 MPa/min in accordance with the Korean standard KS F 2405 (2010).

3. Results

3.1 pH Analysis of the Carbonation Reaction Medium

Figure 3 depicts the pH variation of the reaction medium with the increase in reaction time. In this study, the pH of the reaction medium was measured at the 1 min interval. From the figure, it is observed that irrespective of the reaction conditions, the pH of the reaction medium decreases

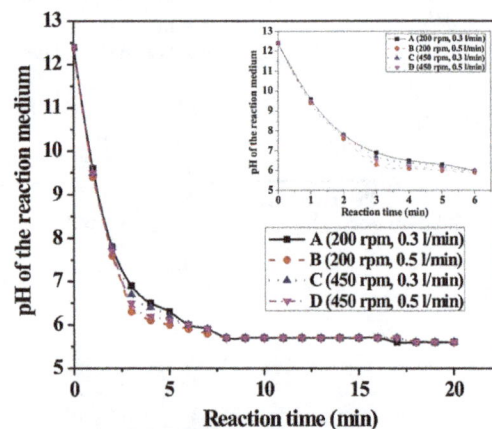

Fig. 3 pH variation of the aqueous carbonation reaction medium as function of the reaction time.

gradually up to 5 min of the reaction, followed by keeping constant with the progress in reaction. Prior to the carbonation reaction started (before CO_2 purged), the pH of the reaction medium is measured to be 12.4, which is measured to be ~ 6.5 after 5 min of the reaction started. Interestingly, the decrement of pH is observed to be different for the different reaction conditions (inset figure of the Fig. 3). The decrement of pH for the reaction condition B is observed to be higher as compared to that of the reaction condition D, C and A. Additionally, the order of pH decrement of the four different reaction conditions is found to be B > D > C > A. Hence, It is indicated that the rapid carbonation reaction occurs in the reaction condition B as compared to that of the other three conditions. The pH of the carbonation reaction medium is reduced to be due to the formation of the carbonic acid, which neutralizes the saturated solution of the slaked lime. During the reaction, the lime solution reacts with the carbonic acid, which in turn leads to produce the calcium carbonate and water. As a consequence, the concentration of OH^- in the reaction system decreases. Eventually, the pH of the reaction system decreases (Olajire 2013). The pH of the reaction medium decreases continuously until the lime molecules are available for the carbonation. After complete neutralization of the lime molecules (after 6 min of the reaction started), the solution becomes CO_2 rich (pH = 5–6), which restrains to reduce the pH.

3.2 Quantification of the CO_2 Consumption

Table 3 represents the moles of reactants and products of the aqueous carbonation reaction. In this study, the experiment was repeated three times to identify the most effective reaction condition. After measuring the mass of the reactant and products, the moles of CO_2 consumed in the carbonation reaction in four different reaction conditions are measured. From the table, the moles of CO_2 (average value) consumed in the four different carbonation reaction conditions such as A, B, C and D are measured to be 0.0107 ± 0.0004, 0.0118 ± 0.0008, 0.0100 ± 0.0004, and 0.0112 ± 0.0011, respectively. It is well known that the 1 mol of lime reacts with 1 mol of CO_2 to produce 1 mol of calcium carbonate (Olajire 2013). Viewing in light of the theoretical prediction, it is assumed that the 0.0117 mol of slaked lime would react with 0.0117 mol of CO_2 to produce 0.0117 mol of calcium carbonate. However, in this investigation, all four reaction conditions do not follow the theoretical prediction conveniently. As presented in Table 3, the reaction conditions B and D nearly follow the theoretical prediction. From the table, it is envisaged that in the reaction condition B and D, 0.0117 mol of slaked lime react with 0.0118 ± 0.0008 and 0.0112 ± 0.0011 mol of CO_2, respectively, and produces 0.0116 ± 0.0003 and 0.0111 ± 0.0002 mol of calcium carbonate, respectively. Additionally, the % of CO_2 consumption in four different reaction conditions such as A, B, C and D is calculated to be 91.2 ± 3.0, 100.1 ± 6.8 85.5 ± 3.1 and 96.0 ± 9.1 %, respectively. Figure 4 represents the moles of slaked lime (calcium hydroxide) and CO_2 used and the moles of calcium carbonate produced in the theoretical carbonation reaction scheme and the four

different carbonation reaction conditions such as A, B, C and D. The result reveals that the reaction condition B consumes higher amounts of CO_2 and produces a greater amount of calcium carbonate as compared to that of the reaction condition D, A and C. Moreover, the reaction condition B follows the theoretical prediction more appropriately. Although, the impact of the reaction condition D is similar to the reaction condition B. However, the reaction condition D (450 RPM agitation speed and 0.5 l/min CO_2 injection speed) would be a more energy consuming process as compared to that of the reaction condition B (200 RPM agitation speed and 0.5 l/min CO_2 injection speed). Therefore, the reaction condition B is considered to be the most suitable for the carbonation of slaked lime, which is used for the production of the carbonated lime in this investigation.

3.3 Thermo Gravimetric Analysis (TGA) of the Carbonated Material

The thermogravimetric analysis of the precipitates obtained from the aqueous carbonation reaction was performed to estimate the extent of carbonated product. In this investigation, the aqueous carbonation reaction in four different conditions was repeated three times to identify the most effective reaction condition. Figure 5 presents the TG thermogram (one replicate plot from each reaction condition) of the carbonated product obtained from the four different reaction conditions. As presented in the figure, the mass losses in the temperature range 400–550 and 600–800 °C are observed to be due to the decomposition of the calcium hydroxide and calcium carbonate, respectively (Chakraborty et al. 2013). From the figure, it is envisaged that irrespective of the reaction conditions used for the carbonation of slaked lime, the thermogram of the carbonated products shows a prominent mass loss in the temperature range 600–830 °C. However, the products obtained from the different reaction conditions show the different mass losses in the same temperature range. From the figure, the average mass loss attributed to be due to the decomposition of the calcium carbonate in four different reaction conditions, i.e., A, B, C and D is measured to be 29.4 ± 2.3, 39.2 ± 2.5, 31.9 ± 1.8 and 35.1 ± 2.3 %, respectively. It indicates that the different carbonation reaction conditions produce different amounts of carbonated product. Additionally, it is observed that the reaction condition B produces a greater extent of calcium carbonate as compared to that of the reaction conditions, i.e., A, C and D. Hence, the reaction condition B is considered to be most appropriate for the aqueous carbonation reaction. Additionally, it is apparent that the aqueous carbonation reaction successfully consumes the CO_2 to produce a solid residue (calcium carbonate). This can be further clarified by the X-ray diffraction analysis as well as FESEM analysis.

3.4 X-ray Diffraction Analysis of the Aqueous Carbonated Product

The X-ray diffraction analysis of the aqueous carbonated lime was performed to identify the chemical phases present in the reaction product. Figure 6 depicts the X-ray diffraction pattern of the products obtained from the carbonation

Table 3 Moles of the reactants and products before and after the carbonization reaction and the summary of the CO_2 consumption in four different reaction conditions.

Chemical constituents	No. of Exp	Moles of the reactants and products before and after the carbonization reaction in four different reaction conditions							
		Condition A		Condition B		Condition C		Condition D	
		Before	After	Before	After	Before	After	Before	After
Moles of $Ca(OH)_2$	(i)	0.0117	–	0.0117	–	0.0117	–	0.0117	–
	(ii)	0.0117	–	0.0117	–	0.0117	–	0.0117	–
	(iii)	0.0117	–	0.0117	–	0.0117	–	0.0117	–
Moles of CO_2	(i)	0.0110	–	0.0127	–	0.0096	–	0.0122	–
	(ii)	0.0103	–	0.0112	–	0.0103	–	0.0101	–
	(iii)	0.0107	–	0.0115	–	0.0101	–	0.0114	–
Average		0.0107 ± 0.0004	–	0.0118 ± 0.0008	–	0.0100 ± 0.0004	–	0.0112 ± 0.0011	–
Moles of $CaCO_3$	(i)	–	0.0095	–	0.0118	–	0.0097	–	0.0113
	(ii)	–	0.0094	–	0.0116	–	0.0098	–	0.0109
	(iii)	–	0.0098	–	0.0113	–	0.0098	–	0.0110
Average		–	0.0096 ± 0.0002	–	0.0116 ± 0.0003	–	0.0098 ± 0.0001	–	0.0111 ± 0.0002
Mole % of CO_2 consumption[a]		91.2 ± 3.0		100.1 ± 6.8		85.5 ± 3.1		96.0 ± 9.1	

[a] Mole percent of CO_2 consumed with respect to the mole of lime $(Ca(OH)_2)$.

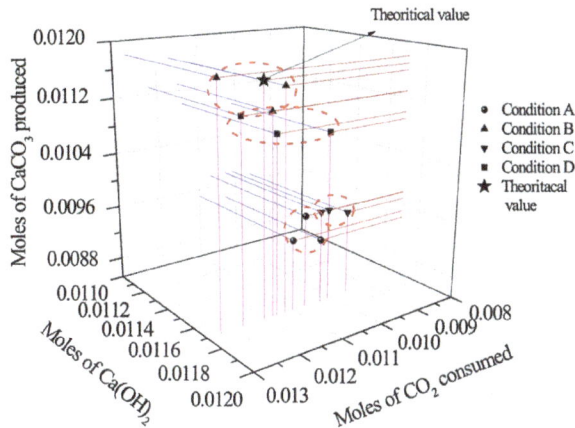

Fig. 4 Moles of slaked lime (calcium hydroxide) and CO_2 used and the moles of calcium carbonate produced in the theoretical carbonation reaction scheme and the four different carbonation reaction conditions such as A, B, C, and D.

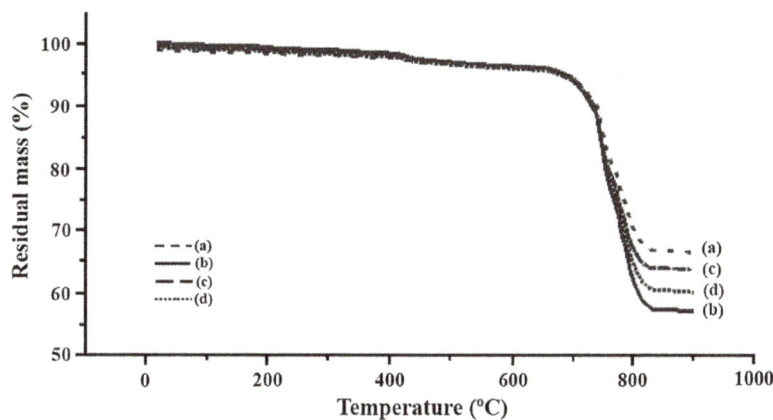

Fig. 6 X-ray diffraction patterns of the products obtained by the aqueous carbonation reaction in four different conditions.

reaction in four different conditions. As presented in the figure, irrespective of the reaction condition used for the aqueous carbonation, the X-ray diffractogram of all samples possesses identical peaks. In the X-ray diffraction pattern, the characteristic peaks at the 2θ equal to 23.02°, 29.41°, 35.97°, 39.4, 43.15, 47.49 and 48.51°, 57.4° appear to be due to the calcium carbonate (calcite) (Stutzman 1996; Jiao et al. 2009; Kontoyannis and Vagenas 2000; Han et al. 2011). Additionally, the very small peak at the 2θ equal to 56.42° and 57.94° appears to be due to the calcium carbonate also (Crowley 2010). Excluding the calcium carbonate peak, no other peak has been observed in the X-ray diffraction pattern. In fact, no peak for the calcium hydroxide (reactant) at the 2θ equal to 18.09°, 28.66°, 34.09°, 47.12°, 50.79° and 54.34° (Stutzman 1996) has been observed. Hence, from the X-ray diffraction analysis, it is demonstrated that the aqueous carbonation reaction is considered to be very effective in producing the calcium

carbonate from the slaked lime. Additionally, it is assumed that the aqueous carbonation of the slaked lime could be an efficient method to consume the environmental CO_2.

3.5 FESEM Analysis

Figure 7 shows the microstructure of the precipitates obtained from the aqueous carbonation reaction in four different conditions. As presented in the figure, the images (Figs. 7a to 7d) obtained from the FESEM analysis are observed to be similar. Hence, it is confirmed that the reaction condition does not affect the microstructure of the carbonated product. Additionally, the microstructure of the carbonated compound obtained from the four different reaction condition is observed to be a crystalline. Generally, the calcium carbonate possesses three polymorphs such as calcite (rhombohedral crystal structure), vaterite (globe or oval shaped) and aragonite (needle-shaped crystal structure) (Han et al. 2011; Stutzman 1996). As evidenced by the XRD

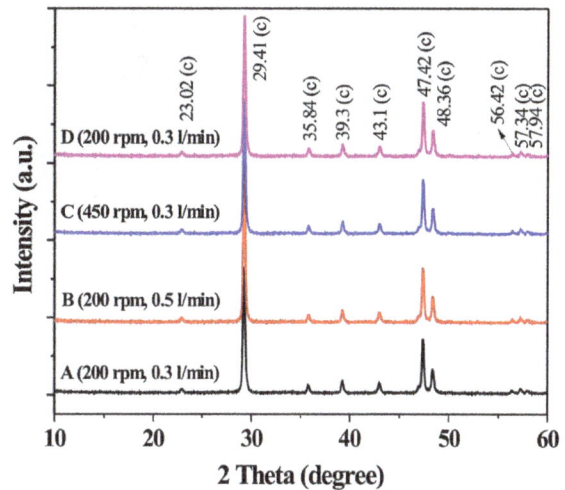

Fig. 5 TG thermograms of the products obtained from the aqueous carbonation reaction in four different conditions. In the figure a, b, c, and d indicate the TG thermogram of the carbonated product obtained from the carbonation reaction condition A, B, C, and D, respectively.

Fig. 7 FESEM images of the products obtained from the carbonation reaction of the slaked lime in four different conditions. **a** Product obtained from the reaction condition A, **b** Product obtained from the reaction condition B, **c** Product obtained from the reaction condition C, **d** Product obtained from the reaction condition D.

Fig. 8 FESEM images of the **a** ordinary Portland cement (Jo et al. 2014c) and **b** the cementitious material produced by hydrothermal synthesis method utilizing aqueous carbonated material.

analysis, irrespective of the reaction condition, the aqueous carbonation of slaked lime produces only calcite. Additionally, no crystal of the oval shaped and needle-shaped has been observed from the FESEM image of the carbonated product. Therefore, the crystal structure of the carbonated product (calcite) obtained from the aqueous carbonation reaction is considered to be rhombohedral.

Figures 8a and 8b depict the FESEM images of the commercially available ordinary Portland cement and the cementitious material produced by the hydrothermal method, respectively. From the Fig. 8a, the particle of the ordinary Portland cement is observed to be irregular shaped. Whilst, the particles of the synthesized cementitious material are observed to be spherical shaped (Fig. 8b). Additionally, an excellent particle size distribution of the synthesized cementitious material has been observed. The TGA, XRD, and FESEM analysis demonstrate that the aqueous carbonation of slaked lime produces rhombohedral crystal of

calcite. However, the microstructure of the alternative cementitious material does not contain any rhombohedral crystal of calcite. Therefore, it is considered that the entire amount of calcium carbonate is consumed in producing the alternative cementitious material. Hence, it is assessed that the hydrothermal method can able to store the carbonated lime as a cementitious material.

Figure 9 presents the microstructure of the 28 days cured control cement mortar (a) as well as the ACM based mortar (b). From Fig. 9b, it is envisaged that the alternative cementitious material can able to produce a compact microstructure after hydration. Comparing the FESEM image of the ACM based mortar with the control mortar, the microstructure of the ACM based mortar is observed to be similar to that of the ordinary Portland cement based mortar. It is a well-known that the development of the microstructure is a crucial factor in enhancing the mechanical strength of the cementitious material. In this study, the cementitious

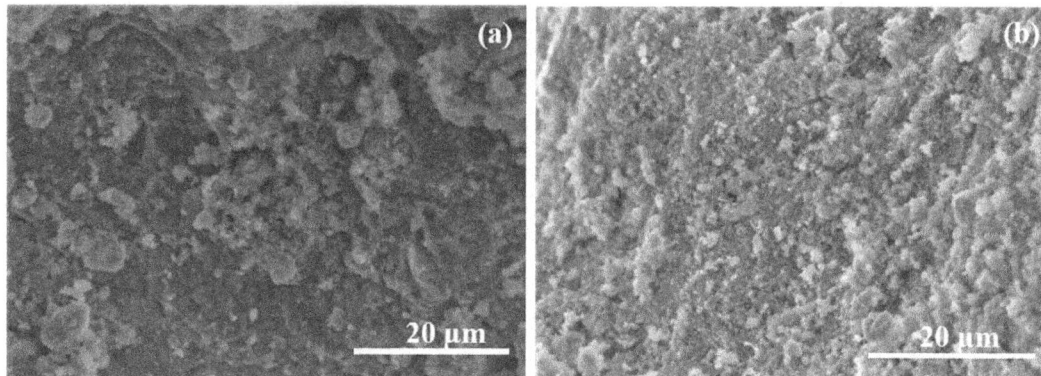

Fig. 9 FESEM images of the 28 days hydrated **a** control cement as well as the **b** ACM based mortar sample.

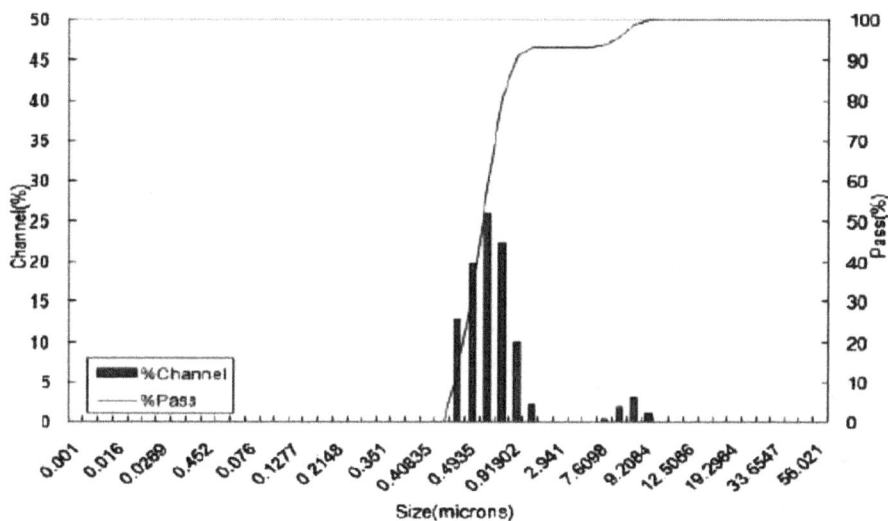

Fig. 10 The particle size distribution curve of the alternative cementitious material produced by the hydrothermal method.

material produced by the hydrothermal method utilizing carbonated lime develops a compact microstructure, which may contribute to improving the strength of the ACM based mortar.

3.6 Particle Size Analysis of the Alternative Cementitious Material

Figure 10 depicts the particle size distribution pattern of the alternative cementitious material. From the figure, it is envisaged that the particle size of the alternative cementitious material belongs in the range between 0.49–9.2 μm. The average particle size of the alternative cementitious material is measured to be 1.4 μm. Whilst, the particle size of the commercially available ordinary Portland cement is reported to be 10–30 μm (Jo et al. 2014c). Therefore, it is apparent that the particle size of the alternative cementitious material produced by the hydrothermal method is significantly smaller as compared to that of the commercially available Portland cement. Additionally, it is ascertained that the hydrothermal synthesis method is able to produce a spherical shaped submicron particle. Additionally, the

specific gravity and fineness modulus of the alternative cementitious material produced by the hydrothermal method are measured to be 2.23 and 68,500 cm^2/g, respectively. Whilst, these are reported to be 3.15 and 2800 cm^2/g, respectively, for ordinary Portland cement (Jo et al. 2014c). As the alternative cementitious material possesses a smaller particle size and greater fineness modulus as compared to that of the OPC, therefore, it is assumed that the alternative cementitious material would expose a greater surface area for the hydration reaction, which may lead to developing the early strength.

3.7 Chemical Composition Analysis

The oxide composition of the alternative cementitious was measured to evaluate the chemical characteristic. The oxide compositions of the alternative cementitious material are observed to be similar to that of the ordinary Portland cement. However, the amounts of the oxide phases are observed to be different. The mass (%) of the oxide phases such as SiO_2, CaO, Al_2O_3, Na_2O, Fe_2O_3 of alternative cementitious material are measured to be 50.84, 9.69, 28.97,

2 and 1.90 %, respectively. Whilst, the same are reported be 20.36, 64.33, 5.77, 0 and 2.84 %, respectively, for ordinary Portland cement (Jo et al. 2014a). Analyzing the oxide composition, it is envisaged that the alternative cementitious material contains a greater amount of SiO_2 and Al_2O_3 as compared to that of the ordinary Portland cement. Therefore, it is expected that the alternative cementitious material may produce a geopolymer compound after hydration reaction (Mayer et al. 2013; Kar et al. 2014). Additionally, it is assumed that the geopolymer material would produce an interpenetrating net structure in the bulk of the mortar, which may lead to developing the strength of the ACM based mortar.

3.8 Setting Time Analysis

Setting time is an important characteristic of the of the hydraulic cement. Usually, the setting time of the cement is measured to know the time required to set and harden the cement sample. In this investigation, the setting time is measured to know the time required for setting and hardening of the alternative cementitious material. The initial and final setting time of the alternative cementitious material is measured to be 242 ± 6 and 528 ± 5 min, respectively. Whereas, the initial and final setting time of the ordinary Portland cement is measured to be 210 ± 5 and 300 ± 8 min, respectively (Jo et al. 2014a). It is reported elsewhere that the initial and final setting time of a standard hydraulic cement should be more than 60 min and less than 10 h, respectively, at the ambient condition (KS L 5201 2013). In this study, the cementitious material produced by the hydrothermal method possesses a similar characteristic. Hence, it is assumed that the alternative cementitious material produced by the hydrothermal method can be considered as an alternative binder for the fabrication of mortar and concrete.

3.9 Compressive Strength

In this investigation, the mortar samples were fabricated using the ACM, sand and varying amounts of water and alkali activator. Table 2 depicts the mix proportion and sample code of the different types of the mortar samples. In this study, six samples from each batch were tested to obtain an average result of the compressive strength. Figure 11 shows the compressive strength of the control as well as ACM based mortar. As presented in the figure, the compressive strength of control mortar (CCM), ACM-M1 (ACM denotes alternative cementitious material and M indicates mortar), ACM-M2, and ACM-M3 is measured to be 34.2 ± 1.2, 25.4 ± 1.5, 28.6 ± 1.2, 33.8 ± 1.3 MPa, respectively. Additionally, it is envisaged that the compressive strength of the ACM based mortar increases gradually with the increase in alkali activator content. Interestingly, the compressive strength of the ACM based mortar fabricated using the maximum amount of alkali activator (ACM-M3) is observed to be comparable with the control mortar. The lower compressive strength of the mortar prepared using less extent of alkali activator is observed to be due to the inadequate hydration of the alternative cementitious material. In

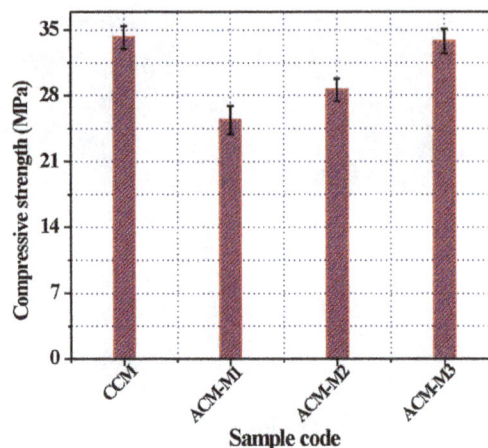

Fig. 11 Compressive strength of the 28 days cured control cement mortar (CCM) and alternative cement based mortar (ACM-M1, ACM-M2, ACM-M3).

fact, due to the lack of driving force to transpire the hydration of the alternative cementitious material in the presence of water (Jo et al. 2014a), the ACM demands high dose of alkali activator. The high dose of alkali activator supplies the driving force for the hydration of the alternative cementitious material, which in turn develops a compact microstructure. Eventually, the strength of the mortar increases. Hence, it is demonstrated that the ACM based mortar possesses a strong ability to gain strength in the presence of alkali activator. Accordingly, it is apparent that the alternative cementitious material could be used as a primary binder for the fabrication of mortar and concrete.

4. Discussion

Viewing in light of the results presented in the preceding sections, it is apparent that the aqueous carbonation of slaked lime can accumulate the CO_2. Additionally, the hydrothermal method successfully produces a cementitious material utilizing the aqueous carbonated lime. In this section, we have tried to establish a reaction scheme for the aqueous the carbonation of the slaked lime and a plausible model for the hydrothermal synthesis of the cementitious material. The Eq. (2) represents a plausible reaction scheme for the carbonation of the slaked lime. Figure 12 depicts the plausible schematic view of the carbonation reaction steps. At the beginning of the carbonation reaction, the slaked lime solution contains the calcium (Ca^{2+}) and hydroxyl ion (OH^-) (Eq. (3)) (Olajire 2013), which in turn increases the pH of the solution (12.4). However, the purging of CO_2 into the reaction medium leads to decrease the pH of the reaction system by dissolving the CO_2 in the aqueous medium (Eq. 4). During this time, the aqueous CO_2 reacts with hydroxyl ions (OH^-) to form bicarbonate ions (Eq. (5)). However, the lifetime of this bicarbonate ion in the alkali medium is very small (Han et al. 2011), which, therefore, reacts with the other hydroxyl ion to form a carbonate ion and water (Eq. (6)). In fact, in this step, the hydroxyl ions

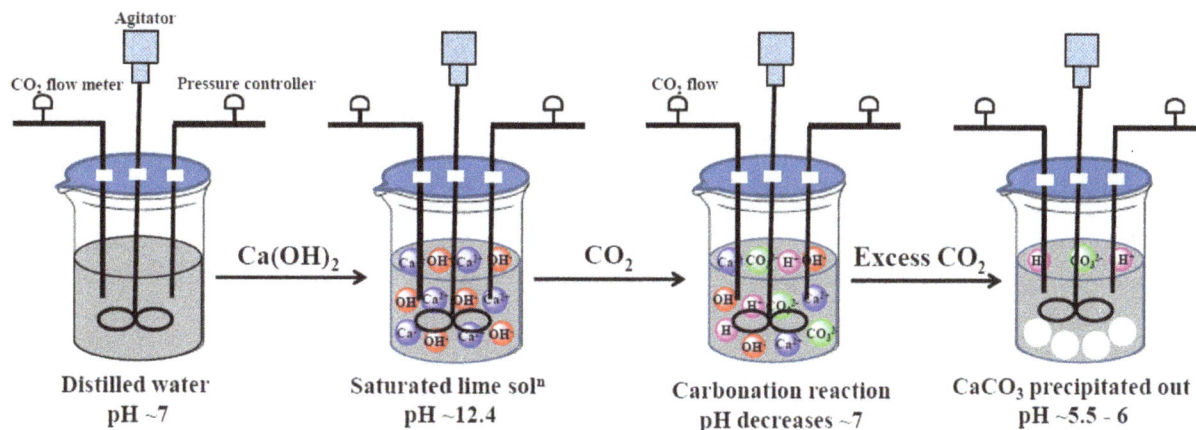

Fig. 12 The plausible schematic view of the carbonation reaction steps.

are consumed to neutralize the bicarbonate ions (Han et al. 2011), which in turn leads to reduce the pH of the medium. Additionally, in this step, due to the production of water, the total volume of water in the reaction system increases, which in turn reduces the OH^- concentration in the per unit volume of water. Eventually, the pH of the reaction medium decreases. Finally, the aqueous medium contains the calcium (Ca^{2+}) and carbonate (CO_3^{2-}) ions. The combination of these two ions forms the calcium carbonate as a white color precipitate (Eq. 7). The excess amount of CO_2 may present in the aqueous medium in the form of carbonic acid, which may lead to decrease the pH of the medium below 6 (Fig. 12).

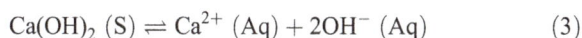

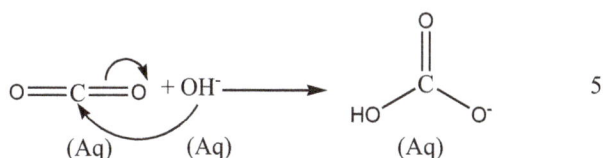

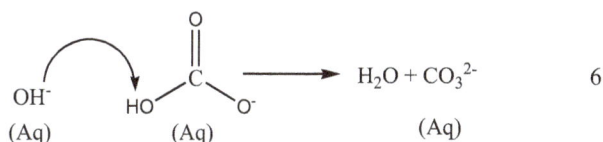

$$Ca(OH)_2 + CO_2 + H_2O \rightarrow CaCO_3 + 2H_2O \tag{2}$$

$$Ca(OH)_2 \ (S) \rightleftharpoons Ca^{2+} \ (Aq) + 2OH^- \ (Aq) \tag{3}$$

$$CO_2 \ (g) \rightleftharpoons CO_2 \ (Aq) \tag{4}$$

The carbonated material produced from the aqueous carbonation was used for the hydrothermal synthesis of the cementitious material. In this study, the alternative cementitious material was produced using aqueous carbonated lime infused with silica and hydrated alumina. Figure 2 represents a plausible model for the hydrothermal synthesis of the cementitious material. Xu and Deventer (2000) stated that a geopolymer compound can be produced from the silica infused with hydrated alumina. Hence, it is assumed that the alternative cementitious material produced in this investigation may contain geomaterial. During the hydrothermal synthesis, the mixture of the silica fume and calcium carbonate may produce a calcium bonded hydrated silica, (Eq. 8). Additionally, the hydrolysis of sodium aluminate may produce a hydrated alumina (Eq. (9)). After aging, the compounds produced by the hydrolysis (Eqs. (8) and (9)) lead to produce the gel. Subsequently, the mixing of these gel compounds may produce the calcium bonded aluminosilicates (Eq. 10). As the alternative cementitious material contains calcium-substituted aluminosilicate geopolymer compound, therefore, the hydration of this cementitious material in the presence of the water is difficult. This could be due to the lack of driving force to initiate the hydration reaction of the alternative cementitious material (Jo et al. 2014a). Nevertheless, the use of alkaline solution (50 % NaOH) may supply the driving force to initiate the hydration reaction and lead to hardening the sample. During the hydration of the ACM in the presence of alkali activator, the mixture of geopolymer compounds may produce an interpenetrating net structure of the hydrated alumino-silicate compounds (Xu and Deventer 2000). Particularly, the hydration of the alternative cementitious material possibly produces the calcium-sodium aluminosilicate hydrate gel (Myers et al. 2013; Kar et al. 2014), which may hold the crystalline component and develop a compact

$$SiO_2 + CaCO_3 + H_2O \longrightarrow (HO)_3Si \!-\!\! O \!-\!\! Si(OH)_3 \; + CaCO_3 \quad +$$

$$\left[-Si(OH)_2 \!-\! O \!-\! Si(OH)_2 - \right]_n \!\!-\! O \!-\! Ca \left[-Si(OH)_2 \!-\! O \!-\! Si(OH)_2 - \right]_n \tag{8}$$

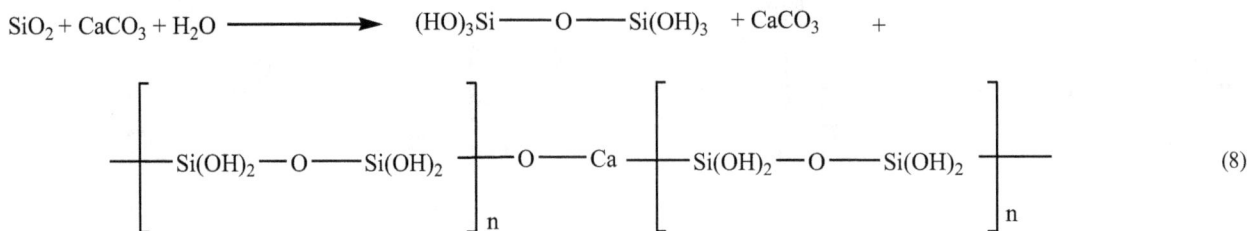

microstructure. This phenomenon may, therefore, contribute to enhancing the strength of the ACM based mortar.

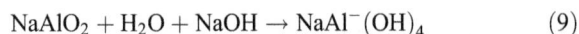

$$NaAlO_2 + H_2O + NaOH \rightarrow NaAl^-(OH)_4 \tag{9}$$

carbonate (0.0116 ± 0.0003 mol). Besides, the alternative cementitious material was synthesized using the carbonated lime produced by the reaction condition B. Based on the particle size analysis, setting time measurement and com-

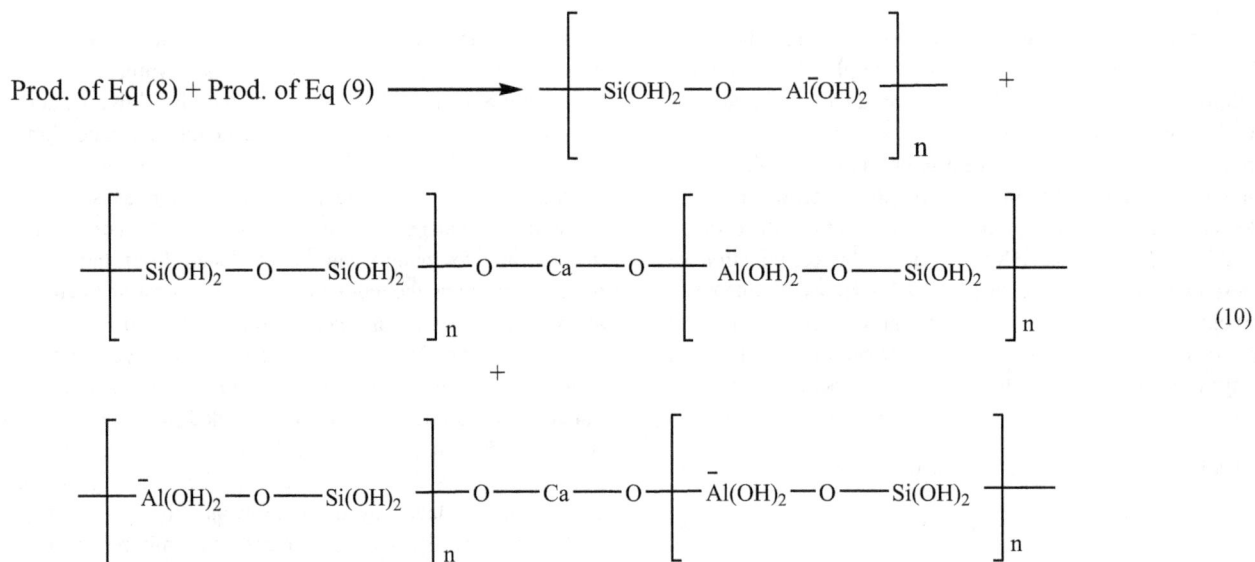

$$\text{Prod. of Eq (8) + Prod. of Eq (9)} \longrightarrow \left[-Si(OH)_2 \!-\! O \!-\! \overline{Al}(OH)_2 - \right]_n \;\; +$$

$$\left[-Si(OH)_2 \!-\! O \!-\! Si(OH)_2 - \right]_n \!\!-\! O \!-\! Ca \!-\! O \left[-\overline{Al}(OH)_2 \!-\! O \!-\! Si(OH)_2 - \right]_n \tag{10}$$

$$+$$

$$\left[-\overline{Al}(OH)_2 \!-\! O \!-\! Si(OH)_2 - \right]_n \!\!-\! O \!-\! Ca \!-\! O \left[-\overline{Al}(OH)_2 \!-\! O \!-\! Si(OH)_2 - \right]_n$$

5. Conclusions

The paper presents a unique technique to produce an using the aqueous carbonated lime, silica fume, and hydrated alumina. Prior to producing the ACM, the aqueous carbonation of the slaked lime was performed in four different conditions to produce the carbonated lime. Analyzing the carbonated product using different analytical tools such as TGA, XRD, and FESEM, it is revealed that the carbonated product is calcium carbonate (calcite). Additionally, it is predicted that the carbonation reaction condition B (agitation speed 200 RPM and CO_2 injection speed 0.5 l/min) consumes a maximum amount carbon dioxide (0.0118 ± 0.0008 mol or 100.1 % with respect to moles of slaked lime) and produces a maximum amount of calcium

pressive strength analysis, it is demonstrated that the ACM can control the physical and mechanical properties of the mortar. The compressive strength of the ACM based mortar (ACM-M3) prepared using the maximum amount of alkali activator is measured to be 33.8 ± 1.3 MPa, whilst the same is measured to be 34.2 ± 1.2 for the control mortar. Interestingly, it is revealed that the compressive strength of the ACM based mortar increases gradually with the increase in alkali activator content. Moreover, a plausible model has been proposed to explain the aqueous carbonation and the synthesis of the ACM. Hence, it is concluded that an alternative cementitious material can be produced by the hydrothermal method utilizing aqueous carbonated lime, which could be used as a primary binder for the fabrication of mortar. Accordingly, the utilization of the aqueous carbonated lime in producing a cementitious material is assumed to be an initial alternative approach to reduce the atmospheric CO_2.

Acknowledgments

Author would like to acknowledge the BK21+, the Government of Korea (Republic of) for their funding to pursue this research program.

References

Amato, I. (2013). Concrete solutions. *Nature News, 494*, 300–301.

Chakraborty, S., Kundu, S. P., Roy, A., Adhikari, B., & Majumder, S. B. (2013). Effect of jute as fiber reinforcement controlling the hydration characteristics of cement matrix. *Industrial and Engineering Chemistry Research, 52*, 1252–1260.

Chen, Z. Y., O'Connor, W. K., & Gerdemann, S. J. (2006). Chemistry of aqueous mineral carbonation for carbon sequestration and explanation of experimental results. *Environmental Progress & Sustainable Energy, 25*(2), 161–166.

Chindaprasirt, P., & Cao, T. (2015). Setting, segregation and bleeding of alkali-activated cement mortar and concrete binders. In F. P. Torgal, J. A. Labrincha, C. Leonelli, A. Palomo, & P. Chindaprasirt (Eds.), *Handbook of alkali-activated cements, mortars and concretes* (pp. 113–131)., Woodhead Publishing series in Civil and Structural Engineering, No.: 54 Cambridge: Woodhead publishing.

Chizmeshya, A. V. G., McKelvy, M. J., Marzke, R., Ito, N., Wolf, G., Bèarat, H., et al. (2007) Investigating geological sequestration reaction processes under in situ process conditions. 32nd International Technical Conference on Coal Utilization & Fuel Systems, 441.

Criado, M., Fernandez-Jimenez, A., & Palomo, A. (2010). Alkali activation of fly ash. Part III: effect of curing conditions on reaction and its graphical description. *Fuel, 89*, 3185–3192.

Crowley, S. F. (2010). Mineralogical and chemical composition of international carbon and oxygen isotope calibration material NBS 19, and reference materials NBS 18, IAEA-CO-1 and IAEA-CO-8. *Geostandards and Geoanalytical Research, 34*(2), 193–206.

Dinakar, P., Sahoo, P. K., & Sriram, G. (2013). Effect of metakaolin content on the properties of high strength concrete. *International Journal of Concrete Structures and Materials, 7*(3), 215–223.

Feely, R. A., Sabine, C. L., Lee, K., Berelson, W., Kleypas, J., Fabry, V. J., et al. (2004). Impact of anthropogenic CO_2 on the $CaCO_3$ system in the oceans. *Science, 305*, 362–366.

Galan, I., Andrade, C., Mora, P., & Sanjuan, M. A. (2010). Sequestration of CO_2 by concrete carbonation. *Environmental Science and Technology, 44*, 3181–3186.

Gerdemann, S. J., O'Connor, W. K., Dahlin, D. C., Penner, L. R., & Rush, H. (2007). Ex-situ aqueous mineral carbonation. *Environmental Science and Technology, 41*, 2587–2593.

Han, S. J., Yoo, M., Kim, D. W., & Wee, J. H. (2011). Carbon dioxide capture using calcium hydroxide aqueous solution as the absorbent. *Energy & Fuels, 25*, 3825–3834.

Hanchen, M., Prigiobbe, V., Baciocchi, R., & Mazzotti, M. (2008). Precipitation in the mg-carbonate system effects of temperature and CO_2 pressure. *Chemical Engineering Science, 63*, 1012–1028.

Huijgen, W. J. J., Comans, R. N. J. (2005). *Carbon dioxide sequestration by mineral carbonation.* Literature review update (2003–2004), ECN-C–05-022. Energy Research Centre of The Netherlands, Petten, Netherlands. Available at: http://www.ecn.nl/docs/library/report/2005/c05022.pdf. Accessed: 23 Dec 2014.

Huijgen, W. J. J., Comans, R. N. J., & Witkamp, G. J. (2007). Cost evaluation of CO_2 sequestration by aqueous mineral carbonation. *Energy Conversion and Management, 48*, 1923–1935.

Huijgen, W. J. J., Witcamp, G. J., & Comans, R. N. J. (2004). Mineral CO_2 sequestration in alkaline solid residues. *Proceedings Materials of 7th International Conference on Greenhouse Gas Control Technologies* (pp. 2415–2418) Vancouver, BC.

Huijgen, W. J. J., Witkamp, G. J., & Comans, R. N. J. (2006). Mechanisms of aqueous wollastonite carbonation as a possible CO_2 sequestration process. *Chemical Engineering Science, 61*, 4242–4251.

Jacobsen, J., Rodrigues, M. S., Telling, M. T. F., Beraldo, A. L., Santos, S. F., Aldridge, L. P., et al. (2013). Nano-scale hydrogen-bond network improves the durability of greener cements. *Scientific Reports, 3*(2667), 1–6. doi: 10.1038/srep02667.

Jeon, D., Jun, Y., Jeong, Y., & Oh, J. E. (2015). Microstructural and strength improvements through the use of Na_2CO_3 in a cementless $Ca(OH)_2$-activated Class F fly ash system. *Cement and Concrete Research, 67*, 215–225.

Jiao, J., Liu, X., Gao, W., Wang, C., Feng, H., Zhao, X., et al. (2009). Two-step synthesis flowerlike calcium carbonate/biopolymer composite materials. *CrystEngComm, 11*, 1886–1891.

Jo, B. W., Chakraborty, S., Jo, J. H., & Lee, Y. S. (2015). Effectiveness of carbonated lime as a raw material in producing a CO_2-stored cementitious material by the hydrothermal method. *Construction and Building Materials, 95*, 556–565.

Jo, B. W., Chakraborty, S., & Kim, K. H. (2014a). Investigation on the effectiveness of chemically synthesized nano cement in controlling the physical and mechanical performances of concrete. *Construction and Building Materials, 70*, 1–8.

Jo, B. W., Chakraborty, S., Kim, K. H., & Lee, Y. S. (2014b). Effectiveness of the top-down nanotechnology in the production of ultrafine cement ($\sim$220 nm). *Journal of Nanomaterials, 57*, 1–9.

Jo, B. W., Chakraborty, S., & Yoon, K. W. (2014c). Synthesis of a cementitious material nanocement using bottom-up nanotechnology concept: An alternative approach to avoid CO_2 Emission during production of cement. *Journal of Nanomaterials, 97*, 1–12.

Juenger, M., Winnefeld, F., Provis, J., & Ideker, J. (2011). Advances in alternative cementitious binders. *Cement and Concrete Research, 41*, 1232–1243.

Kar, A., Ray, I., Halabe, U. B., Unnikrishnan, A., & Dawson-Andoh, B. (2014). Characterizations and quantitative estimation of alkali-activated binder paste from microstructures. *International Journal of Concrete Structures and Materials, 8*(3), 213–228.

Keeling, C. D., Whorf, T. P., Wahlen, M., & VanderPlicht, J. (1995). Inter-annual extremes in the rate of rise of atmospheric carbon dioxide since 1980. *Nature, 75*, 666–670.

Kim, M. S., Jun, Y., Lee, C., & Oh, J. E. (2013). Use of CaO as an activator for producing a price competitive non-cement structural binder using ground granulated blast furnace slag. *Cement and Concrete Research, 54*, 208–214.

Kontoyannis, C. G., & Vagenas, N. V. (2000). Calcium carbonate phase analysis using XRD and FT-Raman spectroscopy. *Analyst., 125*, 251–255.

Kotwal, A. R., Kim, Y. J., Hu, J., & Sriraman, V. (2015). Characterization and early age physical properties of ambient cured geopolymer mortar based on class C fly ash. *International Journal of Concrete Structures and Materials, 9*(1), 35–43.

KSF 2405. (2010). *Testing method for compressive strength of concrete.* Seoul, Korea: Bureau of Korean standard (in Korean).

KSL 5108. (2007). *Testing method for setting time of hydraulic cement by vicat needle.* Seoul, Korea: Bureau of Korean standard (in Korean).

KSL 5201. (2013). *Portland cement.* Seoul, Korea: Bureau of Korean standard (in Korean).

Metz, B., Davidson, O., deConinck, H., Loos, M., & Meyer, L., (Eds.) (2005). *IPCC special report on carbon dioxide capture and storage.* Cambridge University Press, New York, NY: 431. Available at: http://www.ipcc.ch/pdf/special-reports/srccs/srccs_wholereport.pdf. Accessed: 4 Nov 2014.

Myers, R. J., Bernal, S. A., Nicolas, R. S., & Provis, J. L. (2013). Generalized structural description of calcium–sodium aluminosilicate hydrate gels: The cross-linked substituted tobermorite model. *Langmuir, 29*, 5294–5306.

Olajire, A. A. (2013). A review of mineral carbonation technology in sequestration of CO_2. *Journal of Petroleum Science and Engineering, 109*, 364–392.

Phair, J. (2006). Green chemistry for sustainable cement production and use. *Green Chemistry, 8*, 763–780.

Roychand, R., De Silva, S., Law, D., & Setunge, S. (2016). Micro and nano engineered high volume ultrafine fly ash cement composite with and without additives. *International Journal of Concrete Structures and Materials.* doi: 10.1007/s40069-015-0122-7.

Schrabback, J. M. (2010). Concepts for 'green' cement. ICR. Available at: www.sika.com/dms/get//Concepts%20for%20Green%20Cement.pdf. Accessed 11 March 2015.

Siegenthaler, U., & Oeschger, H. (1987). Biospheric CO2 emissions during the past 200 years reconstructed by deconvolution of ice core data. *Tellus, 39B*, 140–154.

Stutzman, P.E. (1996). Guide for X-ray powder diffraction analysis of portland cement clinker. NISTIR 5755. Building and fire research laboratory, National Institute of standards and Technology, U.S Department of Commerce, Gaitheresburg, MD. Available at: http://fire.nist.gov/bfrlpubs/build96/PDF/b96138.pdf. Accessed 4 Nov 2014.

Xu, Hua, & Van Deventer, J. S. J. (2000). The geopolymerisation of alumino-silicate minerals. *International Journal of Mineral Processing, 59*, 247–266.

5

Utilization of Waste Glass Micro-particles in Producing Self-Consolidating Concrete Mixtures

Yasser Sharifi*, Iman Afshoon, Zeinab Firoozjaei, and Amin Momeni

Abstract: The successful completion of the present research would be achieved using ground waste glass (GWG) microparticles in self-consolidating concrete (SCC). Here, the influences of GWG microparticles as cementing material on mechanical and durability response properties of SCC are investigated. The aim of this study is to investigate the hardened mechanical properties, percentage of water absorption, free drying shrinkage, unit weight and Alkali Silica Reaction (ASR) of binary blended concrete with partial replacement of cement by 5, 10, 15, 20, 25 and 30 wt% of GWG microparticles. Besides, slump flow, V-funnel, L-box, J-ring, GTM screen stability, visual stability index (VSI), setting time and air content tests were also performed as workability of fresh concrete indicators. The results show that the workability of fresh concrete was increased by increasing the content of GWG microparticles. The results showed that using GWG microparticles up to maximum replacement of 15 % produces concrete with improved hardened strengths. From the results, when the amount of GWG increased there was a gradual decrease in ASR expansion. Results showed that it is possible to successfully produce SCC with GWG as cementing material in terms of workability, durability and hardened properties.

Keywords: self-consolidating concrete (SCC), ground waste glass (GWG), hardened properties, binary blended concrete, cementing material.

1. Introduction

Use of recycled materials in construction is among the most attractive options because of the large quantity consumptions of the materials, relatively low quality requirements and widespread construction sites. The main applications include a partial replacement for aggregate in asphalt concrete, as fine aggregate in unbond base course, pipe bedding, landfill gas venting systems and gravel backfill for drains (Afshoon and Sharifi 2014; Sharifi et al. 2015; Shi et al. 2005). Due to the ever strict environmental regulations, waste treatment costs and limited availability of disposal sites, the development of new and cost-effective waste management practices has become increasingly significant in recent years (Alp et al. 2008). Waste reuse and recycling are among modern society's environmental priorities, and considerable effort is being devoted to achieve these objectives. Green construction materials play an important role in the sustainable development of the construction industry. Concrete, the most widely used construction material, absorbs natural mineral resources and

should be considered to reduce energy consumption during construction. A sustainable concrete design includes minimizing both the quantity of global carbon dioxide (CO_2) released and the energy consumed to produce concrete and the various components required (Zong et al. 2014). This huge size of production consumes large amounts of energy and is one of the largest contributors to CO_2 release. Accordingly, there is a pressing demand to minimize the quantity of cement used in the concrete industry (Alya et al. 2012).

The Self-consolidating concrete (SCC) was first developed in Japan in the late 1980s (Okamura and Ouchi 2003). Over the last decades, SCC as a new generation of high-performance concrete, has been known as a significant progress in concrete industry and consequently considered as the subject of extensive research studies (Nikbin et al. 2014). SCC's unique property gives it significant constructability, economic and engineering advantages (Uysal 2012; Mahmoud et al. 2013; Lotfy et al. 2015). Moreover, SCC can be pumped to a great distance and increases the speed of construction. Changes to mix design or placing of the material can lead to the modifications of the porous structure and consequently permeability of the material. The fines content of SCC is higher than in normally-vibrated concrete (NVC) and the absence of compacting lowers the risks inherent in the process, either from excessive vibration or from insufficient vibration (Valcuendea et al. 2012). In order to avoid separation of large particles in SCC, viscosity increasing additives or fillers are utilized. An additive to increase the

Department of Civil Engineering, Vali-e-Asr University of Rafsanjan, Rafsanjan, Iran.
*Corresponding Author; E-mail:
yasser_sharifi@yahoo.com; y.sharifi@vru.ac.ir

viscosity is often used when concrete is cast under water and for SCC in tunnels. Mineral admixtures like fly ash, glass filler, limestone powder, silica fume or quartzite fillers may be used in the mixture to increase the viscosity of SCC (Bingol and Tohumcu 2013). The use of mineral additives in concrete or cement is one of the main trends in the development of concrete technology; at the same time it is an important element of a sustainable development strategy. It enables the properties of the concrete to be improved, especially in the aspect of resistance to the aggressive influence of the environment, as well as to obtain significant economic benefits (Ponikiewski and Golaszewski 2014). Apart from the significant effect on hardened concrete properties, the incorporation of mineral additives in concrete is also known to have a considerable effect on its fresh properties. The use of such powders provides greater cohesiveness by improving the grain-size distribution and particle packing. Moreover, their high pozzolanic activity leads to a further particle packing enhancement that is achieved by the pozzolanic products and acts complementary to the physical action (Sfikas et al. 2014).

Glass in general is a highly transparent material formed by melting a mixture of materials such as silica, soda ash, and $CaCO_3$ at high temperatures followed by cooling during which solidification occurs without crystallization. Glass is widely used in our lives through manufactured products such as sheet glass, bottles, glassware, and vacuum tubing (Park et al. 2004). The concept of using waste glass in concrete is not new; early efforts were conducted in the 1960s to use crushed waste glass as a replacement for aggregate. However, these attempts were not satisfactory due to the strong reaction between the alkali in cement and the reactive silica in glass, namely ASR (Alya et al. 2012). In this phenomenon the amorphous silica in glass is susceptible to attack by the alkaline environment and would depolymerize to form a monomer $Si(OH)_4$, which could further react with alkalis such as Na^+, K^+ and Ca^{2+} to form the ASR gel. This ASR gel can absorb water and swell inside the microstructure of concrete, resulting in internal stress. Once the internal stress exceeds the strength of concrete, severe cracking and damage may occur (Du and Tan 2013). When waste glass is collected, different color glass is often intermixed. Mixed color glass cannot be recycled, however, because a mixing of coloring agents results in an unpredictable and uncontrollable color in the new glass. Machines are capable of using optical sensors to efficiently sort large glass pieces by color; however, sorting small glass pieces is not economical and much of this unrecyclable glass cullet is then landfilled. As the economic and environmental consequences of landfilling rise, the incentive to reuse glass cullet has grown. The concrete industry is one of the potential ways of reusing millions tons of glass cullet per year either as aggregate or supplementary cementitious material (SCM) (Mirzahosseini and Riding 2014). Although work on the use of finely ground glass as a pozzolanic material also started as early as 1960s, most of the work in this area is relatively recent, and has been encouraged as a result of continual accumulation of waste glass and its consequent environmental issues (Shayan

and Xu 2006). On the other hand, limited work (one study by this time) has been carried out on the application of ground glass as SCM in SCC (Liu 2011).

A study on the properties of fresh self-consolidating glass concrete (SCGC) investigated by Wang and Huang (2010). They reported that, the slump flow of self-compacting glass concrete (SCGC) increased with higher glass sand replacement. V-funnel and U-test indicated that, when the glass sand replacement increases, the time required to flow and pass through the space between the steel bars increases, mainly because the unit weight is reduced. The air content and unit weight would be raised with glass sand contents decreasing. Recycled glass replacement as fine aggregate in SCC also investigated by Sharifi et al. (2013). Fresh results indicate that the flow-ability characteristics have been increased as the waste glass incorporated to paste volume. Nevertheless, compressive, flexural and splitting tensile strengths of concrete containing waste glass have been shown to decrease when the content of waste glass is increased.

Shayan and Xu (2006) investigated the performance of glass powder (GLP) as a pozzolanic material in concrete. They reported that mixtures containing GLP also performed satisfactorily with respect to drying shrinkage and alkali reactivity, and there were indications that GLP reduces the chloride ion penetrability of the concrete, thereby reducing the risk of chloride induced corrosion of the steel reinforcement in concrete. The results demonstrated that GLP can be incorporated into 40 MPa concrete at dosage rates of 20–30 % to replace cement without harmful effects. The use of GLP provides for considerable value-added utilization of waste glass in concrete and significant reductions in the production of greenhouse gases by the cement industry.

Jain and Neithalath (2010) studied the chloride transport in fly ash and glass powder modified concretes. Rapid chloride permeability (RCP), non-steady-state migration (NSSM) and steady state conduction (SSC) tests are performed on plain and modified concretes. Results shown the glass powder modified concretes demonstrate similar or lower RCP values as compared to the fly ash modified concretes of the same cement replacement level whereas the steady state conductivities are lower for the fly ash modified mixtures. They reported that the NSSM coefficients are lower for the fly ash modified concretes even when the initial conductivities are similar to those of plain or glass powder modified concretes.

Liu (2011) studied incorporating ground glass in SCC. They conclude that, to keep the filling ability constant, the inclusion of ground glass would require an increase in water/powder ratio and a reduction in superplasticizer dosage. These did not change the passing ability, but degraded the consistence retention and hardened properties such as strength but not to a prohibitive extent.

Carpenter and Cramer (1668) also reported that powdered glass was effective in reducing ASR expansion in accelerated mortar bar tests, similar to the effects of fly ash, silica fume and slag.

Waste glass as a supplementary cementitious material in concrete was investigated by Federico and Chidiac (2009).

They reported that the similarity between the ASR and pozzolanic reactions observed for waste glass in concrete suggests that they are closely related and may be simply various stages of one another subject to several controlling factors, including particle size, pore solution, and chemical composition. So far, just the research which has been conducted by Liu (2011) investigated the behavior of SCC containing GWG microparticle substitute as cementing material, and the lack of a comprehensive study on the engineering response of SCC incorporating GWG which partially replaced with cement was evident.

2. Materials

2.1 Cement

Ordinary Portland Cement (OPC) meeting the requirements of ASTM C150 (2009) was used in preparation of concrete mixtures. The chemical and physical properties of cement are given in Table 1. The micro-particle size distribution pattern of the used OPC is also illustrated in Fig. 1.

2.2 Waste Glass

Waste glass (Fig. 2) provided from Rafsanjan-Iran, then in the laboratory is ground by using an electric mill (Fig. 3), for 10 min to a maximum particle size of 100 micron meters (μm) before it was used as a cement replacement material. It was added to the concrete mixtures as a secondary binder replacing up to 30 % by weight of cement. Incorporation of high-volume mineral admixtures reduces heat of hydration as cement content in concrete was reduced and thus rate of hydration reduced. The total heat of hydration produced by the pozzolanic reactions with mineral admixtures is considered as half of the average heat of hydration produced by Portland cement. Due to the reduced heat of hydration, it leads to improvement of rheology properties and reduces

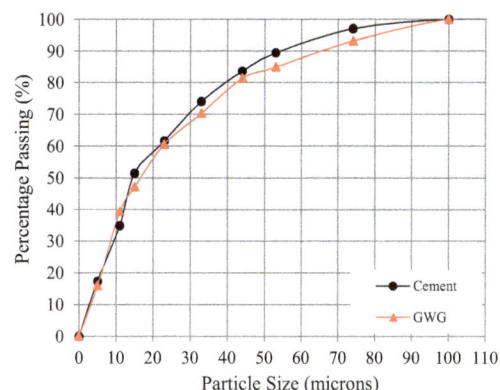

Fig. 1 Particle size distribution for binders.

Fig. 2 Waste glass used in this study: before grounding (a), after grounding (b).

thermally-induced cracking of concrete as well as long term properties of concrete (Nuruddin et al. 2014). The chemical and physical properties of the waste glass is given also in Table 1.

2.3 Aggregates

Locally available sand from natural sources was used in the present experimental investigation. The aggregates used in this experiment are both angular and circler angle together

Table 1 Chemical analysis and physical properties of cement and GWG—compressive strength of cement.

Chemical analysis			Physical properties			
Compound (%)	Cement	Glass		Cement		Glass
SiO_2	21.74	70.5	Water absorption (%)	–		0.17
Al_2O_3	5	2.6	Specific density	3.15		2.50
Fe_2O_3	4	–	Specific surface area (cm^2/g)	2900		2480
CaO	63.04	5.7	Setting time (final) (min)	170		–
MgO	2	2.9	Setting time (initial) (min)	120		–
Na_2O	2.3	16.3	Autoclave expansion (%)	0.1		–
K_2O	1	1.2	Compressive strength (kg/cm^2)	220	3 days	–
SO_3	2.9	0.2		275	7 days	–
LOI	1.3	–		380	28 days	

Fig. 3 Electric mill used in this study.

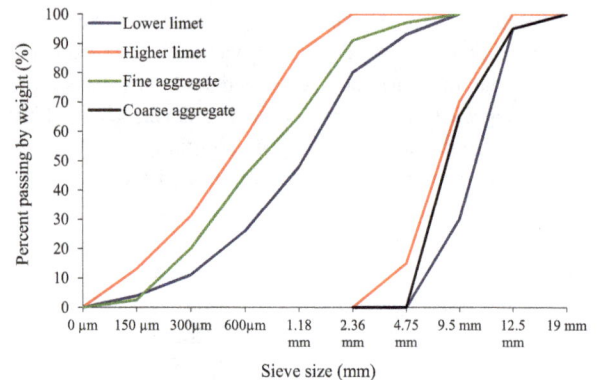

Fig. 4 Particle size distribution of fine and coarse aggregates.

that are from river materials. The fine aggregate was the natural sand free from impurities with minimum and maximum size of 0.3 and 4.75 mm. 19.5 mm nominal maximum size dolomite is used as coarse aggregate and minimum size of coarse aggregate is 4.75 mm. Dolomite powder was used replacing of sand that is smaller than 300x. Table 2 shows the physical properties of coarse and fine aggregates. The aggregate was kept in a condition greater than saturated surface dry (SSD) level. Aggregate particle size distribution was determined in accordance with ASTM C33 (2008) and is presented in Fig. 4. The particle size distribution indicated that it was continuously distributed with 35 % over the size range of 9.5–19.5 mm. The particle size distribution was well-graded with over 47 % of the sand over the size range of 0.3–1.18 mm.

2.4 Water

Regular tap water was used as mixing water, according to the ASTM C94 specifications (2009).

2.5 Super-Plasticizer

In order to improve the workability of high performance concrete, polycarboxylic ether based high range water reducer (HRWR) namely P10-3R was used. Table 3 shows the properties of HRWR according to the ASTM C494 (2010).

3. Mix Proportions

A total of seven SCC mixes were made and their detailed mix proportions are presented in Table 4. These included one control mix (Mix-0) and six mixes (Mix-5 to Mix-30) made by replacing cement with 5, 10, 15, 20, 25 and 30 % of GWG microparticles partially. For all SCCs, the amount of the cementations used was generally maintained at

400 kg/m^3. Coarse and fine aggregates contents were maintained at 700 and 950 kg/m^3, respectively. The W/B ratio was 0.51 for all mixes. Super-plasticizer was also used in the mixtures at the ratio of 1.4–1.1 % of binder materials by weight for providing the desired fluidity of the SCC. In the production of SCCs, the mixing sequence and duration are very important. The batching sequence consisted of homogenizing the fine and coarse aggregates for 30 s in a rotary planetary mixer, then about half of the mixing water introduced. It continued for 1 min. thereafter, cement were added and the mixing was resumed for another minute. Finally, the super-plasticizer with remaining water was introduced, and the concrete was mixed for 3 min and then left for 2 min rest. Eventually, the concrete was mixed for additional 2 min to complete the mixing sequence. Concrete mixes were designed in a way to give a slump flow of 680 ± 30 mm which was achieved by using the super-plasticizer at varying amounts. For this, trial batches were produced for each mixture till the desired slump flow was obtained.

4. Casting, Curing and Testing

4.1 Fresh Concrete Tests

According to European Federation of Producers and Contractors of Specialist Products for Structures (EFNARC) guide for making self-compacting concrete (2005), a concrete mixture can only be classified as SCC if the requirements for filling, passing, and segregation resistivity characteristics are fulfilled. Various tests have been used in present experimental study to investigate the properties of

Table 2 Physical properties of aggregates.

Properties	Fine aggregate	Coarse aggregate
Specific gravity	2.73	2.81
Bulk density (kg/m^3)	1590	1565
Void content (%)	36.71	47.58
Water absorption (%)	1.09	0.32

Table 3 Properties of the super-plasticizer.

Items	Standards quality	Testing results	Regulatory
Density (20 °C) (g/cm^3)	0.938–1.146	1.1 ± 0.02	ASTM-C494
PH (20 °C)	5.4–7.4	7	–
Chlorine (ppm)	≤ 2400	500	ASTM-C494
Color	–	Dark green	–

Table 4 Concrete design mix proportions.

Detail mix	Mix	Aggregates (kg/m^3)		Cementitious material (B) (kg/m^3)		W/B	SP[a] (%)
		Coarse	Fine	GWG	Cement		
0 % GWG + 100 %C	Mix-0	768	1060	0	400	0.51	1.4
5 % GWG + 95 %C	Mix-5	766	1058	20	380	0.51	1.4
10 % GWG + 90 %C	Mix-10	763	1055	40	360	0.51	1.4
15 % GWG + 85 %C	Mix-15	762	1054	60	340	0.51	1.4
20 % GWG + 80 %C	Mix-20	760	1050	80	320	0.51	1.4
25 % GWG + 75 %C	Mix-25	758	1047	100	300	0.51	1.2
30 % GWG + 70 %C	Mix-30	756	1046	120	280	0.51	1.1

[a] Percent by mass of binder (B).

fresh concrete mixes compositions. So far no single method or combination of methods has achieved universal approval and most of them have their adherents. Hence, each mix design should be tested by more than one test method for different workability parameters. Fresh concrete properties of SCCs were determined by using the slump (Diameter, T_{500} and T_{final}), V-funnel (T_0), J-ring (step height, diameter, T_{500} and T_{final}), GTM screen stability and L-box (h_1/h_2, T_{200}, T_{400} and T_{final}) tests according to EFNARC (2005). Besides, the setting time and air content tests were measured for SCCs according to ASTM C403 (2008) and ASTM C231 (2009), respectively.

4.2 Hardened Concrete Tests
4.2.1 Compressive Strength of Concretes
Compressive strength test usually gives an overall picture of the quality of concrete, because strength is directly related to the structure of the hydrated cement paste. The compression test is an important test to determine the strength development of the concrete specimens (Hameed et al. 2012). It was defined as the maximum load sustained divided by the cross-sectional area of the sample. We used three 100 mm cube specimens for determined the compressive strength SCCs. This test were performed on 3, 7, 14, 28, 42, 56, 70 and 91 day old specimens.

4.2.2 Tensile and Flexural Strength of Concretes
For the evaluation of the tensile strength of concrete, there are three well-known methods: (i) the direct tensile strength, (ii) the splitting tensile strength and (iii) the flexural tensile strength (3- or 4-point loading). Due to the high degree of difficulty during execution, direct tensile strength test is rather scarce. It is possible to convert these test results from one to another. However, it is not quite clear whether these conversion factors can still be used for SCC (Craeye et al. 2014).

Splitting tensile strength was performed on three cylindrical molds Ø100 mm × 200 mm, at 3, 7, 14, 28, 42 and 56 day old specimens. After the specified curing period was over, the concrete cylinders were subjected to splitting tensile strength test by measured using a compressive machine with a loading capacity of 300 ton. The tests were carried out triplicately and average splitting tensile strength values were obtained.

The splitting tensile strength was determined using the following equation:

$$F_t = 2P/\pi Ld$$

where F_t is splitting tensile strength, P is the maximum applied load indicated by testing machine, and L and d are the length and diameter of specimen, respectively.

The flexural strength of concrete is conducted on prism of size $100 \times 100 \times 500$ mm. Six concrete prisms were casted for each concrete mix proportions for 28- and 56-days. Four point loading has been used to predict the flexural strength. The average modulus of rupture (flexural strength) was determined using the following expression:

$$F_{cr} = PL/bd^2$$

where F_{cr} is the modulus of rupture; P is the maximum applied load indicated by testing machine; and L, b and d are the average length, width and depth of specimen, respectively.

It is generally agreed that the splitting tensile strength test gives a better evaluation of the concrete's response to tensile stresses than that obtained from the modulus of rupture test (Druta et al. 2014).

4.2.3 Unit Weight and Water Absorption of Concretes

The unit weight (density) of hardened concrete was measured at the age of 28 days. This property predominately depends on aggregate density. Therefore, the replacement of cement may not change the density of the concrete remarkably (Beltran et al. 2014). Water absorption test is used to determine the amount of absorbed water under specified conditions which indicates the degree of porosity of a material (Siddique 2013). The water absorption test was conducted by completely immersing dried specimens (the mentioned specimens dried in the oven for 72 ± 2 h) in water and the amount of absorbed water percentage per mass after a specified time records. Here, it was conducted every day until the day of 10 after initial curing.

4.2.4 Free Drying Shrinkage of Concretes

Drying shrinkage can be defined as the volumetric change due to the drying of concrete. This change in volume of the concrete is not equal to volume of the water lost. The loss of free water occurs first; this causes little to no shrinkage. As the drying of the concrete continues, the adsorbed water held by hydrostatic tension in the small capillaries (<50 nm) is removed. The shrinkage due to this water loss is significantly larger than that associated with the loss of free water. The loss of water produces tensile stresses, which lead the concrete to shrink (Guneyisi et al. 2010). Free drying shrinkage of the SCCs specimens after drying was assessed at 1, 3, 7, 14, 28, 42, 56, 70 and 91 days. For this test, we used two beams of dimensions (50 mm $\times$ 50 mm $\times$ 285 mm). In this test, the specimens were removed from the molds at the age of 24 h. At the end of the 28 day curing period, the specimens were stored in air until the time of testing. After then, the strain was measured at the special times.

4.2.5 Alkali–Silica Reaction Test

The expansion due to the ASR was determined on three prisms with dimensions of $25 \times 25 \times 285$ mm. A zero reading was taken after storing the prisms in distilled water at 80 °C for 24 h. The mortar bars were then transferred and immersed in 1 N NaOH solution at 80 °C until the testing time (Fig. 4). The expansion of the mortar bars was measured within 15 ± 5 s after they were removed from the 80 °C water or alkali storage condition by using a length comparator. The measurements were conducted at the 1, 3, 7, 10, 14, 28 and 42 days. According to ASTM C 1260 the expansion of concrete 16 days after casting is classified as non-detrimental if it is below 0.10 %, as potentially detrimental if it is between 0.10 % and 0.20 % and as detrimental if it is over 0.20 %.

5. Testing Procedure

After curing, the following tests were carried out on the concrete specimens:

- Compressive strength test was conducted on concrete samples with BS 1881: Part 116 (1983), using a loading rate of 2.5 kN/s;
- Cylinder tensile (splitting) strength test was done in accordance with ASTM C496 (2004), using a loading rate of 2.1 kN/s;
- Flexural strength test was conducted in accordance with BS EN 1351:1997 (1997), using a simple beam with four point loading at a loading rate of 0.2 kN/s;
- Unit weight test was conducted in accordance with BS EN-12390-7 (2009);
- The water absorption test was conducted in accordance with BS 1881: Part 122 (1983);
- ASTM C157 (2008) was used for free drying shrinkage;
- ASTM C1260 (2007) was used for the ASR test.

6. Experimental Results and Discussion

6.1 Fresh Concrete Results

The consistency and workability of SCC were evaluated using the slump flow, J-Ring, L-box, V-funnel, GTM screen stability, air content and setting time tests. The typical workability acceptance criteria for SCCs based on EFNARC (2005) are listed in Table 5. Properties of fresh SCCs are summarized in Table 6. As shown in Table 6, utilization of GWG increased workability of the fresh concrete. Similarly, slump flow time was always lower than 5 s for all of the concretes meeting the upper limit of EFNARC (2005). The results presented in Table 6 also indicated that, irrespective of W/B ratio and SP dosage, the V-funnel time shows a distinct tendency to decrease with increasing GWG content. For instance, Mix-0 had a V-funnel time of 8.15 s which decreased to 5.87 s as GWG introduced up to 30 % by mass. From Table 6, the step height of the J-ring test, changed from 10 mm (Mix-0) to 12.5 mm (Mix-20) for SCC with SP = 1.4 %, but this parameter for Mix-25 and Mix-30 mixes measured 14 mm and 16 mm, respectively. It was observed that the J ring flow (slump flow with J ring) increased with increase in GWG content but this values were lower than the control mix (Mix-0). The J-Ring diameter was in the range of 645–664 mm and the difference in height

Table 5 Acceptance criteria for SCC according to EFNARC.

Slump flow test		V-funnel test		L-box test	
Slump flow classes	Slump flow (mm)	Viscosity classes	V-funnel times (s)	Passing ability classes	Blocking ratio
SF1	550–650	VF1	≤ 8	PA1	≥ 0.8 with 2 bars
SF2	660–750	VF2	9–25	PA2	≥ 0.8 with 3 bars

Table 6 Properties of fresh SCC.

	Mix-0	Mix-5	Mix-10	Mix-15	Mix-20	Mix-25	Mix-30
Slump flow							
Diameter (mm)	670	678	690	695	703	676	681
T_{500} (s)	4.96	4.12	3.64	2.79	1.68	2.49	2.02
T-Final (s)	19.31	20.68	21.3	23.35	22.23	22.01	23.9
J-ring							
Step height (mm)	10	10.5	15	13.7	12.5	14	16
Diameter (mm)	655	657	650	660.5	664	655.5	645
T_{500} (s)	5.1	5.17	5.61	4.69	4.38	4.65	5.11
T-final (s)	27.16	26.44	25.33	25.44	24.43	26.36	27.47
L-box							
h_1/h_2	0.84	0.93	0.92	0.96	0.99	0.95	0.97
T_{200} (s)	0.97	0.68	0.78	0.6	0.56	0.61	0.58
T_{400} (s)	1.73	1.42	1.51	1.11	0.94	1.06	0.96
T-final (s)	13.67	12.45	13.55	12.24	10.36	11.54	10.67
V-funnel							
T_0 (S)	8.15	7.78	7.11	6.65	6.18	6.58	5.87
GTM screen stability (%)	8.7	8.2	6.9	7.6	8.1	7.1	5.6
Air content (%)	3.8	3.7	3.1	2	1.2	2.5	3.3
Setting time							
Initial set (min)	416	430	450	457	480	410	395
Final set (min)	540	565	630	670	695	543	600

was less than 20 mm. The blocking ratio (h_2/h_1) should be between 0.8 and 1.00. All mixtures of SCC are within this target range. The results presented in Table 6 also indicated that, in constant W/B ratio, the blocking ratio (h_1/h_2) shows a distinct tendency to increase with increasing GWG content. As presented in Table 6, with increase of GWG % the segregation index (SI) increased up to 10 %. This reason may be due to increase in free water content and decrease in cohesion. The SI for the mixes was lower than 18 % which is as per EFNARC (2005) standards and they were ranked to the SR2 class. In this study air content of all mixes between 1.2 and 3.8 % (Table 6). Increasing GWG content caused air content decreased. This reason may be due to increase in free water content and filling ability of concrete mixes with

GWG. Table 6 shows that increasing the GWG content led to considerable increase of the initial and final sets. This can be attributed to increasing of free water in SCCs and smooth surface texture and low moisture absorption glass. But decrease in SP dosage could lead to the decrease of initial and final setting times of concrete for Mix-25 and Mix-30.

Figure 5's various tests have been used in present experimental study to investigate the fresh properties for mixes compositions.

6.2 Hardened Concrete Results
6.2.1 Compressive Strength
The compressive strength results of SCCs with GWG microparticles at curing periods up to 91 days are presented

Fig. 5 Fresh properties tests (sump flow, J-ring, L-box, GTM and VSI test).

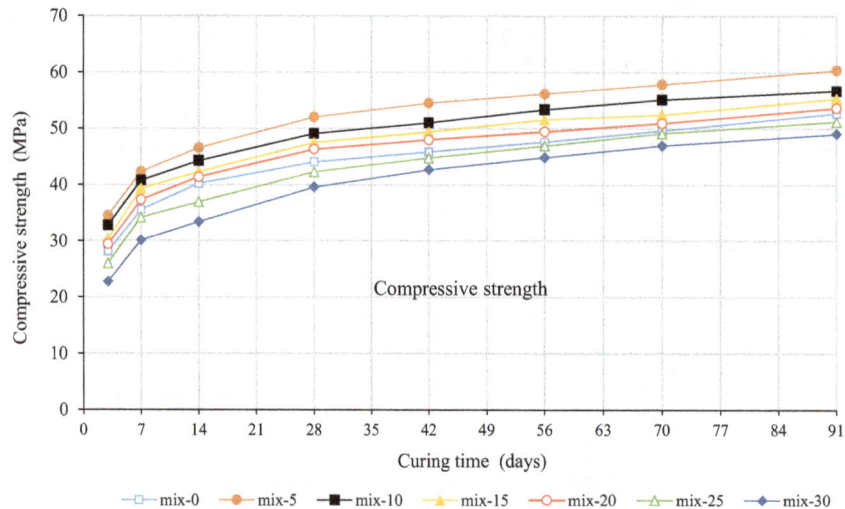

Fig. 6 Compressive strength of SCC with GWG microparticles as ages.

in Fig. 6. The rate of gain in compressive strength was rapid up to 14 days and then slowed down afterward. The 28-day compressive strength varied from 39.56 to 51.96 MPa while the 91-day compressive strength differed from 49.03 to 60.41 MPa for different values of GWG microparticles. As can be seen from Fig. 7, GWG microparticles addition from 0 to 5 % causes the compressive strength to increase 15.7, 18 and 14.67 %, for 14, 28 and 91 days, respectively. It is observed that the maximum compressive strength was achieved for specimen containing 5 % GWG microparticles. Comparison of the results from the 3-, 7-, 14-, 28-, 42-, 56-, 70- and 91-day samples shows that the compressive strength increases with GWG up to 5 % replacement and then it

decreases, although the results of up to 20 % replacement are still higher than those of the plain cementitious composite. This may be attributed to achievement of suitable workability of concrete which in turn causes more in compaction levels and improves compressive strength. This issue is highlighted in SCC mixtures when no compaction method is applied for molding and the compaction only performed by their own concrete weights. Second factor that may be caused an increase in compressive strength is the filler effect of GWG microparticles grains. It has been observed that Mix-25 and Mix-30 which containing 25 and 30 % GWG achieved a slightly decrease in compressive strength about 2.8 and 6.9 % compare to Mix-0, respectively. This may be due to

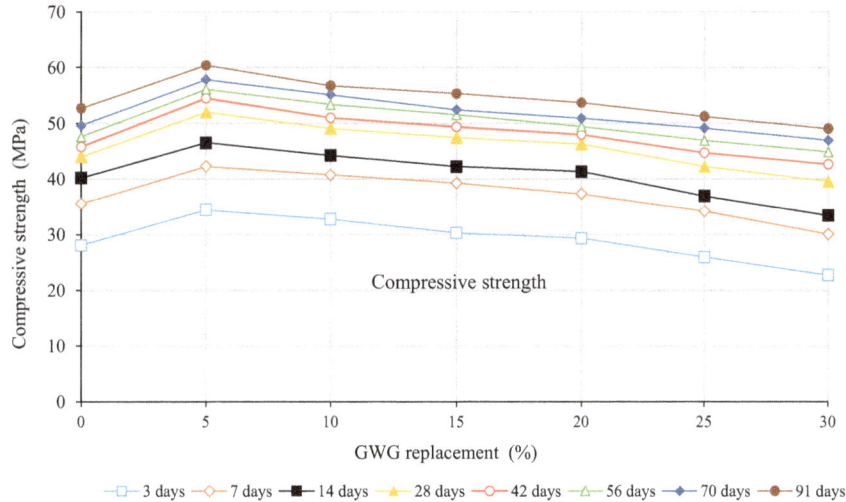

Fig. 7 Compressive strength versus GWG replacement as cementing material.

the fact that the quantity of GWG present in the mix is higher than the amount required to combine with the liberated lime during the process of hydration, thus leading to excess silica leaching out and causing a deficiency in strength as it replaces part of the cementitious material. Therefore, the optimum amount of GWG microparticle may be 20 % as cement replacement, and the maximum strength achieved by 5 % GWG replacement. Dyer and Dhir (2001) expressed that the maximum rate of hydration heat evolution drops as the Portland cement content is reduced and glass included. They noted this can be attributed to that any pozzolanic reaction the glass undergoes will occur at later stages and emits only minor quantities of heat. They also believed that the presence of green glass has little effect on the kinetics of Portland cement hydration. However, the production of portlandite is enhanced relative to the control, and C–S–H gel levels were found to be higher in the pastes containing glass.

Liu (2011) investigated the effect of waste glass inclusion as both cement and sand substitute in SCC. He found that inclusion of waste glass as cement and sand substitute in SCCs decreases the compressive strength. However he expressed that in terms of strength, glass could be a suitable candidate for addition in SCC. Federico and Chidiac (2009) considered the effect of waste glass as cement replacement in conventional concrete and reported that when cement replaced with waste glass between 10 and 20 % the highest strength achieved. Our study corroborates their consequence in which, maximum strength happened in 5 % GWG substitute as cementitious material, although inclusion of 20 % of GWG gives similar compressive strength as the plain cementitious composite.

It should be noted that the particle sizes of GWG has a remarkable influence on the properties of mortar and concrete. Shao et al. (1999) investigated the response mortars containing glass which has been replace with cement partially. They found that the particle size has a clear influence on mortar behavior. Finer glass particles led to an increase in the reactivity of glass with lime, and hence improved

compressive strength and decreased shrinkage. Chen et al. (2006) considered the performance of concretes with various waste E-glass particle amounts. The size distribution of glass particle was from 38 to 300μ and about 40 % of E-glass particle was less than 150μ. Based on the response of hardened concrete, optimum E-glass content was found to be 40–50 % by mass. Oliveira et al. (2008) analyzed the effect of crushed glass of various sizes on mortars. The mixtures were prepared with replacing from 0 to 40 % of the cement by glass. They concluded that mortars prepared with 45–75μ glass particles improve in terms of compressive strength, have a denser cementitious matrix and are less liable to expansive reactions such as ASR. As it was mentioned previously, in the present study the particle distribution of GWG is similar the cement particle distribution (Fig. 1), and the obtained results clarified the real behavior exactly.

6.2.2 Splitting Tensile Strength

Figure 8 shows the splitting tensile strength results which were also found to depend on the GWG microparticles contents. The 28-day splitting tensile strengths of Mix-5, Mix-10 and Mix-15 specimens showed an increase of 4.16, 6.62 and 1.57 % more than 28-day splitting tensile strength of Mix-0, respectively. Generally, the splitting tensile strengths of the concretes containing GWG microparticles increased with an increase in GWG content up to 15 %. As stated previously, this may be attributed to the chemical reaction between glass and calcium hydroxide that was up to the amount of products of hydration, basically calcium silicate hydrate. As expected, the tensile strength trend performance is similar the compressive strength approximately. When GWG content increased more, splitting tensile strengths decreased lower than those of the plain cement concrete (Mix-0). Similarly, this can be attributed to the fact that the quantity of GWG microparticles (pozzolan) exist in the mixture is higher than the required content to combine with the liberated lime during the process of hydration. Therefore, this leading to a deficiency in strength as it replaces part of the cementitious material but does not contribute to strength (Fig. 9).

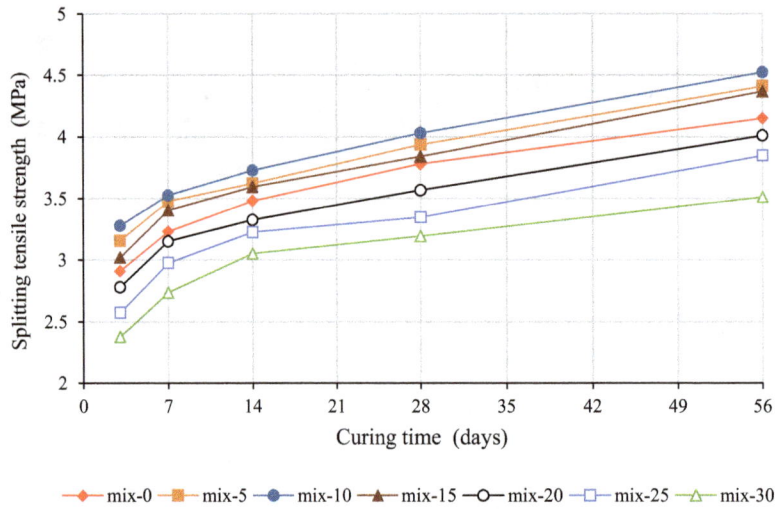

Fig. 8 Split tensile strength of SCC with GWG microparticles as ages.

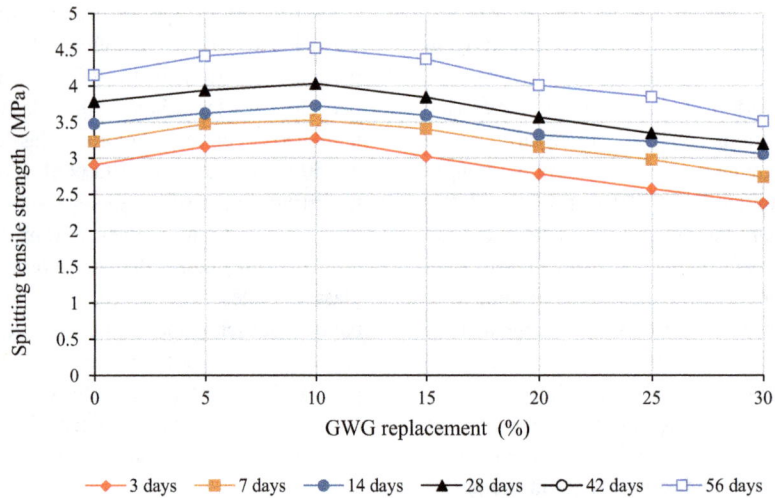

Fig. 9 Split tensile strength versus GWG replacement as cementing material.

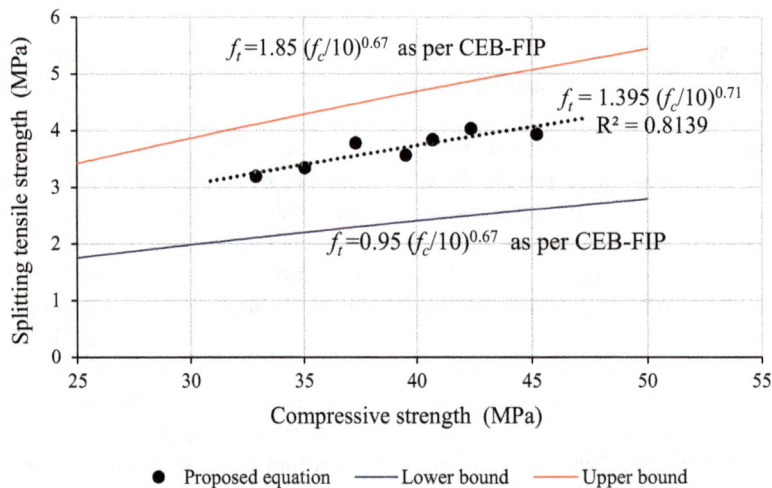

Fig. 10 Variation of split tensile strength versus compressive strength.

The average tensile strength was within the permissible values in accordance with the design specifications based CEB-FIP (1990) code provisions for conventional concrete. For design purposes, the tensile strength can be empirically taken as $0.45\sqrt{F_{cu}}$, where F_{cu} is the compressive strength (Jackson and Dhir 1996). The splitting tensile strength of the SCC mixtures versus corresponding 28-day cylindrical compressive strength was presented in Fig. 10. It should be noted that the 100-mm cube compressive strength was converted to cylindrical strength by applying suitable conversation factor (Domone 1997). Accordingly, it can be seen that splitting tensile strength values of SCC with GWG lies in the range of bound value suggested by CEB-FIP code for normal concrete (1990). However, the mean relationship proposed by CEB-FIP model generally provided higher tensile strength value. This finding may be confirmed by the results of Parra et al. (2011), reported that at more advanced ages (28 and 90 days), there is a clearly higher tensile strength in normal vibration concrete than in SCC for the same compressive strength.

6.2.3 Flexural Strength

The flexural strengths determined at 28 and 56 days as function of time in day are shown in Fig. 11. From Fig. 11 it can be seen that flexural strength increases by less than 20 for 10 % GWG replacement, but at 30 % the reduction was as much as 14 % in the case of GWG replacement (at 28 days). When compared to that of the control mixture increasing amounts of GWG (5–15 %), generally increase the flexural strength. This implies that the material skeleton influences the flexural strength. Higher replacements of GWG also have resulted in decrease in strength. The decrease in the flexural strength of SCCs was about 8.36, 11.9 and 19.33 %, respectively, at 20, 25 and 30 %, GWG replacement in comparison with the control mixture (56 days) (Fig. 12). The experimental modulus of rupture compare with the empirical expression suggested in the design specifications that is $F_{cr} = 0.75\sqrt{F_{cu}}$ where F_{cu} is the cube compressive strength (Jackson and Dhir 1996). Wang (2011) stated that the compressive strength, flexural strength and the tensile strength of the cement mortar with waste LCD glass powder decrease with increasing substitution amounts, while the property of the waste LCD glass powder with 10 % substitution amount was similar to that of the control group, with respect to compressive strength and flexural strength. But here the 20 % substitution amount gives similar strength as the control mix in term of hardened strength approximately (Mix-0).

6.2.4 Unit weight

The unit weight for different mixes with GWG percentages varying from 0 to 30 % is shown in Fig. 13. Unit weight for the hardened specimens ranged between 2359 and 2395 kg/m^3. Mix-15 with 2395 kg/m^3 showed the highest unit weight value among the mixtures. As GWG amount increased up to 10 %, unit weight of concrete increased accordingly. Mix-20, Mix-25 and Mix 30 were 0.37, 0.62 and 1.14 % lower than that of the control concrete (mix-0), respectively. Utilizing pozzolanic materials (fly ash, slag, glass and etc.) in suitable quantities improved the workability of concrete and lead

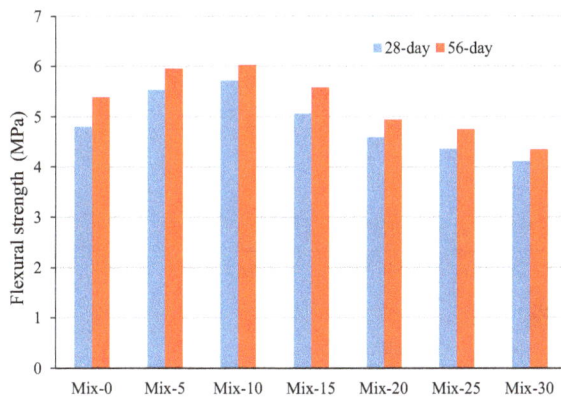

Fig. 11 Flexural strength of SCC with GWG microparticles in different ages.

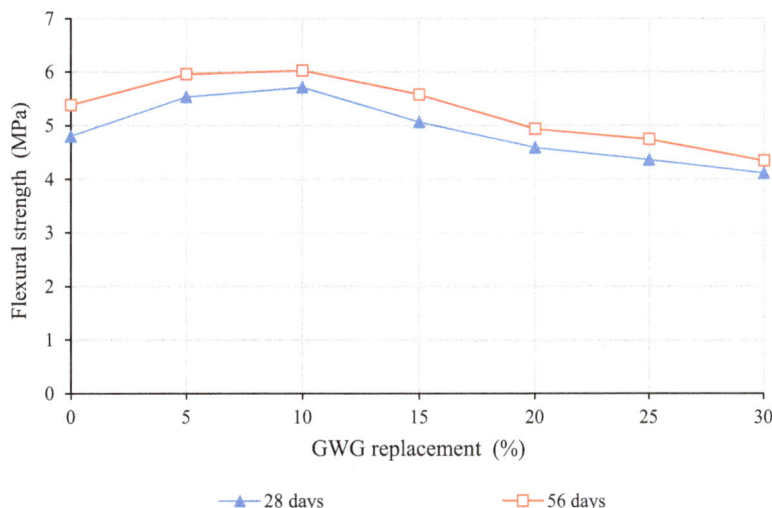

Fig. 12 Flexural strength versus GWG microparticles replacement as cementing material.

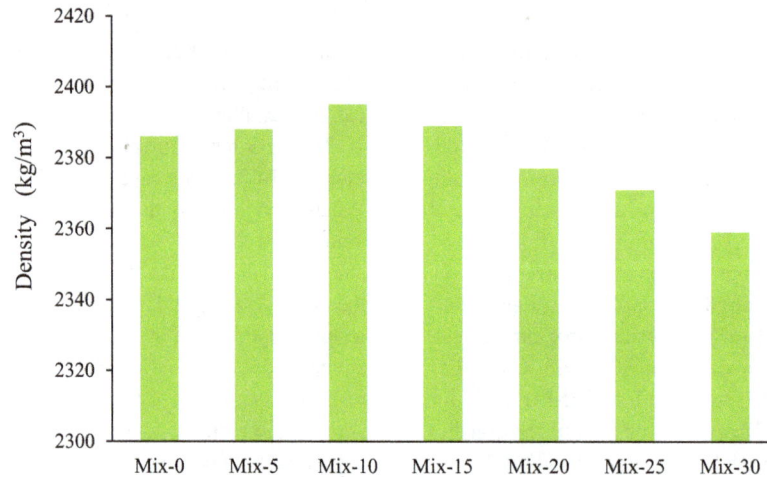

Fig. 13 Unit weights of SCCs incorporating partial level of GWG microparticles.

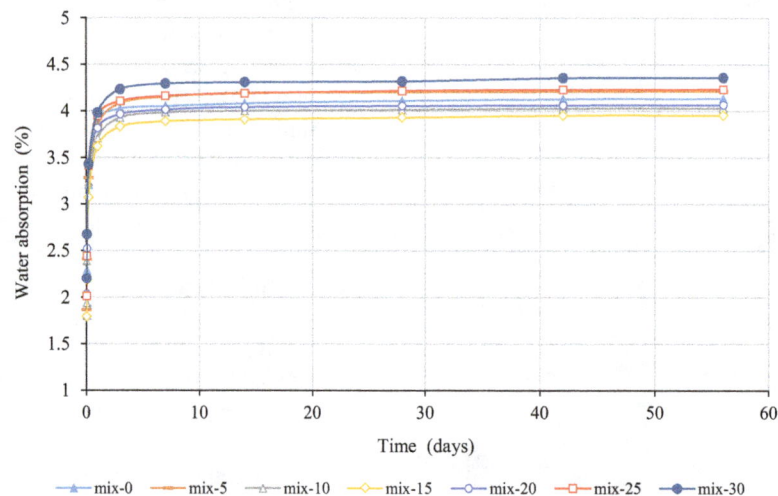

Fig. 14 Water absorption of SCCs incorporating partial level of GWG microparticles as ages.

to sufficient compaction (especially for SCC), hence more compressive strength and density. Besides, the pozzolanic grains fill the interface between aggregate and paste similar filler. Therefore, using GWG microparticles properly (up to 15 %), the density of concrete increases and vice versa porosity decreases. While the quantity of pozzolanic materials increases from the required amount for hydration process the silica leaching out; hydration product decreased and pores increase and consequently densities decrease. The other reason which made the unit weights of concretes containing GWG substitute decrease, is the less specific density of glass compare to cement. Therefore, with an increase in glass content, the weight of concrete mixes decrease because the specific gravity of glass is lower than that of cement. Generally, when the weak interface zone between aggregate and paste of the concrete is reduced, the strength of the concrete may be increased.

6.2.5 Water Absorption

One of the most important issue that relates to durability is water absorption. For service life prediction and long term

behavior of concrete the aforementioned test should be investigated deeply. Liu (2011) believes that water absorption phenomena can be one of the most important factor for predicting the concrete deterioration subjected to freezing and thawing cycling and carbonation. Figure 14 shows the water absorption results of SCCs. Note that the replacement of Portland cement with GWG microparticles more than 20 %, slightly increased water absorption. But the results showed that up to 20 % GWG replacement the water absorption is less or could be compared with the control mix. This can be attributed to hydration process which is continuing at the later days in reason of the pozzolanic effects of remaining GWG powder. As one knows absorption is the amount and rate of water absorbed into the concrete pores by capillary suction. Therefore, as the capillary pores decreased by more hydration process, the absorption would be also decreased as ages consequently. In this reason when GWG content increases (i.e., Mix-30 %) the hydration products decreases and the water absorption increased slightly.

Figure 14 shows that the use of GWG significantly decreases the rate of water absorption of SCC with increasing curing ages. It is evident that in the early ages (up to 6 days) the water absorption is increased significantly, but it slow down afterward. This may be due to the pozzolanic activity of GWG during the curing time. All concrete mixtures under investigation showed the water absorption lower than 4.5 %. Moreover, the water absorptions were in the range of 1.8–2.2 % at 30 min, in the range of 4.1–4.31 % at 28 days and in the range of 4.13–4.36 % at 56 days. From the water absorption results, use of up to 20 % volume ratio of GWG microparticles in SCC may not influence the durability more than cement. Totally from the water absorption results, use of GWG as cement substitute in SCC may not incur severe durability problems. Liu (2011) reported that at each age, the sorptivity values are similar, which shows that the current glass content does not significantly increase the water absorption. Figure 15 shows some specimens before absorption measurement.

6.2.6 Free Drying Shrinkage

Drying shrinkage defined as the time-dependent strain measured at constant temperature in an unloaded and unrestrained specimen (Tam et al. 2012). Free shrinkage tests can

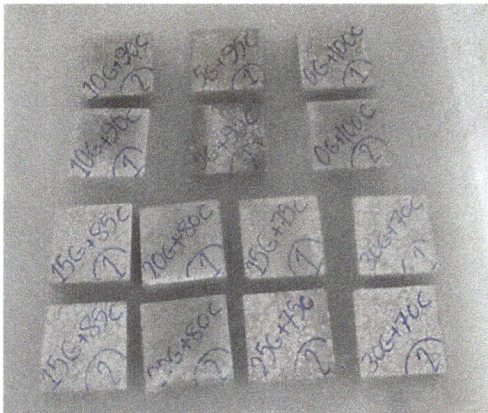

Fig. 15 Specimens before absorption measurement.

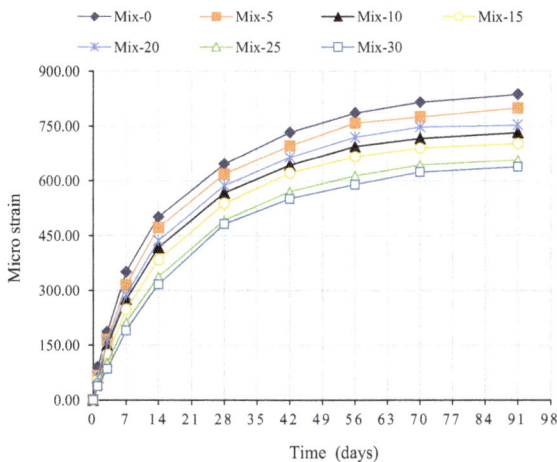

Fig. 16 Drying shrinkage of SCCs incorporating partial level of GWG microparticles as ages.

provide necessary information on how the drying shrinkage stresses develop although they cannot offer sufficient information on the behavior of concrete structures (Wiegrink et al. 1996). As shown in Fig. 16, between 1 and 14 days, all mixes have the higher shrinkage rates. But, after this period time the shrinkage rates was decreased. A clear distinction was observed for different concretes with different GWG microparticles content, especially after about 14 days. Generally, the shrinkage strains were somewhat comparable at very early ages while there was a considerable distinction at later ages of the drying period. Tam et al. (2012) reported that this is mainly because of the loss of physically absorbed water from Calcium Silicate Hydrate (C–S–H), resulting in a shrinkage strain. There are various factors influencing the drying shrinkage of hardened concrete. Water content is probably the largest single factor influencing the shrinkage of paste and concrete (Feldman and Hauang 1985a, b). Moreover, high paste content and the increasing use of admixtures may affect the shrinkage of concrete.

Based on the Australian Standard AS3600 provisions the acceptable value for shrinkage required a value below 0.075 % at the age 56 days. From Fig. 16, it is evident that all of the concrete maintained below the acceptable value except the control mix. Therefore, this clarified that inclusion of GWG microparticles substitute in SCC improved the shrinkage performance. Mix-0 which contained no GWG showed shrinkage values larger than those of any of the other mixes. This may be attributed to existing more free water that do not combined in hydration process.

Mix-20 has 3.55 % increase in the shrinkage strains compare to Mix-10 at 28 days. Mix-25 and Mix-30 have the lowest shrinkage strains among of all mixes, due to the combined effects of the decrease in the superplasticizer dosage and the increase in the GWG content that do not let to exist more free water in the paste. From the results (Fig. 16), the 91 days drying shrinkage of the concretes with 5, 10, 15, 20, 25 and 30 % GWG were 4.46, 12.47, 16.08, 9.99, 21.44 and 23.75 % lower than that of the control concrete (Mix-0), respectively. The shrinkage measuring apparatus used to determine the free drying shrinkage strain for SCCs is the concrete length comparator is shown in Fig. 17.

Fig. 17 Concrete length comparator.

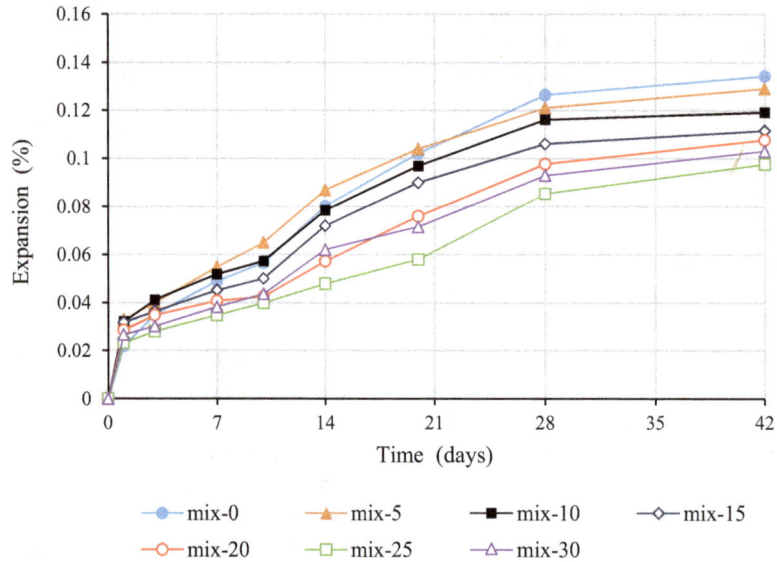

Fig. 18 ASR expansion of SCCs incorporating partial level of GWG microparticles as ages.

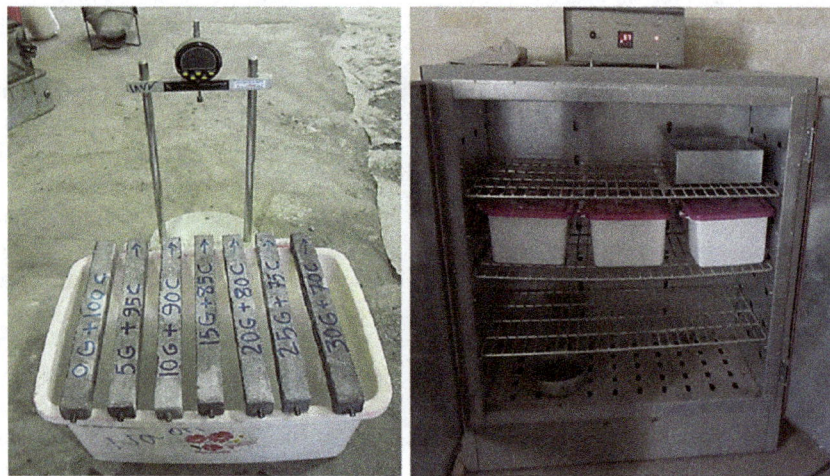

Fig. 19 Mortar bars stored in a 1 N NaOH solution at 80 °C.

6.2.7 ASR

The ASR expansion of SCC mixtures at different ages are shown in Fig. 18. Each value presented is the average of two measurements. The ASR expansion was great up to 14 days. The suppressing effect of GWG microparticles on ASR expansion is very clear that the expansion at 14 days was less than 0.09 % for all concrete mixtures, which is slightly lower than 0.1 %, as specified by ASTM C 1260 for innocuous reaction. As it is obvious glass contains high level of alkaline and it is able to be leached out and an expansion due to alkali aggregate occurred. A large number studies have confirmed that glass particles will not generate deleterious expansion themselves once they are smaller than 300μ (Meyer and Baxter 1997; Shi et al. 2004). The ASR expansion decreased with decreasing particle size of glass. This might be due to some glass containing high content of active silica can be classified as a reactive

aggregate or a pozzolanic material (Shi et al. 2005; Zhu et al. 2009). Here, it is very clear that the expansion of mortar bar decreases as the GWG microparticles replacement level increases. One possible reason may be due to the highly reactive GWG microparticles which react with lime and form C–S–H gel which retains the alkalis in the C–S–H.

From Fig. 18, when the amount of GWG microparticles increased from 0 to 5, 10, 15, 20, 25 and 30 %, there was a gradual decrease in ASR expansion. Expansions at all ages however did not exceed 0.20 %, the allowable limits by ASTM 1260 (2007). Liu (2011) investigated the expansions in the ASR mortar bar tests of the mixes with and without glass. He reported that ASR expansion are similar to those of the control mix and all can be considered innocuous. This indicates glass incurs no more ASR risks than the cement in concrete.

In brief no deleterious ASR expansion was occurred in the concrete specimens test, indicating that ASR would not be a problem in the presence of GWG microparticles. As previously stated this may have arisen because the pozzolanic reaction of GWG microparticles with cement appeared to enhance the binding of alkali, making it unavailable for reaction in other ways. After 42 days of curing, the surface of all mortar bars remained quite smooth and no cracks were observed. Mortar bars stored in a 1 N NaOH solution at 80 °C are presented in Fig. 19.

7. Concluding Remarks

This paper presents the experimental results of a study on the feasible use of GWG microparticles as cementing material for the production of SCC. Based on the results of this study, the following conclusions can be presented:

1. The compressive, splitting tensile and flexural strengths of the GWG-SCCs increased with an increase in GWG microparticles content up to 20 %, vice versa as the GWG content increased more than 20 % the strengths have been decreased. The concrete containing 5 % GWG microparticles resulted in the highest strength properties. There was more than 18 % improvement in the compressive strength of GWG-SCCs with 5 % GWG microparticles substitution in comparison with the control mixture, but mixture containing 20 % GWG was comparable to the control mixture. The variation of compressive strength is ranged 0, 18.01, 11.49, 7.72, 5.08, −3.99 and −10.15 % from Mix-0 (control specimen) to Mix-30 containing 30 % GWG, respectively.
2. The relationship between compressive strength and tensile splitting tensile strength located between the upper and lower bound per CIB-FIP code provisions.
3. The unit weight of the GWG-SCCs increased up to 15 % GWG microparticles replacement, and for the mixes containing more than 15 %, it has been decreased. The unit weight and strengths behavior trends are similar.
4. Totally it was seen a decrease in the surface water absorption as GWG microparticles quantity increased up to 20 %. The variation of water absorption of GWG-SCCs in 28 age is ranged 0, 2, −3, −4 %, −1, 2 and 4 % from Mix-0 (control specimen) to Mix-30 containing 30 % GWG, respectively.
5. The drying shrinkage of the GWG-SCCs decreased with an increase in the GWG microparticles content. The variation of drying shrinkage of GWG-SCCs in age 28 is decreasing in range 0, 4, 7, 8, 10, 27 and 45 % from Mix-0 (control specimen) to Mix-30 containing 30 % GWG, respectively.
6. The expansions in the ASR mortar bar tests of the mixes with GWG microparticles are similar to those of the control mix and all can be considered innocuous. This indicates GWG microparticles incur no more ASR risks than the cement in concrete.
7. From this study it may be concluded that the use of GWG microparticles as cement substitution improves ASR, drying shrinkage and workability characteristics, while the hardened behavior increases up to 20 % replacement. The GWG microparticles in the range of 0–20 % may replace cement in concrete mixture. The economical SCC could be achieved with sufficient strength as the conventional concrete. Based on the materials used in this study, the results suggested that it is technically feasible to utilize GWG microparticles as a part of paste content in the production of SCC based on the present test results.
8. The improvement in the engineering and bulk properties of concrete mixtures incorporating GWG microparticles as cement indicates that GWG microparticles can be used beneficially as cementing material of SCC, however, additional experimental results are needed for micro properties of this type of concrete.

Acknowledgments

The authors are pleased to acknowledge the Roads and Urban Development Company of Kerman Province support. The experimental work of the present study was undertaken at the Institute for Concrete Research (ICR) of Rafsanjan University, Iran.

References

Afshoon, I., & Sharifi, Y. (2014). Ground copper slag as a supplementary cementing material and its influence on the fresh properties of self-consolidating concrete. *The IES Journal Part A: Civil & Structural Engineering, 7*(4), 229–242.

Alp, I., Deveci, H., & Sungun, H. (2008). Utilization of flotation wastes of copper slag as raw material in cement production. *Journal of Hazardous Materials, 159*, 390–395.

Alya, M., Hashmi, M. S. J., Olabi, A. G., Messeiry, M., Abadir, E. F., & Hussain, A. I. (2012). Effect of colloidal nano-silica on the mechanical and physical behavior of waste-glass cement mortar. *Materials and Design, 33*, 127–135.

ASTM C496/C496M-04. (2004). *Standard test method for splitting tensile strength of cylindrical concrete specimens.* West Conshohocken, PA: ASTM International.

ASTM C1260-07. (2007). *Standard specification for potential alkali reactivity of aggregates (mortar-bar method)*. West Conshohocken, PA: ASTM International.

ASTM C33/C33M-08. (2008). *Standard specification for concrete aggregates*. West Conshohocken, PA: ASTM International.

ASTM C403/C403M-08. (2008). *Standard specification for time of setting of concrete mixtures by penetration resistance*. West Conshohocken, PA: ASTM International.

ASTM C157/C157M-08. (2008). *Standard specification for length change of hardened hydraulic-cement mortar and concrete*. West Conshohocken, PA: ASTM International.

ASTM C150/C150M-09. (2009). *Standard specification for Portland cement*. West Conshohocken, PA: ASTM International.

ASTM C94/C94M-09. (2009). *Standard specification for ready—mixed concrete*. West Conshohocken, PA: ASTM International.

ASTM C231/C231M-09. (2009). *Standard specification for air content of freshly mixed concrete by the pressure method*. West Conshohocken, PA: ASTM International.

ASTM C494/C494M-10. (2010). *Standard specification for chemical admixtures for concrete*. West Conshohocken, PA: ASTM International.

Beltran, M. G., Barbudo, A., Agrela, F., Galvn, A. P., & Jimenez, J. R. (2014). Effect of cement addition on the properties of recycled concretes to reach control concretes strengths. *Journal of Cleaner Production, 79*, 124–133.

Bingol, A. F., & Tohumcu, I. (2013). Effects of different curing regimes on the compressive strength properties of self - compacting concrete incorporating fly ash and silica fume. *Materials and Design, 51*, 12–18.

BS 1881: Part 116. (1983). *Testing concrete: method for determination of compressive strength of concrete cubes*. British Standard Institution (BSI).

BS 1881: Part 122. (1983). *Testing concrete: Method for determination of water absorption*. British Standard Institution (BSI).

BS EN 12390:2009. (2009). *Part 7. Testing hardened concrete: Density of hardened concrete*.

Carpenter, A. J., & Cramer, S. M. (1999). *Mitigation of ASR in pavement patch concrete that incorporates highly reactive fine aggregate* (pp. 60–67). Transportation Research Record 1668, Paper No. 99-1087.

CEB-FIB model code 1990. (1993). *Committee Euro-International du Beton*. Thomas Telford, London, UK.

Chen, C. H., Huang, R., Wu, J. K., & Yang, C. C. (2006). Waste E-glass particles used in cementitious mixtures. *Cement and Concrete Research, 36*, 449–456.

Craeye, B., Itterbeeck, P. V., Desnerck, P., Boel, V., & De Schutter, G. (2014). Modulus of elasticity and tensile strength of self-compacting concrete: Survey of experimental data and structural design codes. *Cement & Concrete Composites, 54*, 53–61.

Domone, P. L. (1997). A review of the hardened mechanical properties of self-compacting concrete. *Cement and Concrete Composites, 29*, 1–12.

Druta, C., Wang, L., & Stephen, Lane D. (2014). Tensile strength and paste–aggregate bonding characteristics of self-consolidating concrete. *Construction and Building Materials, 55*, 89–96.

Du, H., & Tan, K. H. (2013). Use of waste glass as sand in mortar: Part II—Alkali–silica reaction and mitigation methods. *Cement & Concrete Composites, 35*, 118–126.

Dyer, T. D., & Dhir, R. K. (2001). Chemical reactions of glass cullet used as cement component. *Journal of Materials in Civil Engineering, 13*(6), 412–417.

EFNARC. (2005). *The European guidelines for self-compacting concrete: specification, production and use*. The Self-Compacting Concrete European Project Group.

EN 1351:1997. (1997). *Determination of flexural strength of autoclaved aerated concrete*.

Federico, L. M., & Chidiac, S. E. (2009). Waste glass as a supplementary cementitious material in concrete—critical review of treatment methods. *Cement & Concrete Composites, 31*, 606–610.

Feldman, R. F., & Hauang, C. Y. (1985a). Properties of Portland cement–silica-fume pastes I: Porosity and surface properties. *Cement and Concrete Research, 15*(5), 765–774.

Feldman, R. F., & Hauang, C. Y. (1985b). Properties of Portland cement–silica-fume pastes II: Mechanical properties. *Cement and Concrete Research, 15*(6), 943–952.

Guneyisi, E., Gesoglu, M., & Ozbay, E. (2010). Strength and drying shrinkage properties of self-compacting concretes incorporating multi-system blended mineral admixtures. *Construction and Building Materials, 24*, 1878–1887.

Hameed, M. S., Sekar, A. S. S., & Saraswathy, V. (2012). Strength and permeability characteristics study of self-compacting concrete using crusher rock dust and marble sludge powder. *The Arabian Journal for Science and Engineering, 37*, 561–574.

Jackson, N., & Dhir, K. R. (1996). *Civil engineering materials* (5th ed.). London, UK: Macmillan Press Ltd.

Jain, J. A., & Neithalath, N. (2010). Chloride transport in fly ash and glass powder modified concretes—influence of test methods on microstructure. *Cement & Concrete Composites, 32*, 148–156.

Liu, M. (2011). Incorporating ground glass in self-compacting concrete. *Construction and Building Materials, 25*, 919–925.

Lotfy, A., Hossain, K. M. A., & Lachemi, M. (2015). Lightweight self-consolidating concrete with expanded shale aggregates: Modelling and optimization. *International Journal of Concrete Structures and Materials, 9*(2), 185–206.

Mahmoud, E., Ibrahim, A., El-Chabib, H., & Patibandla, V. C. (2013). Self-consolidating concrete incorporating high volume of fly ash, slag, and recycled asphalt pavement. *International Journal of Concrete Structures and Materials, 7*(2), 155–163.

Meyer, C., & Baxter, S. (1997). *Use of recycled glass for concrete masonry blocks final report 97-15*. Albany, NY: New York State Energy Research and Development Authority.

Mirzahosseini, M., & Riding, K. A. (2014). Effect of curing temperature and glass type on the pozzolanic reactivity of glass powder. *Cement and Concrete Research, 58*, 103–111.

Nikbin, I. M., Beygi, M. H. A., Kazemi, M. T., Amiri, J. V., Rabbanifar, S., Rahmani, E., & Rahimi, S. (2014). A comprehensive investigation into the effect of water to cement ratio and powder content on mechanical properties of self-compacting concrete. *Construction and Building Materials, 57*, 69–80.

Nuruddin, M. F., Chang, K. Y., & Azmee, N. M. (2014). Workability and compressive strength of ductile self-compacting concrete (DSCC) with various cement replacement materials. *Construction and Building Materials, 55*, 153–157.

Okamura, H., & Ouchi, M. (2003). Self-compacting concrete. *Journal of Advanced Concrete Technology, 1*(1), 5–15.

Oliveira, L. A., Castro-Gomes, J. P., & Santos, P. (2008). Mechanical and durability properties of concrete with ground waste glass sand. In *11th international conference on durability of building materials and components*, Istanbul, Turkey.

Park, S. B., Lee, B. C., & Kim, J. H. (2004). Studies on mechanical properties of concrete containing waste glass aggregate. *Cement and Concrete Research, 34*, 2181–2189.

Parra, C., Valcuende, M., & Gomez, F. (2011). Splitting tensile strength and modulus of elasticity of self-compacting concrete. *Construction and Building Materials, 22*, 201–207.

Ponikiewski, T., & Golaszewski, J. (2014). The influence of high-calcium fly ash on the properties of fresh and hardened self-compacting concrete and high performance self compacting concrete. *Journal of Cleaner Production, 72*, 212–221.

Sfikas, I. P., Badogiannis, E. G., & Trezos, K. G. (2014). Rheology and mechanical characteristics of self-compacting concrete mixtures containing metakaolin. *Construction and Building Materials, 64*, 121–129.

Shao, Y., Lefort, T., Moras, S., & Rodriguez, D. (1999). Studies on concrete containing ground waste glass. *Concrete and Cement Research, 30*(1), 91–100.

Sharifi, Y., Hoshiar, M., & Aghebati, B. (2013). Recycled glass replacement as fine aggregate in self-compacting concrete. *Frontiers of Structural and Civil Engineering, 7*(4), 419–428.

Sharifi, Y., Afshoon, I., & Firoozjaie, Z. (2015). Fresh properties of self-compacting concrete containing ground waste

glass microparticles as cementing material. *Journal of Advanced Concrete Technology, 13*(2), 50–66.

Shayan, A., & Xu, A. (2006). Performance of glass powder as a pozzolanic material in concrete: A field trial on concrete slabs. *Cement and Concrete Research, 36*, 457–468.

Shi, C., Wu, Y., Shao, Y., & Riefler, C. (2004). Alkali-aggregate reaction of concrete containing ground glass powder. In *Proceedings of the 12th international conference on AAR in concrete* (pp. 789–795).

Shi, C., Wu, Y., Riefler, C., & Wang, H. (2005). Characteristics and pozzolanic reactivity of glass powders. *Cement and Concrete Research, 35*(5), 987–993.

Siddique, R. (2013). Compressive strength, water absorption, sorptivity, abrasion resistance and permeability of self-compacting concrete containing coal bottom ash. *Construction and Building Materials, 47*, 1444–1450.

Tam, C. M., Tam, V. W. Y., & Ng, K. M. (2012). Assessing drying shrinkage and water permeability of reactive powder concrete produced in Hong Kong. *Construction and Building Materials, 26*, 79–89.

Uysal, M. (2012). Self-compacting concrete incorporating filler additives: Performance at high temperatures. *Construction and Building Materials, 26*, 701–706.

Valcuendea, M., Parra, C., Marco, E., Garrido, A., Martínez, E., & Cánoves, J. (2012). Influence of limestone filler and viscosity-modifying admixture on the porous structure of self-compacting concrete. *Construction and Building Materials, 28*, 122–128.

Wang, H. Y. (2011). The effect of the proportion of thin film transistor–liquid crystal display (TFT–LCD) optical waste glass as a partial substitute for cement in cement mortar. *Construction and Building Materials, 25*, 791–797.

Wang, H. Y., & Huang, W. L. (2010). A study on the properties of fresh self-consolidating glass concrete (SCGC). *Construction and Building Materials, 24*, 619–624.

Wiegrink, K., Marikunte, S. M., & Shah, S. P. (1996). Shrinkage cracking of high-strength concrete. *ACI Materials Journal, 93*(5), 409–415.

Zhu, H. Y., Chen, W., Zhou, W., & Byars, E. A. (2009). Expansion behaviour of glass aggregates in different testing for alkali-silica reactivity. *Materials and Structures, 42*(4), 485–494.

Zong, L., Fei, Z., & Zhang, S. (2014). Permeability of recycled aggregate concrete containing fly ash and clay brick waste. *Journal of Cleaner Production, 70*, 175–182.

6

Determination of Double-K Fracture Parameters of Concrete Using Split-Tension Cube: A Revised Procedure

Shashi Ranjan Pandey[1], Shailendra Kumar[2],*, and A. K. L. Srivastava[1]

Abstract: This paper presents a revised procedure for computation of double-K fracture parameters of concrete split-tension cube specimen using weight function of the centrally cracked plate of finite strip with a finite width. This is an improvement over the previous work of the authors in which the determination of double-K fracture parameters of concrete for split-tension cube test using weight function of the centrally cracked plate of infinite strip with a finite width was presented. In a recent research, it was pointed out that there are great differences between a finite strip and an infinite strip regarding their weight function and the solution of infinite strip can be utilized in the split-tension specimens when the notch size is very small. In the present work, improved version of LEFM formulas for stress intensity factor, crack mouth opening displacement and crack opening displacement profile presented in the recent research work are incorporated. The results of the double-K fracture parameters obtained using revised procedure and the previous work of the authors is compared. The double-K fracture parameters of split-tension cube specimen are also compared with those obtained for standard three point bend test specimen. The input data required for determining double-K fracture parameters for both the specimen geometries for laboratory size specimens are obtained using well known version of the Fictitious Crack Model.

Keywords: split-tension cube test, three point bend test, concrete fracture, double-K fracture parameters, weight function, cohesive stress, size-effect.

Abbreviations

CBM	Crack band model
CCM	Cohesive crack model
CT	Compact tension
DGFM	Double-G fracture model
DKFM	Double-K fracture model
ECM	Effective crack model
FCM	Fictitious crack model
FPZ	Fracture process zone
LEFM	Linear elastic fracture mechanics
SEM	Size effect model
SIF	Stress intensity factor
STC	Split tension cube
TPBT	Three point bend test
TPFM	Two parameter fracture model
WST	Wedge splitting test

List of Notations

a_o	Initial crack length
A_i	Regression coefficients
a_c	Effective crack length at peak (critical) load
B	Width of beam
B_i	Regression coefficients
c_1, c_2	Material constants for nonlinear softening function
CMOD	Crack mouth opening displacement
$CMOD_c$	Crack mouth opening displacement at critical load
CTOD	Crack tip opening displacement
$CTOD_c$	Crack tip opening displacement at critical load
D	Depth or characteristic dimension of specimen
E	Modulus of elasticity of concrete
f_t	Uniaxial tensile strength of concrete
G_F	Fracture energy of concrete
$G(x,a)$	Weight function
H	Height or total depth (2D) for split tension cube specimen
$k(\alpha, \beta)$	Non-dimensional function for KI or geometry factor
K_I	Stress intensity factor
K_{IC}^{ini}	Initial cracking toughness
K_{IC}^{un}	Unstable fracture toughness

[1]Department of Civil Engineering, National Institute of Technology, Jamshedpur, Jharkhand 831014, India.

[2]Department of Civil Engineering, Institute of Technology, Guru Ghasidas Vishwavidyalaya (A Central University), Bilaspur, CG 495009, India.

*Corresponding Author; E-mail: shailendrakmr@yahoo.co.in

K_{IC}^C	Cohesive toughness
$m(x,a)$	Universal weight function
M_1, M_2, M_3	Parameters of weight function
P_u	Maximum applied load or critical load
S	Span of beam
t	Half of the width of distributed load for split tension cube specimen
$V(\alpha, \beta)$	Dimensionless function for CMOD
w_c	Maximum crack opening displacement at the crack-tip for which the cohesive stress becomes equals to zero
α	Ratio of crack length to depth of specimen (a/D)
β	Ratio of load-distributed to height of specimen ($2t/h = t/D$) for split tension cube specimen
σ	Cohesive stress
υ	The Poisson's ratio
$\sigma_s(\text{CTOD}_c)$	Cohesive stress at the tip of initial notch corresponding to CTOD$_c$

1. Introduction

It is well known that fracture parameters of *quasibrittle* material like concrete cannot be determined by directly applying the concepts of linear elastic fracture mechanics (LEFM) because of the existence of large and variable size of fracture process zone (FPZ) ahead of a crack-tip. In order to account for and characterize FPZ in the analysis, several non-linear fracture mechanics models have been developed which primarily involve cohesive crack model (CCM) or fictitious crack model (FCM) (Hillerborg et al. 1976; Modeer 1979; Petersson 1981; Carpinteri 1989; Planas and Elices 1991; Zi and Bažant 2003; Roesler et al. 2007; Park et al. 2008; Zhao et al. 2008; Kwon et al. 2008, Cusatis and Schauffert 2009, Elices et al. 2009; Kumar and Barai 2008b, 2009b) and crack band model (CBM) (Bažant and Oh 1983), two parameter fracture model (TPFM) (Jenq and Shah 1985), size effect model (SEM) (Bažant et al. 1986), effective crack model (ECM) (Nallathambi and Karihaloo 1986), K_R-curve method based on cohesive force distribution (Xu and Reinhardt 1998, 1999a), double-K fracture model (DKFM) (Xu and Reinhardt 1999a, b, c) and double-G fracture model (DGFM) (Xu and Zhang 2008).

In recent time, much of research and studies (Xu and Reinhardt 1999a, b, c, 2000; Zhao and Xu 2002; Zhang et al. 2007; Xu and Zhu 2009; Kumar and Barai 2008a, 2009a, 2010; Kumar 2010; Zhang and Xu 2011; Kumar and Pandey 2012; Hu and Lu 2012; Murthy et al. 2012; Hu et al. 2012; Ince 2012; Kumar et al. 2013; Choubey et al. 2014; Kumar et al. 2014) have been carried out to determine and characterize the fracture parameters of concrete using double-K fracture model for which the reasons are obvious (Kumar et al. 2013). The double-K fracture model is characterized by two material parameters: initial cracking toughness K_{IC}^{ini} and unstable fracture toughness K_{IC}^{un}. The initiation toughness is defined as the inherent toughness of the materials, which

holds for loading at crack initiation when material behaves elastically and micro cracking is concentrated to a small-scale in the absence of main crack growth. It is directly calculated by knowing the initial cracking load and initial notch length using LEFM formula. The total toughness at the critical condition is known as unstable toughness K_{IC}^{un} which is regarded as one of the material fracture parameters at the onset of the unstable crack propagation and it can be obtained by knowing peak load and corresponding effective crack length using the same LEFM formula. Recently, Kumar and Pandey (2012) presented the formulation and determination of double-K fracture parameters using split-tension cube test specimen using weight function method in which the LEFM formulas for stress intensity factor (SIF), crack mouth opening displacement (CMOD) and crack opening displacement (COD) profile derived by Ince (2010) and the universal weight function of Wu et al. (2003) were adopted. The authors (Kumar and Pandey 2012) mentioned that there are several advantages of using split-tension cube (STC) test specimen over the testing of other specimens like three point bend test (TPBT), compact tension (CT) and wedge splitting test (WST) specimens. However, there should be a limitation that the notch can be only produced at the time of casting of concrete cubes (pre-cast notch) in the split tension cube specimen. The authors also presented the results of the initial cracking toughness, cohesive toughness and unstable fracture toughness obtained using split tension cube test specimen and they were compared with those obtained using standard compact tension specimen. From the study it was concluded that the double-K fracture parameters as obtained using split-tension cube test are in good agreement and consistent with those as calculated using standard compact tension specimen. However, the results of fracture parameters are influenced by the distributed-load width during the loading of split-tension cube specimen and it was observed that the values of unstable fracture toughness and cohesive toughness increase with increase in the distributed-load width whereas the initial cracking toughness is not significantly affected by the distributed-load width. In the formulation, the authors (Kumar and Pandey 2012) used the weight function of the centrally cracked infinite strip with a finite width specimen (Tada et al. 2000) and the equivalent four terms of universal weight function (Wu et al. 2003) for computing the value of cohesive toughness and consequently determining the initial cracking toughness. Later, Ince (2012) put forward a method for determination of double-K fracture parameters using weight function for split—tension specimens such as splitting tests on cubical, cylindrical and diagonal cubic concrete samples. The author pointed out that there are great differences between a finite strip and an infinite strip regarding their weight function and the solution of infinite strip can be utilized in the split-tension specimens when the notch size is very small. It was concluded that the central cracked plate can be considered as an infinite strip when the length/width (l/D) ratio of a plate is equal or greater than 3 (Isida 1971, Tada et al. 2000). In case of a cube-split tension test specimen the value of the length/characteristic dimension (l/D) ratio is taken to be 1 for which

Ince (2012) derived the four term universal weight function using boundary element method and finite element method. The author also presented the improved version of LEFM formulas for stress intensity factor, CMOD and COD profile over the previously derived LEFM equations by the same author (Ince 2010) for split tension cube test specimen. In view of the above development, it was felt necessary to carry out a comparative study on the double-K fracture parameters computed using the procedure outlined by Kumar and Pandey (2012) and using the weight function of the centrally cracked plate of finite strip with a finite width incorporating the improved version of LEFM formulas for stress intensity factor, CMOD and COD profile derived by Ince (2012).

The paper presents the revised procedure for determination of double-K fracture model using *weight function method* for the split-tension cube specimen of concrete considering improved LEFM formulas for stress intensity factor, CMOD and COD profile and the weight function of the centrally cracked plate of finite strip with a finite width derived by Ince (2012). The results of the fracture parameters obtained using revised procedure and the previous work of Kumar and Pandey (2012) are compared. Further, the double-K fracture parameters of split-tension cube specimen are also compared with those obtained for standard three point bend test specimen. The input data required for determining for split-tension cube test and three point bend test for laboratory size specimens are obtained using well known version of the fictitious crack model.

2. Dimensions of Test Specimens

For present investigation, the standard test geometries, dimensions and loading conditions for STC and TPBT specimens are considered as shown in Fig. 1. The symbols in Fig. 1(a): a_o, D, h and t are half of the initial notch-length, characteristic dimension as specimen size ($D = h/2$), height or total depth and half of the width of distributed load respectively for STC geometry. RILEM Technical Committee 50-FMC (1985) has recommended the guidelines for determination of fracture energy of cementitious materials using standard three-point bend test on notched beam. This method has been widely used for determination of fracture energy of concrete with certain modification in the experimental setup (Lee and Lopez 2014). In present study, standard three—point bend test (RILEM Technical Committee 50-FMC 1985) is considered for which the symbols: B, D and S in Fig. 1b are the width, depth and span respectively with $S/D = 4$.

3. Determination of Double-K Fracture Parameters for STC Specimen

3.1 Assumptions

Linear asymptotic superposition assumption is considered to introduce LEFM for calculating the double-K fracture parameters. The hypotheses of the assumption are given below:

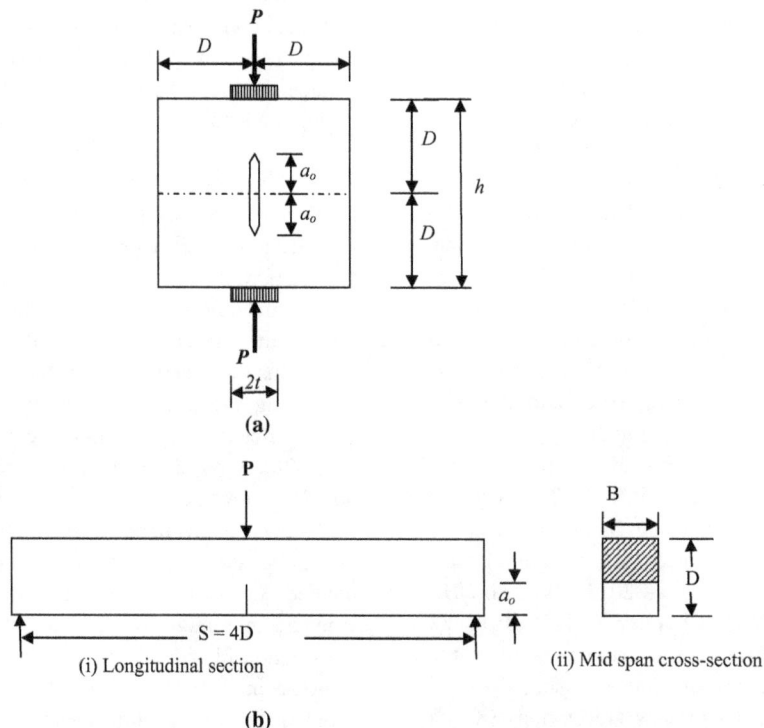

Fig. 1 Dimensions and loading schemes for STC and TPBT test specimens. **a** Split tension cube test specimen, **b** Dimensions and loading schemes of TPBT.

1. the nonlinear characteristic of the load-crack mouth opening displacement (P-CMOD) curve is caused by fictitious crack extension in front of a stress-free crack; and
2. an effective crack consists of an equivalent-elastic stress-free crack and equivalent-elastic fictitious crack extension.

A detailed explanation of the hypotheses may be seen elsewhere (Xu and Reinhardt 1999b).

3.2 Effective Crack Extension

For the applied load (Fig. 1) on the STC specimen, the critical value of CMOD (CMOD$_c$) is measured across the crack faces at the centre of specimen. The P-CMOD curve up to peak load for this test geometry should be known a priori for determining the value of effective crack extension during the crack propagation. Using linear asymptotic superposition assumption, the equivalent-elastic crack length a_c corresponding to maximum load P_u is solved using the *revised* LEFM formulae (Ince 2012). Hence, the CMOD is expressed as:

$$CMOD = \frac{\pi D \sigma_N}{E} \alpha V(\alpha, \beta) \tag{1}$$

$$V(\alpha, \beta) = B_0(\beta) + B_1(\beta)\alpha + B_2(\beta)\alpha^2 + B_3(\beta)\alpha^3 + B_4(\beta)\alpha^4 + + B_5(\beta)\alpha^5 \tag{2}$$

In which $\alpha = a/D$, β is the relative load-distributed width and expressed as $\beta = 2t/h = t/D$, $V(\alpha,\beta)$ is dimensionless geometric function, coefficients B_i ($i = 0$ to 5) are the function of β as given in Table 1. Equation (2) is valid for $0.1 \leq \alpha \leq 0.9$ within 0.3 % accuracy for $0 \leq \beta \leq 0.2$. The modulus of elasticity of concrete (E) obtained using cylinder test is taken as a constant value for a particular concrete mix. Ince (2012) used boundary element numerical method to improve the LEFM formulas over the previous LEFM

formulas (Ince 2010) for the split tension cube specimens which was based on centrally cracked infinite strip with a finite width specimen. Equation (2) and Table 1 used in the present study are extracted by Ince (2012) from the numerical results based on centrally cracked finite strip with a finite width specimen. Since the values of coefficients B_i (Table 1) are given (Ince 2012) at discrete intervals, these coefficients can be determined by linear interpolation at any value of β for the given range $0 \leq \beta \leq 0.2$.

Also, the nominal stress for STC test specimen in Eq. (1) can be written using the following formula (Timoshenko and Goodier 1970).

$$\sigma_N = \frac{2P}{\pi Bh} \tag{3}$$

At critical condition that is at maximum load P_u the half of crack length a becomes equal to a_c and σ_N to σ_{Nu} in which σ_{Nu} is the maximum nominal stress. Karihaloo and Nallathambi (1991) concluded that almost the same value of E might be obtained from P-CMOD curve, load–deflection curve and compressive cylinder test. Hence, in case that is not known the value of E determined using compressive cylinder tests may be used to obtain the critical crack length of the specimen.

3.3 Calculation of Double-K Fracture Parameters

A linearly varying cohesive stress distribution is assumed in the fictitious crack zone, which gives rise to cohesion toughness as a part of total toughness of the cracked body. Superposition method is used in order to calculate the stress intensity factor (SIF) at the tip of effective crack length K_I. According to this method, total stress intensity factor K_I is taken as the summation of stress intensity factor caused due to external load K_I^P and stress intensity factor contributed by cohesive stress K_I^{C-} as shown in

Table 1 The values of coefficients A_i and B_i for split- tension cube specimen (Ince 2012).

Coefficient	$\beta = t/D$					
	0.0	0.067	0.1	0.133	0.167	0.2
A_0	0.842	0.995	1.050	1.060	1.036	0.995
A_1	2.861	-0.147	-1.366	-1.815	-1.655	-1.219
A_2	-17.384	1.847	9.772	12.762	11.794	8.986
A_3	53.695	-0.480	-23.296	-32.385	-30.268	-22.774
A_4	-70.864	-1.908	27.794	40.275	38.365	29.263
A_5	35.033	2.429	-12.082	-18.691	-18.479	-14.669
B_0	1.159	1.192	1.211	1.216	1.208	1.188
B_1	1.974	1.160	0.582	0.175	-0.047	-0.133
B_2	-11.204	-5.970	-2.239	0.397	1.834	2.379
B_3	37.233	22.364	11.650	3.942	-0.417	-2.252
B_4	-48.035	-29.008	-15.160	-5.051	0.803	3.389
B_5	23.823	14.741	8.015	2.972	-0.093	-1.597

(a)

(b)

(c)

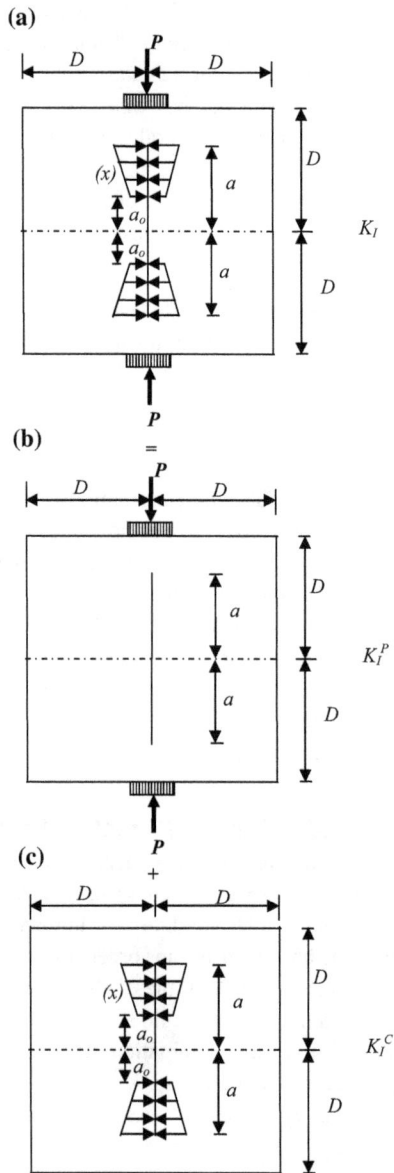

Fig. 2 Calculation of SIF using superposition method.

Fig. 2. The value of K_I is expressed in the following expression:

$$K_I = K_I^P + K_I^C \tag{4}$$

After determining the critical effective crack extension at unstable condition of loading, the unstable fracture toughness K_{IC}^{un} is determined using the revised LEFM formulae (Ince 2012) for which the stress intensity factor is expressed as:

$$K_I = \sigma_N \sqrt{D} \sqrt{\pi\alpha}\, k(\alpha, \beta) \tag{5}$$

$$k(\alpha, \beta) = A_0(\beta) + A_1(\beta)\alpha + A_2(\beta)\alpha^2 + A_3(\beta)\alpha^3 \\ + A_4(\beta)\alpha^4 + A_5(\beta)\alpha^5 \tag{6}$$

where $k(\alpha, \beta)$ is a geometric factor and coefficients A_i ($i = 0\text{--}5$) are the function of β as summarized in Table 1.

Equation (6) yields results within 0.7 % accuracy for $0.1 \le \alpha \le 0.9$ and $0 \le \beta \le 0.2$. Within the range of $0 \le \beta \le 0.2$, any value of coefficients A_i can be determined by linear interpolation. The unstable fracture toughness K_{IC}^{un} is calculated using Eq. (5) at maximum load P_u when a becomes equal to a_c and σ_N to σ_{Nu}.

If the crack initiation load P_{ini} is known from experiment, the initiation toughness K_{IC}^{ini} is calculated using Eq. (5) in which P is equal to P_{ini} and a is equal to a_o. Alternatively, it can be determined analytically by applying the following relation.

$$K_{IC}^{ini} = K_{IC}^{un} - K_{IC}^C \tag{7}$$

Equation (7) is known as inverse method for determining the initiation toughness.

4. Determination of SIF Due to Cohesive Stress

4.1 Cohesive Stress Distribution

The cohesive stress acting in the fracture process zone on STC test specimen is idealized as series of pair normal forces subjected symmetrically to central cracked specimen of finite strip and a finite width as shown in Fig. 3. The linearly varying distribution of cohesive stress is also shown in Fig. 4.

A centrally cracked specimen with finite strip of a finite width plate subjected to pair of normal forces as shown in Fig. 3 takes into consideration for a split tension test cube specimen where the value of the length/characteristic dimension (l/D) ratio becomes 1. The SIF due to cohesive stress distribution as shown in Fig. 4 becomes to cohesive toughness K_{IC}^C of the material at the critical loading condition with negative value because of closing stress in fictitious fracture zone. However, the absolute value of K_{IC}^C is taken as a contribution of the total fracture toughness (Xu and Reinhardt 1999b) at the critical condition.

At this loading condition, the crack-tip opening displacement (CTOD) is termed as critical crack-tip opening displacement ($CTOD_c$). In Fig. 4, the $\sigma_s(CTOD_c)$ is cohesive stress at the tip of initial notch where CTOD is equal to $CTOD_c$ and then $\sigma(x)$ can be expressed as:

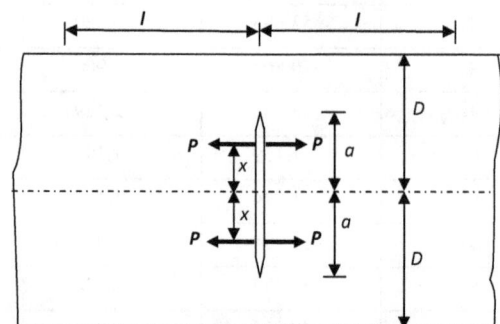

Fig. 3 Central cracked specimen with finite strip of a finite width plate subjected to pair of normal forces.

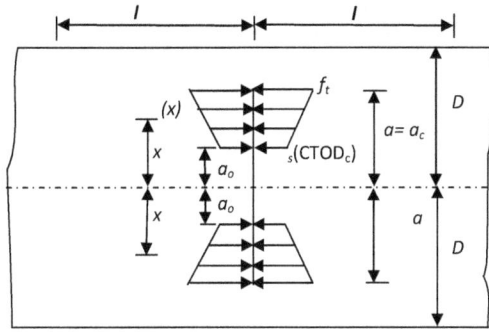

Fig. 4 Distribution of cohesive stress in the fictitious crack zone at critical load.

$$\sigma(x) = \sigma_s(CTOD_c) + \frac{x - a_o}{a - a_o}[f_t - \sigma_s(CTOD_c)]$$
$$for \ 0 \le CTOD \le CTOD_c \tag{8}$$

The value of $\sigma_s(CTOD_c)$ is calculated using softening functions of concrete. In the present work, the nonlinear softening function (Reinhardt et al. 1986) is used for the computation which can be expressed as:

$$\sigma(w)$$
$$= f_t \left\{ \left[1 + \left(\frac{c_1 w}{w_c} \right)^3 \right] \exp \left(\frac{-c_2 w}{w_c} \right) - \frac{w}{w_c}(1 + c_1^3) \exp(-c_2) \right\} \tag{9}$$

The value of total fracture energy of concrete G_F is expressed as:

$$G_F = w_c f_t \left\{ \frac{1}{c_2} \left[1 + 6 \left(\frac{c_1}{c_2} \right)^3 \right] - \left[1 + c_1^3 \left(1 + \frac{3}{c_2} + \frac{6}{c_2^2} + \frac{6}{c_2^3} \right) \right] \right.$$
$$\left. \frac{\exp(-c_2)}{c_2} - \left(\frac{1 + c_1^3}{2} \right) \exp(-c_2) \right\} \tag{10}$$

In which, $\sigma(w)$ is the cohesive stress at crack opening displacement w at the crack-tip and c_1 and c_2 are the material constants. Also, $w = w_c$ for $f_t = 0$, i.e., w_c is the maximum crack opening displacement at the crack-tip at which the cohesive stress becomes to be zero. The value of w_c is computed using Eq. (10) for a given set of values c_1, c_2 and G_F. For normal concrete the value of c_1 and c_2 is taken as 3 and 7, respectively.

4.2 Determination of CTOD$_c$

For a given value of critical crack mouth opening displacement CMOD$_c$, the crack opening displacement within the crack length $COD(x)$ is computed using the revised expression (Ince 2012) as given below.

$$COD(x) = CMOD_c$$
$$\times \left\{ \left(1 - \frac{x}{a} \right)^2 + \left[1.967 - 0.454(1 + \beta)^{6.363} \alpha^{1.984} \left(\frac{x}{a} \right)^{1.913} \right] \right.$$
$$\left. \times \left[\frac{x}{a} - \left(\frac{x}{a} \right)^2 \right]^2 \right\}^{1/2} \tag{11}$$

The accuracy of Eq. (11) is greater than 4 % for $0.1 \le \alpha \le 0.9$ and any value of β and is greater than 2.5 % for $0.1 \le \alpha \le 0.6$ and any value of β. The value of x is taken as a_o and a as a_c for evaluation of CTOD$_c$ using Eq. (11).

4.3 Calculation of Cohesive Toughness Using Weight Function Approach

According to weight function approach (Bueckner 1970, Rice 1972), the SIF for mode –I loading is given by following expression.

$$K_I = \int_0^a \sigma(x).m(x,a)dx \tag{12}$$

where $\sigma(x)$ is the distribution of stress along the crack line x in the uncracked body, the term $m(x,a)$ is known as weight function, a is the crack length and dx is the infinitesimal length along the crack surface. The four term universal form of weight function (Glinka and Shen 1991, Kumar and Barai 2008a, 2009a, 2010) is written as:

$$m(x,a) = \frac{2}{\sqrt{2\pi(a - x)}}$$
$$\times \left[1 + M_1(1 - x/a)^{1/2} + M_2(1 - x/a) + M_3(1 - x/a)^{3/2} \right] \tag{13}$$

For centrally through cracked specimen of infinite strip and a finite width subjected to pairs of normal forces symmetrically (Fig. 3), the weight function as given by Tada et al. (2000) is expressed as:

$$G(x,a) = \frac{2}{\sqrt{2D}} \left\{ 1 + 0.297 \sqrt{1 - \left(\frac{x}{a} \right)^2} \left[1 - \cos \left(\frac{\pi a}{2D} \right) \right] \right\}$$
$$F \left(\frac{a}{D}, \frac{x}{a} \right) F \left(\frac{a}{D}, \frac{x}{a} \right) = \sqrt{\tan \left(\frac{\pi a}{2D} \right)}$$
$$\times \left[1 - \left(\frac{\cos \frac{\pi a}{2D}}{\cos \frac{\pi x}{2D}} \right)^2 \right]^{-1/2}$$

Equation (14) as equivalently expressed in terms of universal weight function $m(x,a)$ of Eq. (13) by Wu et al. (2003) was used by Kumar and Pandey (2012) in the previous formulation. In the present investigation the weight function parameters M_1, M_2 and M_3 derived by Ince (2012) for the split tension cube specimen are used. According to Ince (2012) the parameters of four term weight function for a centrally through cracked specimen of finite strip and a finite width subjected to pairs of normal forces (Fig. 3) can be obtained as:

$$M_i = m_{i0} + m_{i1}\alpha + m_{i2}\alpha^2 + m_{i3}\alpha^3 + m_{i4}\alpha^4 + m_{i5}\alpha^5$$
$$+ m_{i6}\alpha^6 + m_{i7}\alpha^7 \tag{15}$$

where $\alpha = a/D$ and m_{ij} ($i = 1$–3 and $j = 0$–7) are the coefficients of the polynomial Eq. (15) as presented in

Table 2 Coefficients m_{ij} ($j = 0$–7) of the four term universal weight function parameters M_1, M_2 and M_3 (Ince 2012).

M_i	0	1	2	3	4	5	6	7
1	0.070	0.407	−5.405	49.393	−199.837	384.617	−359.928	132.792
2	−0.089	−2.017	24.839	−86.042	207.787	−243.596	114.431	
3	0.432	2.581	−31.022	134.511	-329.531	437.642	−292.768	69.925

Table 2. The sixth degree polynomial ($m_{i7} = 0$) is used for M_2. Equation (15) is valid for $0 \leq a/D \leq 0.9$ and $0 \leq x/a$ 1 (exactly 0.993). The accuracy of Eqs. (13) and (15) is greater than 3 % for all the split—tension cube specimens.

Once the weight function parameters are determined, Eq. (12) is used to calculate the SIF at critical condition (cohesive toughness) due to trapezoidal cohesive stress distribution as shown in Fig. 4. The value of $\sigma(x)$ in Eq. (12) is replaced by Eq. (8), hence the closed form expression of K_{IC}^C can be obtained in the following form.

$$
K_{IC}^C = \frac{2}{\sqrt{2\pi a}} \left\{ A_1 a \left[2s^{1/2} + M_1 s + \frac{2}{3} M_2 s^{3/2} + \frac{M_3}{2} s^2 \right. \right.
$$
$$
\left. + \frac{2}{5} M_4 s^{5/2} \right] + A_2 a^2 \left[\frac{4}{3} s^{3/2} + \frac{M_1}{2} s^2 + \frac{4}{15} M_2 s^{5/2} \right.
$$
$$
\left. \left. + \frac{4}{35} M_4 s^{7/2} + \frac{M_3}{6} \left\{ 1 - (a_o/a)^3 - 3sa_o/a \right\} \right] \right\}
$$

(16)

where, $A_1 = \sigma_s(CTOD_c)$, $A_2 = \frac{f_t - \sigma_s(CTOD_c)}{a - a_o}$ and $s = (1 - a_o/a)$, also $a = a_c$ at $P = P_u$. After computing the value of K_{IC}^C using Eq. (16), initiation toughness can be evaluated using Eq. (7).

5. Fictitious Crack Model and Material Properties for Double-K Fracture Model

The cohesive crack model (Modeer 1979; Petersson 1981; Carpinteri 1989; Planas and Elices 1991; Zi and Bažant

2003; Roesler et al. 2007, Park et al. 2008, Zhao et al. 2008, Kwon et al. 2008; Cusatis and Schauffert 2009; Elices et al. 2009; Kumar and Barai 2008b, b) is developed for STC and TPBT specimens to determine the input data such as P_u and $CMOD_c$ for these specimens. Three material properties such as modulus of elasticity E, uniaxial tensile strength f_t, and fracture energy G_F are required to model FCM. In this method, the governing equation of COD along the potential fracture line is written. The influence coefficients of the COD equation are determined using linear elastic finite element method. Four noded isoparametric plane elements are used in finite element calculation. The COD vector is partitioned according to the enhanced algorithm introduced by Planas and Elices (1991). Finally, the system of nonlinear simultaneous equation is developed and solved using Newton–Raphson method. For standard STC and TPBT specimens with $B = 100$ mm having size range $D = 200$-500 mm, the finite element analysis is carried out for which the one-quarter of STC and half of TPBT specimens are discretized due to symmetry as shown in Fig. 5 considering 80 numbers of equal isoparametric plane elements along the characteristic dimension D. In the discretization, both the specimens are divided into three bands perpendicular to characteristic dimension D such as $D/4$, $D/4$ and $D/2$ in case of STC specimen and $0.25D$, $0.75D$ and D in case of TPBT specimen as shown in Fig. 5. This arrangement facilitates to obtain finer mesh size near the potential fracture line. For STC specimen, the number of divisions is taken as 20, 5 and 5 in the bands $D/4$, $D/4$ and $D/2$ respectively whereas it is 20, 10 and 5 in the bands $0.25D$, $0.75D$ and D respectively for TPBT specimen. Ten nodes from top along the potential

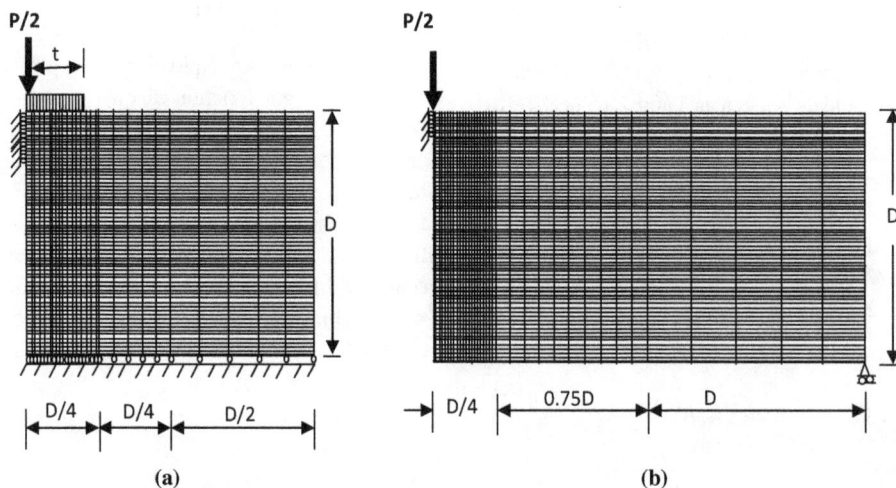

Fig. 5 Finite element discretization of test geometries. **a** Split tension cube test specimen, **b** Three point bend test specimen.

Table 3 Values of P_u and $CMOD_c$ obtained from FCM for TPBT and STC specimens for different specimen sizes.

D (mm)	a_o/D	P_u (kN)					$CMOD_c$ (mm)				
		For TPBT	For STC specimen				For TPBT	For STC specimen			
			Value of β for STC					Value of β for STC			
			0.0	0.05	0.1	0.15		0.0	0.05	0.1	0.15
500	0.3	10.73	20.66	20.81	21.24	21.96	0.0822	0.0426	0.0427	0.0442	0.0449
400	0.3	9.47	17.56	17.69	18.074	18.72	0.0720	0.0379	0.0380	0.0385	0.0403
300	0.3	7.94	14.15	14.27	14.604	15.18	0.0624	0.0316	0.0318	0.0323	0.0340
200	0.3	6.05	10.33	10.43	10.724	11.23	0.0510	0.0243	0.0251	0.0259	0.0288

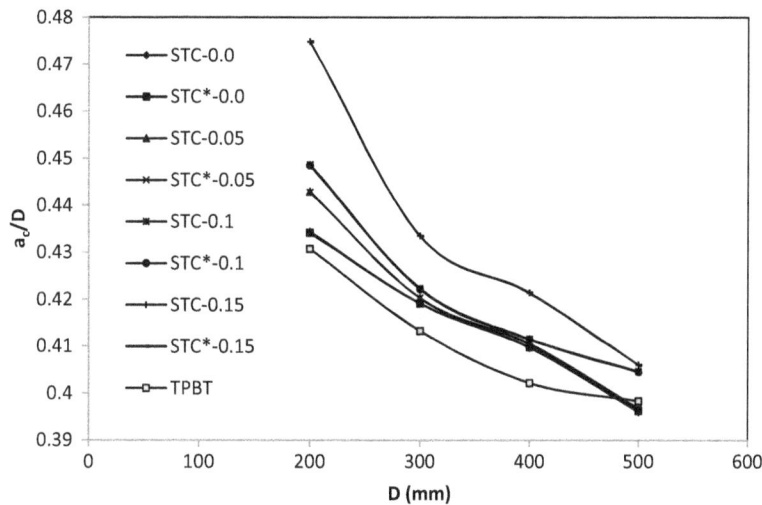

Fig. 6 Comparison in the values of a_c/D for STC obtained using the previous method (Kumar and Pandey 2012) and the present revised method.

fracture line are restrained against horizontal movement and all the nodes at the bottom perpendicular to fracture line are restrained against vertical movement in case of STC specimen. For the TPBT specimen, three nodes from top along the potential fracture line are restrained in horizontal direction. The concrete mix with material properties: $v = 0.18$, $f_t = 3.21$ MPa, $E = 30$ GPa, and $G_F = 103$ N/m along with nonlinear stress-displacement softening relation with $c_1 = 3$ and $c_2 = 7$ are used as the input parameters of FCM.

From simulation of FCM, the results of peak load P_u versus $CMOD_c$ for TPBT specimen at a constant a_o/D ratio of 0.3 are presented in Table 3. Similar results of peak load P_u and the corresponding $CMOD_c$ at different load distributed widths ($\beta = 0.0$, 0.05, 0.1 and 0.15) for STC specimens of varying sizes (200–500 mm) at a constant a_o/D ratio of 0.3 are also presented in Table 3.

6. Results and Discussion

The input parameters such as P_u, $CMOD_c$, E and softening function of concrete are required from the tests for determining double-K fracture parameters of concrete using weight function analytical method. In the present study, the values of E, f_t, nonlinear softening function (Eq. (9)) as

mentioned in Sect. 5 and the values of P_u-$CMOD_c$ for STC and TPBT specimens obtained from FCM are used to determine double-K fracture parameters. The *weight function method* with four terms is applied to calculate double-K fracture parameters in which the value of critical crack extension a_c is obtained using improved Eq. (1) for STC specimen. For given values of a_c and $CMOD_c$, the values of $CTOD_c$ are determined using revised Eq. (11). The values of a_c and $CTOD_c$ are also determined using corresponding equations presented in the previous work of Kumar and Pandey (2012) which were based on LEFM equations given by Ince (2010). The values of a_c and $CTOD_c$ for TPBT specimen are determined as mentioned elsewhere (Kumar and Barai 2008a, 2010). All the values of a_c and $CTOD_c$ determined as above are plotted in Figs. 6 and 7, respectively. For determining the value of K_{IC}^C using weight function method, first of all the four parameters M_1, M_2 and M_3 of four terms weight function are computed using Eq. (15) and Table 2, then closed form expression (Eq. (16)) is used to obtain the value of K_{IC}^C and finally the K_{IC}^{ini} is determined using inverse procedure (Eq. (7)). For TPBT specimen, double-K fracture parameters are determined in a similar manner using four terms *weight function method* as mentioned elsewhere (Kumar and Barai 2008a, 2010). Thus the values of K_{IC}^{un}, K_{IC}^C and K_{IC}^{ini} as obtained for STC for different

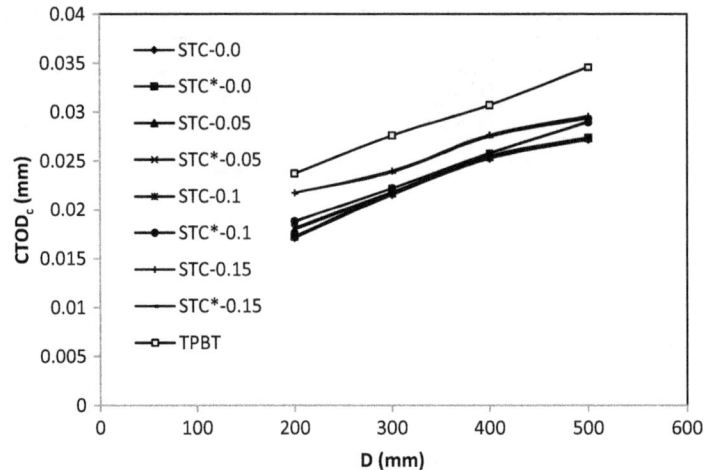

Fig. 7 Comparison in the values of $CTOD_c$ for STC obtained using the previous method (Kumar and Pandey 2012) and the present revised method.

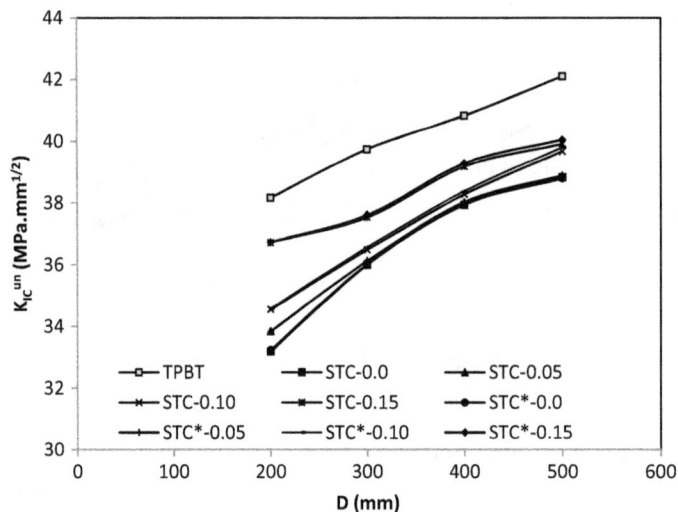

Fig. 8 Comparison of unstable fracture toughness for STC obtained using the previous method (Kumar and Pandey 2012) and the present revised method.

distributed-load widths ($0 \leq \beta \leq 0.15$) and TPBT specimens for specimen size $200 \leq D \leq 500$ mm at a_o/D ratio of 0.3 are plotted in Figs. 8, 9 and 10 respectively. The legends in Figs. 6, 7, 8, 9 and 10 marked with star (*) indicate the fracture parameters of the specimens determined using revised equations presented in this work whereas those legends with no mark with star (*) show the respective parameters of the specimens determined using the equations presented by Kumar and Pandey (2012).

From Figs. 6 and 7 it can be seen that the revised formulae and the previous LEFM equations (Kumar and Pandey 2012) yield the same results of critical values of effective crack length and crack tip opening displacement. These values for split tension cube specimen and three point bend test specimen also depend upon the size of the specimens and show similar pattern. The value a_c/D decreases with the increase in specimen size whereas $CTOD_c$ increases with the increase in specimen size. From Fig. 6 it can be seen that for STC specimen these parameters also depend on distributed-load

width and the a_c/D ratio shows maximum deviation for STC specimen with $\beta = 0.15$ from those obtained for TPBT for a given specimen size. This deviation is more for the lower specimen size and seems to be converging at higher specimen size. The a_c/D values for STC specimen are on higher side as compared with those of TPBT specimen for all values of distributed-load width ($0 \leq \beta \leq 0.15$) considered in the study. On an average, these values for STC for all values of β ($0 \leq \beta \leq 0.15$) are more than those for TPBT specimen by approximately 4.6 % and 0.43 % for $D = 200$ mm and $D = 500$ mm, respectively.

From Fig. 7 it can be observed that for STC specimen the value of $CTOD_c$ depends on distributed-load width and the value of $CTOD_c$ shows maximum deviation for STC specimen having $\beta = 0$ from those obtained for TPBT for a given specimen size. The $CTOD_c$ values for STC specimen are in lower side as compared with those of TPBT specimen for all values of distributed-load width ($0 \leq \beta \leq 0.15$). On an average, these values for STC for all values of β

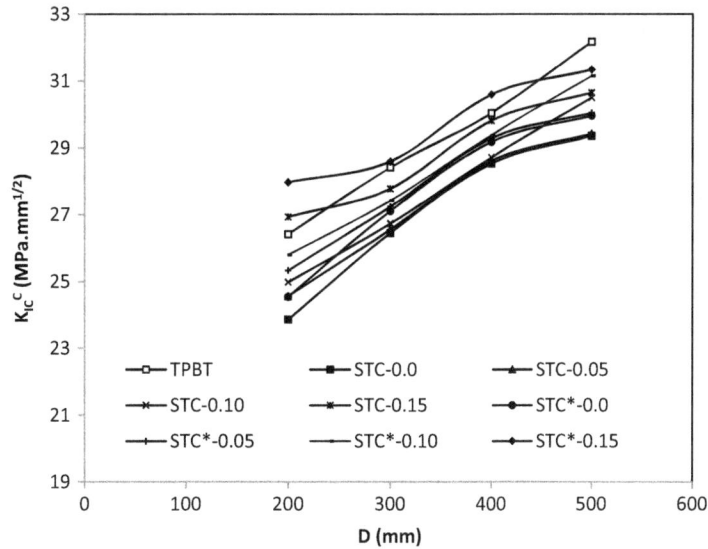

Fig. 9 Comparison of cohesive toughness for STC obtained using the previous method (Kumar and Pandey 2012) and the present revised method.

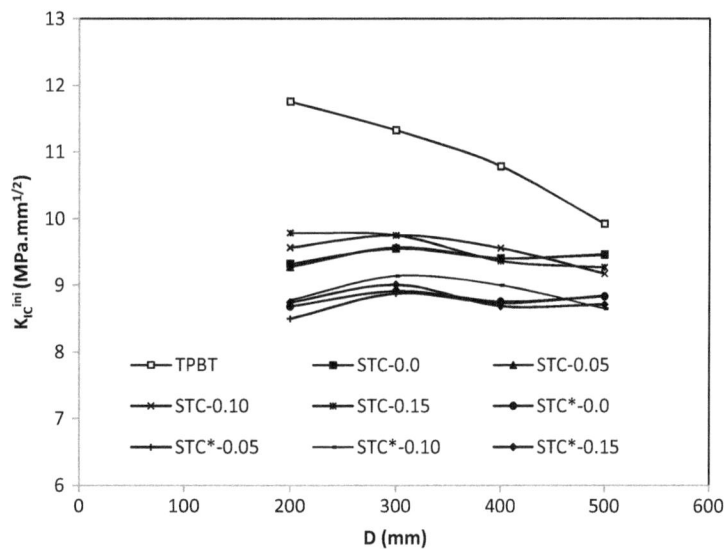

Fig. 10 Comparison of initial cracking toughness for STC obtained using (Kumar and Pandey 2012) and the present revised method.

$(0 \leq \beta \leq 0.15)$ are less than those for TPBT specimen by approximately 19.9 and 18.3 % for $D = 200$ mm and $D = 500$ mm respectively.

It can be observed from Fig. 8 that the values of K_{IC}^{un} determined using LEFM equations presented elsewhere (Kumar and Pandey 2012) and the revised LEFM equations in this work are the almost same for specimen sizes $(D = 200–500$ mm) for all values of β $(0 \leq \beta \leq 0.15)$. It is also seen from the figure that the unstable fracture toughness obtained from STC specimen is compatible with that of TPBT specimen. The value of K_{IC}^{un} for STC is the lowest for distributed-load width $\beta = 0$ and is the highest for $\beta = 0.15$ which is in close agreement with that obtained from TPBT specimen for all sizes of specimens. The values of K_{IC}^{un} are 36.70 and 39.91 MPa mm$^{1/2}$ for STC

with $\beta = 0.15$ and 38.16 and 42.10 MPa mm$^{1/2}$ for TPBT specimens for specimen size 200 mm and 500 mm respectively. It seems from Fig. 8 that there is relatively more difference in results of unstable fracture toughness between STC with $\beta = 0$ and TPBT. Therefore, in case STC specimen is adopted to replace TPBT to test unstable fracture toughness of concrete, the STC with $\beta = 0.15$ can be considered to be reasonable. That means the unstable fracture toughness of concrete can be determined using STC specimen.

The value of cohesive toughness obtained using equations presented elsewhere (Kumar and Pandey 2012) and the revised procedure in this work, varies with the value of β for STC specimen. The values of cohesive toughness for STC and TPBT specimens shown in Fig. 9 also show that these values either obtained using STC specimen or TPBT specimen are in consistent with each other.

The effect of finite strip in the present revised work over the infinite strip (previous work of Kumar and Pandey (2012)) of finite width cracked specimen on the cohesive toughness values for the $0 \leq \beta \leq 0.15$ is clearly observed from Fig. 9. It can be seen that for all values of distributed load width, the values of K_{IC}^C obtained considering the finite strip plate are slightly on higher side than those obtained considering the infinite strip plate.

For STC specimen with infinite strip and $\beta = 0$, the values of K_{IC}^C are found to be 29.36 MPa mm$^{1/2}$ and 23.87 MPa mm$^{1/2}$ for $D = 500$ mm and 200 mm respectively whereas those values are obtained as 29.96 MPa mm$^{1/2}$ and 24.55 MPa mm$^{1/2}$ for finite strip for $D = 500$ mm and 200 mm respectively. Similarly, for STC specimen with infinite strip and $\beta = 0.15$, the value of K_{IC}^C are found to be 30.64 MPa mm$^{1/2}$ and 26.93 MPa mm$^{1/2}$ for $D = 500$ mm and 200 mm respectively whereas those values are obtained as 31.34 MPa mm$^{1/2}$ and 27.98 MPa mm$^{1/2}$ for finite strip for $D = 500$ mm and 200 mm respectively. On an average for all values of β, the K_{IC}^C as obtained using finite strip is 2.14 and 3.29 % more than those obtained using infinite strip of plate for $D = 500$ mm and 200 mm respectively. Also, the values of K_{IC}^C as determined using finite strip of STC is 4.82 and 1.86 % less than those obtained using three point bend test specimen for $D = 500$ mm and 200 mm respectively. It is also observed from Fig. 9 that the size effect on the K_{IC}^C values for STC specimen is less significant than that presented for three point bend test.

It can be observed from Fig. 10 that for STC specimen with infinite strip and $\beta = 0$, the values of K_{IC}^{ini} are found to be 9.46 MPa mm$^{1/2}$ and 8.32 MPa mm$^{1/2}$ for $D = 500$ mm and 200 mm respectively whereas those values are obtained as 8.83 MPa mm$^{1/2}$ and 8.68 MPa mm$^{1/2}$ for finite strip for $D = 500$ mm and 200 mm respectively. Similarly, for STC specimen with infinite strip and $\beta = 0.15$, the values of K_{IC}^{ini} are found to be 9.26 MPa mm$^{1/2}$ and 9.78 MPa mm$^{1/2}$ for $D = 500$ mm and 200 mm respectively whereas those values are obtained as 8.71 MPa mm$^{1/2}$ and 8.74 MPa mm$^{1/2}$ for finite strip for $D = 500$ mm and 200 mm respectively. On an average for all values of β, the K_{IC}^{ini} obtained using finite strip is 6.21 and 5.96 % lower than those obtained using infinite strip of plate for $D = 500$ mm and 200 mm respectively. Also, the values of K_{IC}^{ini} as determined using finite strip of STC is 11.70 and 26.10 % less than those obtained using three point bend test specimen for $D = 500$ mm and 200 mm, respectively. According to the present trend, it seems that the difference in the value of K_{IC}^{ini} obtained between the STC and TPBT specimens may further increase for smaller size specimens such as 150 mm or 100 mm. As per the common convention, this difference should not be more than ±25 % in the fracture test which is a matter of further investigation. It is also seen from Fig. 10 that the size effect on the K_{IC}^{ini} values for STC specimen is less significant than that presented for three point bend test.

7. Conclusions

A revised formulation for determination of double-K fracture parameters using weight function method for split-tension cube test is presented in the paper. In the revised procedure, the weight function of the centrally cracked plate of finite strip with a finite width is used which is an improvement over the previous work of the authors. From the present study considering the specimen sizes ($D = 200$–500 mm) and distributed-load width ($0 \leq \beta \leq 0.15$) of split-tension cube test the following conclusions can be drawn.

- Use of weight function for split-tension cube test considering a centrally cracked plate of finite width with the finite strip or the infinite strip yields the same results of critical values of effective crack length, critical value of crack tip opening displacement and unstable fracture toughness of concrete.

- For all values of distributed load width ($0 \leq \beta \leq 0.15$), the values of cohesive toughness obtained considering the finite strip plate is slightly higher than those obtained considering the infinite strip plate. On an average cohesive toughness obtained using finite strip is 2.14 % and 3.29 % more than those obtained using infinite strip of plate for $D = 500$ mm and 200 mm, respectively

- Consequently, on an average for all values distributed load width ($0 \leq \beta \leq 0.15$), the initial cracking toughness determined using finite strip is 6.21 and 5.96 % lower than those obtained using infinite strip of finite width plate for $D = 500$ mm and 200 mm respectively.

- The value of unstable fracture toughness determined using finite strip of split-tension cube specimen is the lowest for distributed-load width $\beta = 0$ and is the highest for $\beta = 0.15$ which is in close agreement with that obtained from three point bed test for all sizes of specimens. Also, on an average for all values of the distributed-load width, the values of cohesive toughness determined using finite strip of split-tension cube specimen is 4.82 and 1.86 % less than those obtained using three point bend test specimen for $D = 500$ mm and 200 mm respectively. Further, on an average for all values of distributed-load width, the values of initial cracking toughness determined using finite strip of split-tension cube specimen is 11.70 and 26.10 % less than those obtained using three point bend test specimen for $D = 500$ mm and 200 mm, respectively.

References

Bažant, Z. P., Kim, J.-K., & Pfeiffer, P. A. (1986). Determination of fracture properties from size effect tests. *Journal of Structural Engineering ASCE, 112*(2), 289–307.

Bažant, Z. P., & Oh, B. H. (1983). Crack band theory for fracture of concrete. *Materials and Structures, 16*(93), 155–177.

Bueckner, H. F. (1970). A novel principle for the computation of stress intensity factors. *Zeitschrift für Angewandte Mathematik und Mechanik, 50*, 529–546.

Carpinteri, A. (1989). Cusp catastrophe interpretation of fracture instability. *Journal of the Mechanics and Physics of Solids, 37*(5), 567–582.

Choubey, R. K., Kumar, S., & Rao, M. C. (2014). Effect of shear-span/depth ratio on cohesive crack and double-K fracture parameters. *International Journal of Construction, 2*(3), 229–247.

Cusatis, G., & Schauffert, E. A. (2009). Cohesive crack analysis of size effect. *Engineering Fracture Mechanics, 76*, 2163–2173.

Elices, M., Rocco, C., & Roselló, C. (2009). Cohesive crack modeling of a simple concrete: Experimental and numerical results. *Engineering Fracture Mechanics, 76*, 1398–1410.

Glinka, G., & Shen, G. (1991). Universal features of weight functions for cracks in Mode I. *Engineering Fracture Mechanics, 40*, 1135–1146.

Hillerborg, A., Modeer, M., & Petersson, P. E. (1976). Analysis of crack formation and crack growth in concrete by means of fracture mechanics and finite elements. *Cement and Concrete Research, 6*, 773–782.

Hu, S., & Lu, J. (2012). Experimental research and analysis on double-K fracture parameters of concrete. *Advanced Science Letters, 12*(1), 192–195.

Hu, S., Mi, Z., & Lu, J. (2012). Effect of crack-depth ratio on double-K fracture parameters of reinforced concrete. *Applied Mechanics and Materials, 226–228*, 937–941.

Ince, R. (2010). Determination of concrete fracture parameters based on two-parameter and size effect models using split-tension cubes. *Engineering Fracture Mechanics, 77*, 2233–2250.

Ince, R. (2012). Determination of the fracture parameters of the Double-K model using weight functions of split-tension specimens. *Engineering Fracture Mechanics, 96*, 416–432.

Isida, M. (1971). Effect of width and length on stress intensity factor of internally cracked plates under various boundary conditions. *International Journal of Fracture, 7*, 301–316.

Jenq, Y. S., & Shah, S. P. (1985). Two parameter fracture model for concrete. *Journal of Engineering Mechanics, 111*(10), 1227–1241.

Karihaloo, B. L., & Nallathambi, P. (1991). Notched beam test: Mode I fracture toughness. In S. P. Shah & A. Carpinteri (Eds.), *Fracture mechanics test methods for concrete, report of RILEM Technical Committee 89-FMT* (pp. 1–86). London, UK: Chamman & Hall.

Kumar, S. (2010). Behaviour of fracture parameters for crack propagation in concrete. Ph.D. Thesis submitted to Indian Institute of Technology, Kharagpur, India.

Kumar, S., & Barai, S. V. (2008a). Influence of specimen geometry on determination of double-K fracture parameters of concrete: A comparative study. *International Journal of Fracture, 149*, 47–66.

Kumar, S., & Barai, S. V. (2008b). Cohesive crack model for the study of nonlinear fracture behaviour of concrete. *Journal of the Institution of Engineers (India), 89*, 7–15.

Kumar, S., & Barai, S. V. (2009a). Determining double-K fracture parameters of concrete for compact tension and wedge splitting tests using weight function. *Engineering Fracture Mechanics, 76*, 935–948.

Kumar, S., & Barai, S. V. (2009b). Effect of softening function on the cohesive crack fracture parameters of concrete CT specimen. *Sadhana-Academy Proceedings in Engineering Sciences, 36*(6), 987–1015.

Kumar, S., & Barai, S. V. (2010). Determining the double-K fracture parameters for three-point bending notched concrete beams using weight function. *Fatigue & Fracture of Engineering Materials & Structures, 33*(10), 645–660.

Kumar, S., & Pandey, S. R. (2012). Determination of double-K fracture parameters of concrete using split-tension cube test. *Computers and Concrete, 9*(1), 1–19.

Kumar, S., Pandey, S. R., & Srivastava, A. K. L. (2013). Analytical methods for determination of double-K fracture parameters of concrete. *Advances in Concrete Construction, 1*(4), 319–340.

Kumar, S., Pandey, S. R., & Srivastava, A. K. L. (2014). Determination of double-K fracture parameters of concrete using peak load method. *Engineering Fracture Mechanics, 131*, 471–484.

Kwon, S. H., Zhao, Z., & Shah, S. P. (2008). Effect of specimen size on fracture energy and softening curve of concrete: Part II. Inverse analysis and softening curve. *Cement Concrete Res, 38*, 1061–1069.

Lee, J., & Lopez, M. M. (2014). An experimental study on fracture energy of plain concrete. *International Journal of Concrete Structures and Materials, 8*(2), 129–139.

Modeer, M. (1979). A fracture mechanics approach to failure analyses of concrete materials. Report TVBM-1001, Division of Building Materials. University of Lund, Sweden.

Murthy, A. R., Iyer, N. R., & Prasad, B. K. R. (2012). Evaluation of fracture parameters by Double-G, Double-K models and crack extension resistance for high strength and ultra high strength concrete beams. *Computers Materials & Continua, 31*(3), 229–252.

Nallathambi, P., & Karihaloo, B. L. (1986). Determination of specimen-size independent fracture toughness of plain concrete. *Magazine of Concrete Research, 135*, 67–76.

Park, K., Paulino, G. H., & Roesler, J. R. (2008). Determination of the kink point in the bilinear softening model for concrete. *Engineering Fracture Mechanics, 7*, 3806–3818.

Petersson, P. E. (1981). Crack growth and development of fracture zone in plain concrete and similar materials. Report No. TVBM-100, Lund Institute of Technology, Sweden.

Planas, J., & Elices, M. (1991). Nonlinear fracture of cohesive material. *International Journal of Fracture, 51*, 139–157.

Reinhardt, H. W., Cornelissen, H. A. W., & Hordijk, D. A. (1986). Tensile tests and failure analysis of concrete. *Journal of Structural Engineering, 112*(11), 2462–2477.

Rice, J. R. (1972). Some remarks on elastic crack-tip stress fields. *International Journal of Solids and Structures, 8*, 751–758.

RILEM Draft Recommendation (TC50-FMC). (1985). Determination of fracture energy of mortar and concrete by means of three-point bend test on notched beams. *Materials and Structures, 18*(4), 287–290.

Roesler, J., Paulino, G. H., Park, K., & Gaedicke, C. (2007). Concrete fracture prediction using bilinear softening. *Cement Concrete Composites, 29*, 300–312.

Tada, H., Paris, P. C., & Irwin, G. R. (2000). *Stress analysis of cracks handbook* (3rd ed.). New York, NY: ASME Press.

Timoshenko, S. P., & Goodier, J. N. (1970). *Theory of elasticity* (3rd ed.). New York, NY: McGraw Hill.

Wu, Z., Jakubczak, H., Glinka, G., Molski, K., & Nilsson, L. (2003). Determination of stress intensity factors for cracks in complex stress fields. *Archive of Mechanical Engineering, 50*(1), s41–s67.

Xu, S., & Reinhardt, H. W. (1998). Crack extension resistance and fracture properties of quasi-brittle materials like concrete based on the complete process of fracture. *International Journal of Fracture, 92*, 71–99.

Xu, S., & Reinhardt, H. W. (1999a). Determination of double-K criterion for crack propagation in quasi-brittle materials, Part I: Experimental investigation of crack propagation. *International Journal of Fracture, 98*, 111–149.

Xu, S., & Reinhardt, H. W. (1999b). Determination of double-K criterion for crack propagation in quasi-brittle materials, Part II: Analytical evaluating and practical measuring methods for three-point bending notched beams. *International Journal of Fracture, 98*, 151–177.

Xu, S., & Reinhardt, H. W. (1999c). Determination of double-K criterion for crack propagation in quasi-brittle materials, Part III: Compact tension specimens and wedge splitting specimens. *International Journal of Fracture, 98*, 179–193.

Xu, S., & Reinhardt, H. W. (2000). A simplified method for determining double-K fracture meter parameters for three-point bending tests. *International Journal of Fracture, 104*, 181–209.

Xu, S., & Zhang, X. (2008). Determination of fracture parameters for crack propagation in concrete using an energy approach. *Engineering Fracture Mechanics, 75*, 4292–4308.

Xu, S., & Zhu, Y. (2009). Experimental determination of fracture parameters for crack propagation in hardening cement paste and mortar. *International Journal of Fracture, 157*, 33–43.

Zhang, X., & Xu, S. (2011). A comparative study on five approaches to evaluate double-K fracture toughness parameters of concrete and size effect analysis. *Engineering Fracture Mechanics, 78*, 2115–2138.

Zhang, X., Xu, S., & Zheng, S. (2007). Experimental measurement of double-K fracture parameters of concrete with small-size aggregates. *Frontiers of Architecture and Civil Engineering in China, 1*(4), 448–457.

Zhao, Z., Kwon, S. H., & Shah, S. P. (2008). Effect of specimen size on fracture energy and softening curve of concrete: Part I. Experiments and fracture energy. *Cement Concrete Res, 38*, 1049–1060.

Zhao, Y., & Xu, S. (2002). The influence of span/depth ratio on the double-K fracture parameters of concrete. *Journal of China Three Gorges University (Natural Sciences), 24*(1), 35–41.

Zi, G., & Bažant, Z. P. (2003). Eignvalue method for computing size effect of cohesive cracks with residual stress, with application to kink-bands in composites. *International Journal of Engineering Science, 41*, 1519–1534.

7

Laboratory Simulation of Corrosion Damage in Reinforced Concrete

S. Altoubat[1],*, M. Maalej[1], and F. U. A. Shaikh[2]

Abstract: This paper reports the results of an experimental program involving several small-scale columns which were constructed to simulate corrosion damage in the field using two accelerated corrosion techniques namely, constant voltage and constant current. A total of six columns were cast for this experiment. For one pair of regular RC columns, corrosion was accelerated using constant voltage and for another pair, corrosion was accelerated using constant current. The remaining pair of regular RC columns was used as control. In the experiment, all the columns were subjected to cyclic wetting and drying using sodium chloride (NaCl) solution. The currents were monitored on an hourly interval and cracks were visually checked throughout the test program. After the specimens had suffered sufficient percentage steel loss, all the columns including the control were tested to failure in compression. The test results generated show that accelerated corrosion using impressed constant current produces more corrosion damage than that using constant voltage. The results suggest that the constant current approach can be better used to simulate corrosion damage of reinforced concrete structures and to assess the effectiveness of various materials, repair strategies and admixtures to resist corrosion damage.

Keywords: corrosion, damage, potential, current, reinforced concrete.

1. Introduction

Chloride-induced steel corrosion is one of the major worldwide deterioration problems for steel reinforced concrete structures. Especially with the use of de-icing salts in cold climate regions, key bridge components such as the supporting columns are vulnerable to corrosion attack. Other sources of chloride contamination in structures may be saline contamination of aggregate component in concrete or the water used in concrete batching or from direct or indirect exposure to marine environments.

Much effort has been directed towards the design of new concrete mixes to reduce corrosion or establishment of sound techniques for the repair of existing structures (Bonacci and Maalej 2000). In order to evaluate the effectiveness of repair materials and methods in addressing reinforcement corrosion, accelerated corrosion techniques are often used.

Accelerated corrosion tests conducted in the laboratory are used to predict corrosion behaviour when service history is lacking and time or budget constraints prohibit field-testing (Davis 2000). While the desired service lives of many

structures can range from 10 to 75 years, the typical time allowed for testing of candidate materials to support a materials selection process is measured in weeks or months. The purpose of any accelerated laboratory test is to provide reliable information on the performance of candidate materials or coatings in service. It is also typical that accelerated laboratory corrosion tests increase the severity of the environmental conditions, for example, by exposing the materials to more concentrated solutions, or increased periods of wetting. The environment used might simulate a humid tropical area, seaside air with high salt content, or one of many others.

Several research studies have used accelerated corrosion tests to study the corrosion resistance of reinforced concrete structural members incorporated new materials and/or designs. While several studies had employed a constant voltage accelerated corrosion regimes (Pellegrini-Cervantes et al. 2013; Deb and Pradhan 2013; Kishore Kumar et al. 2012; Lee et al. 2000; Al-Swaidani and Aliyan 2015), several others had also used constant current accelerated corrosion regimes (Talakokula et al. 2013; El Maaddawy et al. 2005; Kashani et al. 2013; El Maaddawy and Soudki 2003; Pritzl et al. 2014). One of the studies (El Maaddawy and Soudki 2003) focused on the effectiveness of the impressed (constant) current technique using current densities in the range of 100–500 $\mu A/cm^2$ to simulate corrosion of steel reinforcement in concrete. However, no studies could be found which had considered the effectiveness of the using constant voltage technique versus the constant current technique in simulating corrosion damage in reinforced concrete. In this paper, an experimental program is designed

---

[1]Department of Civil & Environmental Engineering, College of Engineering, University of Sharjah, Sharjah, UAE.
*Corresponding Author; E-mail: saltoubat@sharjah.ac.ae
[2]Department of Civil Engineering, Curtin University, Perth, Australia.

to study which of the two accelerated corrosion techniques (using constant voltage and constant current) produces more corrosion damage for similar amount of induced corrosion level (steel loss).

2. Research Significance

The success of any laboratory corrosion experiments depends largely on the ability of simulating field damage through an accelerated corrosion regime. In the field, significant corrosion damage is generated as a result of small percentage steel loss. In the laboratory, however, only minimal corrosion damage is generated as a result of high percentage steel loss. An ideal accelerated corrosion regime will not only produce the necessary damage within the required time frame, but produces damage that is consistent with that found in the field. This paper presents the evaluation of two different accelerated corrosion regimes: constant voltage and constant current to identify which of the two techniques provides a closer damage simulation to that in the field. As research work on corrosion is still being conducted to date using both techniques, the outcome of the present study will impact future research work in this field.

3. Test Program

3.1 Experimental Plan
The overall experimental program consists of two parts. The first part of the program is to allow test specimens to corrode to around 8 % steel loss using two methods of accelerated corrosion regimes, that is using constant voltage and constant current. The second part of the program is to structurally test the corroded specimens to failure by compression to determine the maximum compression load. The results will then be discussed in relation to the objectives of this study.

The test specimens used in this study were intended to simulate possible existing conditions in reinforced concrete bridge columns in which the concrete would be subjected to salt contamination. The test specimens were immersed in (3 % by weight of water) sodium chloride solution to simulate chloride contamination of the cover by the use of de-icing salt spray. A total of six circular columns were constructed for this study.

3.2 Design of Specimen
The cross sectional details and overall dimensions of a typical column specimen are shown in Figs. 1 and 2, respectively. The largest possible dimension of the column specimen was selected such that it allows for ease of handling during the various stages of the experiment. Furthermore, the same column size was selected for all specimens in this comparative study. Incorporated into each column were six 13 mm diameter deformed reinforcing bars, oriented in the longitudinal direction of the column, and 6 mm diameter spiral at a spacing of 44 mm. A hollow stainless steel pipe

with 20 mm diameter and 2 mm thickness was placed longitudinally in the centre of the column to act as a cathode. Holes of 2 mm in diameter were drilled on four sides at 20 mm spacing along the length of the hollow pipe within the test region to increase oxygen exposure. The reinforcing bars were also pre-drilled and tapped at the protruding end to facilitate the electrical connection from the power supply to the specimen.

To eliminate any possible end effects, the upper and lower 165 mm sections of the reinforcement cage were coated with epoxy to discourage corrosion.

3.3 Specimen Fabrication
Positioning of the six reinforcement bars and the centre cathode bar was done with the help of a wooden base, where bars were slotted in pre-drilled holes. The specimens were cast in the inverted position such that each bar protruded out of one end of the specimen by 15 mm when in the upright position. The protrusions of the bars were meant to allow connection of the electric circuit from the power supply to the specimens. After the main bars were put in place, spiral reinforcements were then tied to the main bars. Subsequently, epoxy coating was applied to both ends of the region of each specimen to prevent corrosion to these regions. Spacer blocks were also put along the spiral reinforcements to ensure required concrete cover. Large pipes of 208 mm diameter were also cut which formed the moulds of the specimens. A slit was made along the longitudinal direction of each pipe to facilitate the removal of the formwork upon demoulding. The slit was then closed by tying wires around the pipe. After that, the pipes were placed onto the wooden base and sealed along the circumference using silicone. Silicone was also applied along the slit to close up the gap.

One end of the stainless steel cathode bar which was going to be embedded in the concrete was taped to prevent concrete from flowing in during the pouring process. A plastic tubing of approximately the same diameter and length as the cathode bar was also inserted into the latter to prevent the cement and water from seeping through the numerous small holes drilled along the length of the cathode bar. The plastic tubing was meant to be removed after the concrete had hardened. Once that was done, the specimens were ready to be cast. After casting, the specimens were wrapped in plastic sheets and were subsequently cured outdoors using wet gunnysacks for 28 days.

3.4 Materials
Ordinary Portland cement was used for all concrete columns in this experimental program with 10 mm maximum coarse aggregate size. The concrete mix proportions for cubic meter of concrete is as follows: cement (OPC) = 300 kg; fine aggregate = 925 kg; coarse aggregate (10 mm) = 1000 kg, water = 185 kg, and admixture (Daracem 100) = 928 ml. The measured average 28-day concrete compressive cube strength was 29.02 MPa. The yield strengths of the 13 mm diameter round deformed bars and 6 mm diameter spiral reinforcing bars were 462 and 285 MPa, respectively.

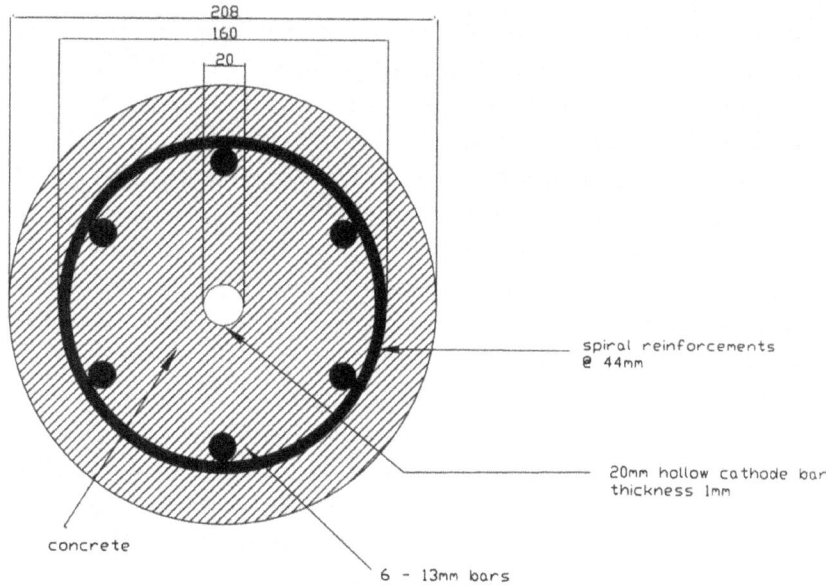

Fig. 1 Cross sectional details of the specimen.

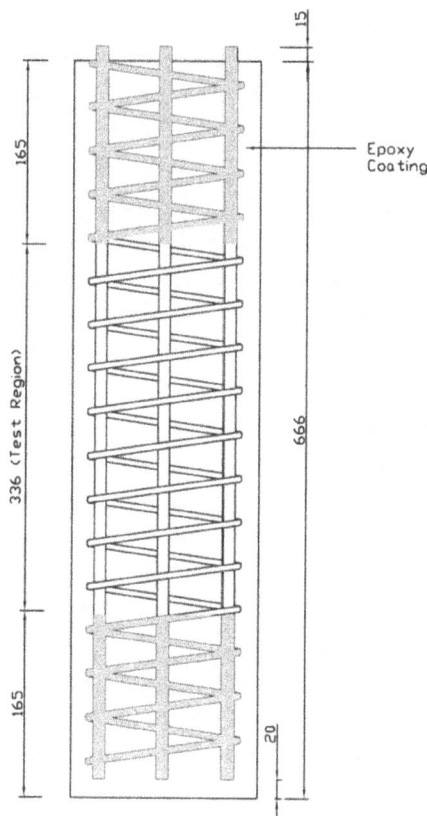

Fig. 2 Overall dimensions of a typical specimen.

3.5 Accelerated Corrosion Testing

Six concrete columns were tested in this study. Two out of the six concrete columns (labeled C1 and C2) were used as control for structural testing purpose and were not subjected to any accelerated corrosion. The remaining four columns were subjected to accelerated corrosion. Two concrete columns (labeled C3 and C4) were subjected to accelerated corrosion using constant voltage and two other concrete

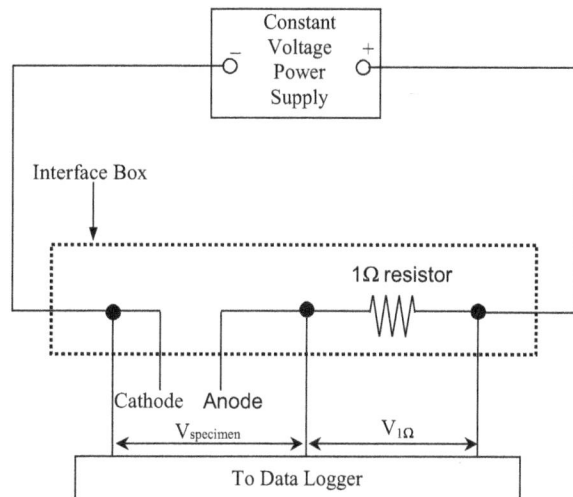

Fig. 3 Experimental setup for constant voltage.

columns (labeled C5 and C6) were subjected to accelerated corrosion using constant current.

All the columns (except the control) were subjected to cyclic wetting and drying using three percent sodium chloride (NaCl) solution by weight of water. The cycle is 1 day wet and two and a half days dry. During the wet cycle, the sodium chloride solution was filled up to the boundary of the upper test region and during the dry cycle, sodium chloride solution was pumped out to below the base level of the specimens. Matsuoka (1987) found that a cyclic wetting and drying test is more useful than continuous wetting or immersion in simulating actual field damage within a short time frame.

3.6 Constant Voltage Setup

Figure 3 shows a schematic diagram of the circuit setup for the accelerated corrosion test using constant voltage. The circuit basically consists of a one-ohm (1 Ω) resistor and the specimen connected in series with the constant voltage power supply. The maximum voltage that the power supply can provide is 32 volts.

The 1 Ω resistor and the wires connecting the specimen to the circuit are contained in a specially designed interface box. Wires were also connected from the top of the interface box to the data logger to measure the output voltage readings across the specimen and the 1 Ω resistor.

3.7 Constant Current Setup

The experimental setup for accelerated corrosion under constant current is similar to that used for constant voltage except that the power supply is replaced by a constant current power supply.

3.8 Testing Procedure

First, the circuits for constant voltage and constant current were set up. Then, the circuits for the constant voltage were closed and the voltages across the 1 Ω resistor were recorded. According to Ohm's Law (V = IR), the current flowing in the circuit is effectively the same value as the voltage reading since the resistor has a resistance of 1 Ω. Next, the currents for the constant current circuits were adjusted such that at the start of the accelerated corrosion, all concrete columns were subjected to the same magnitude of corrosion current. The specimens were subsequently allowed to corrode and the current readings were taken on an hourly interval throughout the experiment until the desired percentage steel loss (i.e. 8 %) had been achieved. By knowing the voltage readings across the specimens, it was also possible to monitor the resistances of all the specimens throughout the experiment since the currents flowing in the various circuits were already known.

To accelerate the corrosion process, the specimens were subjected to cyclic wetting and drying. The cyclic wetting and drying will not only provide the necessary moisture for the corrosion reaction, but also a constant fresh supply of chloride ions to the reinforcement bars.

Pore saturation has two opposing effects (Liu and Weyers 1998): As pore saturation increases, the rate of corrosion increases too since the resistivity of the specimen decreases. However, the rate of oxygen diffusion, which is essential to corrosion, decreases. So initially, due to an improved conductivity, the corrosion rate will increase, but as time goes by, the corrosion rate decreases rapidly to a low value due to the strongly obstructed supply of oxygen. Therefore, to balance these opposing forces, it is important to find the optimum cycle which will be the most effective in producing corrosion damage in the specimens.

Based on a previous study by Lee et al. (2000), it was found that a cycle of 1 day wet and two and a half days dry produces the most corrosion damage to the specimens. So, in the present experiment, the 1 day (1 D) wet and two and a half days (2.5 D) dry cycle was adopted.

Throughout the experimental program, the corrosion current, circumferential expansion of the specimens and cracking were monitored on a periodic basis. From these, the amount of steel loss Δw calculated based on Faraday's Law

using Eq. (1), extent of damage caused by the formation of expansive corrosion products and cracking pattern can be determined.

$$\Delta w = \frac{MIt}{zF} \qquad (1)$$

where Δw = mass of steel consumed due to corrosion (grams), I is the current (amperes), t is the time (s), F is the 96,500 (amp s), z is the ionic charge (two for Fe), M is the atomic weight of metal (56 g for Fe).

It should be noted that other researchers have found that steel loss calculated using Faraday's Law tends to overestimate the actual steel loss (Aiello 1996). When the specimens are subjected to corrosion, the specimens will expand due to the formation of the corrosion products which are expansive in nature. To measure the circumferential expansion of each column, a specially fabricated mechanical expansion collar was placed at mid-height of each column. Each collar was made from 12 mm wide stainless steel sheets and included a small opening held closed by two stainless steel springs. A micrometer was used to measure the change in the opening size to the nearest one-hundredth of a millimeter, to determine the circumferential expansion of the column at mid-height.

After the concrete columns had achieved 8 % steel loss, the experiment was stopped and preparation works were done to the columns prior to structural testing. The protruding longitudinal reinforcement bars were cut off to level with the concrete surface.

In order for the columns to fail within the test region (since only the test region is allowed to corrode), both end regions of the columns were wrapped in glass fibre reinforced polymer wrap to strengthen the both ends.

All columns were tested under load control using an Instron Actuator IV testing machine with an ultimate capacity of 2000 kN to determine the maximum compression load. Two strain gauges were mounted horizontally along the circumference at the mid height of the test region, 180° apart to measure the circumferential strain. Located next to each horizontal strain gauge was another strain gauge mounted longitudinally to measure the longitudinal strain. In addition to that, two linear variable differential transducers (LVDT) were also placed 180° apart at the upper test region to measure the axial deformation. The gauge length of each LVDT was 50 mm. Figure 4 shows the setup prior to the start of the structural test.

4. Results and Discussion

4.1 Effect of Accelerated Steel Loss, Cracking and Circumferential Expansion

The following section presents the results obtained after around 48 days of accelerated corrosion for the four column specimens, excluding the controls, after the specimens had undergone sufficient corrosion. Steel loss, circumferential expansion and visual observation of cracking were used as

Fig. 4 Experimental setup for a typical column.

primary indicators of damage. The percentage steel losses at the end of the accelerated corrosion experiment for the constant voltage and constant current specimens were 8.13 and 8.10 %, respectively.

For the constant voltage specimens, an initial voltage of 12 volts was applied. The average output current reading was noted (0.5 A) and was subsequently applied to the constant current concrete specimens. However, after about 3 days of accelerated corrosion and upon computing the current densities, the applied constant voltage and applied contact current were reduced to 9.6 volts and 0.4 A, respectively to limit the current densities in the specimens to about 250 $\mu A/cm^2$. The currents passing through the four specimens undergoing accelerated corrosion over time are shown in Figs. 5 and 6.

The accumulated steel loss for the four specimens undergoing accelerated corrosion over time is shown in Fig. 7. Steel losses were calculated using Faraday's Law as indicated earlier. From the rate of steel loss over time plot shown in Fig. 8, it can be seen that the rate of steel losses for the constant current specimens are approximately constant throughout the experiment. However, for the constant

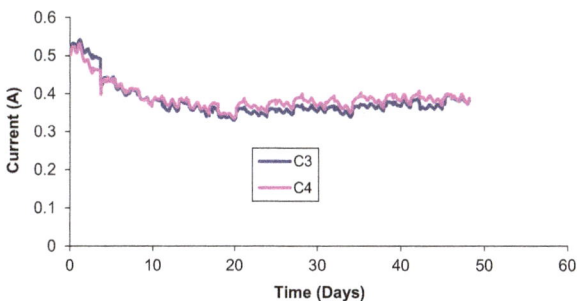

Fig. 5 Variation of current output with time for C3 and C4 (constant voltage).

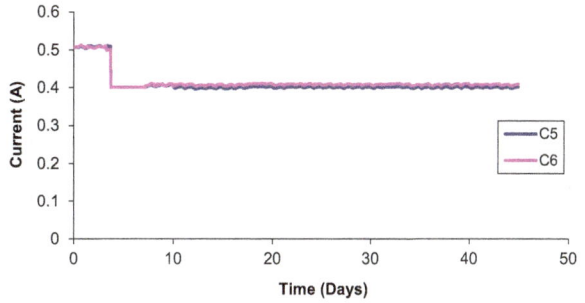

Fig. 6 Variation of current output with time for C5 and C6 (constant current).

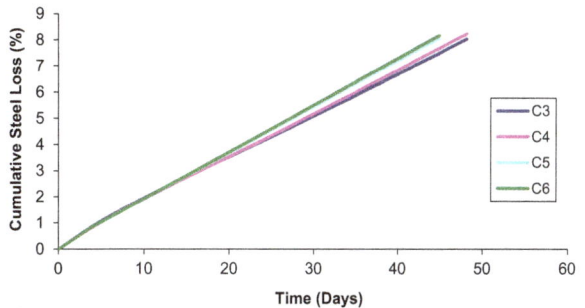

Fig. 7 Variation of cumulative steel loss with time for all specimens.

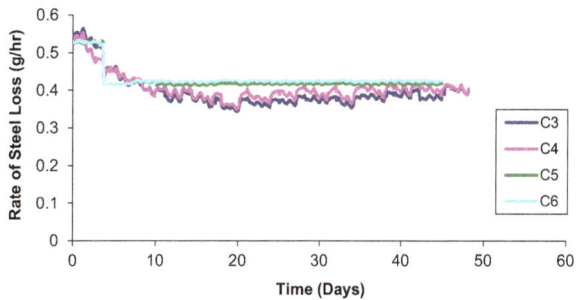

Fig. 8 Variation of the rate of steel loss with time for all specimens.

voltage specimens, there are fluctuations in the rate of steel losses. During the wet cycle, pore volumes get saturated, and the rate of corrosion increases as the resistivity of the specimen decreases. Consequently, more steel is being lost during the wet cycle which resulted in a peak in the rate of steel loss. Conversely, during the dry cycle, as water gets drained out, the pore volumes get less saturated and resistivity of the specimen increases, resulting in a lower current reading which is proportional to the rate of steel loss. Consequently, the steel loss during the dry cycles drops.

Circumferential expansion of each specimen undergoing accelerated corrosion was determined using a specially fabricated mechanical expansion collar fitted at mid height of the column. The circumferential strains at mid height for the various specimens over time are shown in Fig. 9. The graph shows that there was little circumferential expansion for the first 14 days. Subsequently, the constant current specimens (C5 and C6) began to show significant increases

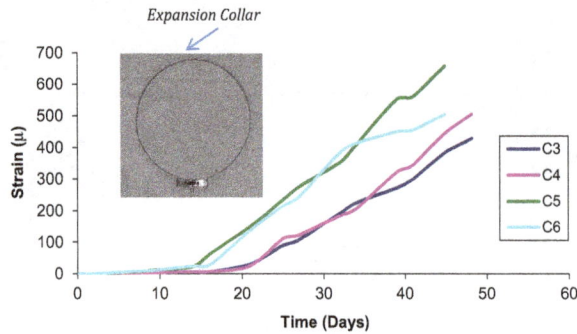

Fig. 9 Variation of the circumferential expansion with time for all specimens as measured with mechanical expansion collar.

in the circumferential strain. The constant voltage specimens (C3 and C4) took around 21 days for significant circumferential expansion to take place. At any given time beyond the first 14 days, the expansion was more for the constant current specimens than for the constant voltage specimens as seen in Fig. 9. Since, circumferential expansion is one of the primary indicators of damage, it can be said that the constant current specimens suffered more damage than constant voltage specimens, although the percentage steel losses were comparable (8.10 % for constant current versus 8.13 % for constant voltage).

The specimens were constantly checked for any visual cracks on the surfaces, especially within the test regions. For both constant current concrete specimens C5 and C6, visible cracks were observed after approximately 15 days of accelerated corrosion. But for the constant voltage specimens, no visible cracks were observed until after approximately 19 days.

For all specimens, cracking began initially as short, vertical and discontinuous in the longitudinal direction near the upper test region. However, as time went by, the cracks began to propagate longitudinally in both directions and some of the vertical cracks joined together. Existing cracks were also seen to be widening. The number and location of each crack was sketched, approximately to scale, on paper as they appeared on the specimen. Typical crack patterns are shown in Figs. 10, 11, for the constant voltage and constant current specimens, respectively.

Horizontal cracks were also observed at the upper test region for both constant current concrete specimens towards

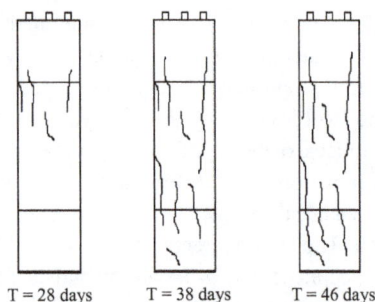

Fig. 11 Typical cracking patterns for constant current specimens (C5 in this figure).

the end of the experiment. However, this was not observed in the constant voltage specimens. Also, there were signs of delamination at the top concrete face where the reinforcement bars protruded for electrical connection for both constant current specimens.

From the observation of cracking patterns of the concrete specimens, it is clear that there were more cracks that had formed in the constant current specimens than in the constant voltage specimens. The cracks observed in the constant current specimens were also longer and the maximum crack widths measured were also larger than those in the constant voltage specimens as shown in Table 1. Photos of developed cracks at the end of the accelerated corrosion regimes are shown in Fig. 12.

One observation from the result is that, in Fig. 8, the general trend lines for the constant voltage specimens are such that the rate of steel loss decreased to a certain value (at around 20 days after the start of the experiment) and subsequently, started to slowly increase. This increase in the rate of steel loss can be explained by the marked increase in the circumference expansion in the specimens themselves. It was observed that the time for significant circumferential expansion of both constant voltage specimens to occur (as shown in Fig. 9) was around 21 days, which was quite close to the time when the steel loss increased. That coincided with the time when first visual cracking of the specimens was being observed (i.e. at 19 days). This means that, the first appearance of the cracks provided easier access for the chloride ions to reach the steel bars and attack them. As more chloride ions made their way to the steel bars through the cracks, the steel bars were more vulnerable to corrosion attack than before and consequently, the steel loss was increased.

4.2   Effect   of   Accelerated   Corrosion on Structural Response

Structural testing was performed on all the specimens (including the control specimens) after wrapping both ends of each column with GFRP sheets to determine the maximum compression load. After all the specimens C1 to C6 had undergone structural testing, the results were compiled and analyzed. Upon preliminary analysis of the results, it was decided that the results for both specimens C1 (control) and C6 (constant current) to be disregarded. The basis for

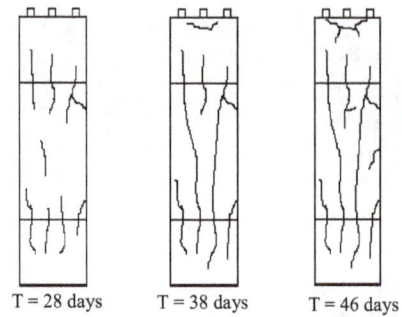

Fig. 10 Typical cracking patterns for constant voltage specimens (C3 in this figure).

Table 1 Test summary results of C2, C3, C4 and C5 and maximum crack width for columns C3 to C6.

Specimen	Steel loss percentage	Maximum load (kN)	Strength reduction (%)[#]	Corresponding axial deformation (mm)	Max crack width (mm)
C2 (control)	–	1100	–	3.8	–
C3 (constant voltage)	8.02	1043	5.2	1.4	0.38
C4 (constant voltage)	8.23	1045	5.0	1.6	0.42
C5 (constant current)	8.04	996	9.5	1.0	0.54
C6 (constant current)	8.15	–	–	–	0.46

[#] Strength reductions were calculated with reference to the control specimen.

Fig. 12 **a** Cracking at the end of constant voltage corrosion process. **b** Cracking at the end of constant current corrosion process.

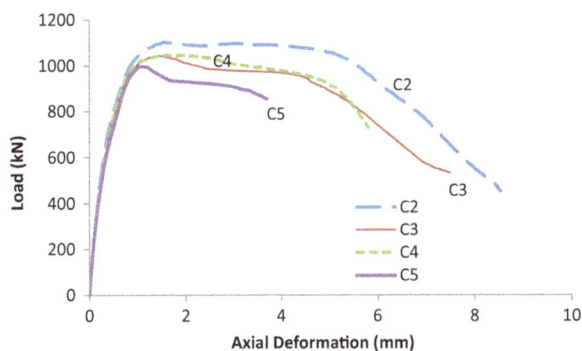

Fig. 13 Plots of axial load versus axial deformation for all column specimens.

Fig. 14 Axial load versus axial deformation curve for specimen C2.

disregarding the results for these specimens was that the loadings for both specimens were not concentric. From the structural test data, the reduction in compression load for both the constant current and constant voltage specimens was compared. Since the amount of steel loss in both types of specimens were approximately the same, the group of specimens that yielded higher reduction in maximum compression load would be the ones that suffered greater corrosion damage.

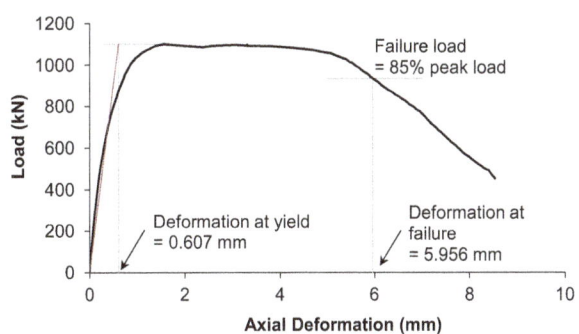

A summary of the structural test results is also shown in Table 1 for comparison purpose. From this table, it is clear that the reduction in the maximum compression load for the constant current specimen is more than those for the constant voltage specimens. The performances of the constant current and constant voltage specimens were also compared with that of the control in the superimposed plot shown in Fig. 13. Since the concrete specimens which had undergone accelerated corrosion experienced similar percentage steel

Table 2 Ductility ratios for specimens C2, C3, C4 and C5.

Specimen	Axial deformation (mm)		Ductility ratio
	At yield	At failure	
C2 (control)	0.607	5.956	9.81
C3 (constant voltage)	0.61	5.019	8.23
C4 (constant voltage)	0.61	5.218	8.55
C5 (constant current)	0.61	3.790	6.21

loss of approximately 8 %, it can be concluded that the constant current specimen (with 9.5 % strength reduction) experienced greater strength reduction per percentage steel loss than constant voltage specimens (with an average of 5.1 % strength reduction).

The performances of the concrete columns (including the control) were also evaluated in terms of ductility. The parameter used in the evaluation was the ductility ratio which is defined as the ratio of deformation at failure to the deformation at yield. It was calculated based on a previous study containing similar analysis (Sheikh et al. 1997). The higher the ductility ratio, the more ductile is the specimen.

To find out the deformation at yield, a straight line, which passed through the point of peak load, was drawn parallel to the minor axis as shown in Fig. 14. Next, another straight line was drawn from the origin, joining the point on the load-deformation curve corresponding to 65 % of the peak load. This line was then extrapolated to intersect the earlier line. The corresponding deformation at the point of intersection was taken as the deformation at yield. Failure load was taken to be 85 % of the peak load. The deformation at failure was obtained directly from the point corresponding to 85 % of the peak load from the graph. Table 2 shows the results of the ductility ratios for specimens C2, C3, C4, and C5.

From the calculated values of the ductility ratios, it can be seen that the corroded specimens with constant current suffered a loss of ductility when compared to the control specimen. This is indicative of greater corrosion damage in the constant current specimen.

5. Conclusions

Six small-scale circular columns were cast for the purpose of comparing two types of accelerated corrosion techniques, namely constant voltage and constant current. The result of this study can be summarized as follows:

1. With respect to corrosion damage, it was observed that during the corrosion process, there were more longitudinal cracks in the constant current specimens than in the constant voltage specimens. Also, the maximum crack widths for the constant current specimens were greater than those of the constant voltage specimens. Furthermore, the constant current specimens also experienced greater circumferential expansion than the constant voltage specimens. Considering all of these visual indicators of damage, and considering that both

methods produced similar steel loss, it can be said that the constant current specimens showed greater signs of damage than the constant voltage specimens. Hence, it can be concluded that accelerated corrosion using constant current is recommended over the constant voltage in studying the effectiveness of various materials, repair strategies and admixtures to resist corrosion damage.

2. With respect to structural response, it was found that there was a greater reduction in the load carrying capacity of the constant current specimen than in the constant voltage specimens when compared to that of the control. In addition, the constant current specimen behaved in a less ductile manner than the constant voltage specimens. Therefore, it can be concluded that accelerated corrosion using constant current also produces greater structural damage to the concrete columns than using constant potential.

In light of the above results, and when considering the use of electro-chemical methods to simulate corrosion in reinforced concrete, it would be recommended to employ a constant current technique rather than a constant voltage technique. In terms of target corrosion current, earlier Laboratory tests had shown that the use of current densities in the range of 100–500 μm/cm^2, produces theoretical mass losses of the steel reinforcement (based on Faraday's Law) that are in good agreement with measured mass losses based on gravimetric weight loss measurement procedures.

Acknowledgments

Support for this research was provided in part by the Sustainable Construction Materials and Structural Systems (SCMASS) research group at the University of Sharjah. This support is gratefully acknowledged.

References

Aiello, J. (1996). The effect of mechanical restraint and mix design on the rate of corrosion in concrete. MEng. Thesis, Department of Civil Engineering, University of Toronto, Toronto, Canada.

Al-Swaidani, A. M., & Aliyan, S. D. (2015). Effect of adding scoria as cement replacement on durability-related properties. *International Journal of Concrete Structures and Materials, 9*(2), 241–254.

Bonacci, J. F., & Maalej, M. (2000). Externally-bonded FRP for service-life extension of RC infrastructure. *ASCE Journal of Infrastructure Systems, 6*(1), 41–51.

Davis, J. R. (2000). *Corrosion: Understanding the basics, materials park.* Novelty, OH: ASM International.

Deb, S., & Pradhan, B. (2013). A study on corrosion performance of steel in concrete under accelerated condition. In Proceedings of the International Conference on Structural Engineering Construction and Management, Kandy, Sri Lanka, December 13th–15th 2013.

El Maaddawy, T. A., & Soudki, K. A. (2003). Effectiveness of impressed current technique to simulate corrosion of steel reinforcement in concrete. *Journal of Materials in Civil Engineering, 15*(1), 41–47.

El Maaddawy, T. A., Soudki, K. A., & Topper, T. (2005). Analytical model to predict nonlinear flexural behavior of corroded reinforced concrete beams. *ACI Structural Journal, 102*(4), 550–559.

Kashani, M. M., Crewe, A. J., & Alexander, N. A. (2013). Nonlinear stress–strain behaviour of corrosion-damaged reinforcing bars including inelastic buckling. *Engineering Structures, 48*, 417–429.

Kumar, M. K., Rao, P. S., Swamy, B. L. P., & Mouli, C. C. (2012). Corrosion resistance performance of fly ash blended cement concretes. *International Journal of Research in Engineering and Technology, 1*(3), 448–454.

Lee, C., Bonacci, J. F., Thomas, M. D. A., Maalej, M., Khajehpour, S., Hearn, N., Pantazopoulou, S., & Sheikh, S. (2000). Accelerated corrosion and repair of reinforced concrete columns using CFRP sheets. *Canadian Journal of Civil Engineering, 27*(5), 941–948.

Liu, T., & Weyers, R. W. (1998). Modeling the dynamic corrosion process in chloride contaminated concrete structures. *Cement and Concrete Research, 28*(3), 365–379.

Matsuoka, K. (1987). Monitoring of corrosion of reinforcing bar in concrete. In CORROSION/87 Symposium on Corrosion of Metals in Concrete. Houston, TX: National Association of Corrosion Engineers.

Pellegrini-Cervantes, M. J., et al. (2013). Corrosion resistance, porosity and strength of blended portland cement mortar containing rice husk ash and nano-SiO$_2$. *International Journal of Electrochemical Science, 8*, 10697–10710.

Pritzl, M. D., Tabatabai, H., & Ghorbanpoor, A. (2014). Laboratory evaluation of select methods of corrosion prevention in reinforced concrete bridges. *International Journal of Concrete Structures and Materials, 8*(3), 201–212.

Sheikh, S., Pantazopolou, S., Bonacci, J. F., Thomas, M. D. A., & Hearn, N. (1997). Repair of delaminated circular pier columns by ACM. Ontario Joint Transportation Research Report, Ministry of Transportation Ontario (MTO).

Talakokula, V., Bhalla, S., & Gupta, A. (2013). Corrosion assessment of reinforced concrete structures based on equivalent structural parameters using electro-mechanical impedance technique. *Journal of Intelligent Material Systems and Structures.* doi:10.1177/1045389X13498317.

8

The Use of Advanced Optical Measurement Methods for the Mechanical Analysis of Shear Deficient Prestressed Concrete Members

K. De Wilder[1],*, G. De Roeck[2], and L. Vandewalle[1]

Abstract: This paper investigates on the use of advanced optical measurement methods, i.e. 3D coordinate measurement machines (3D CMM) and stereo-vision digital image correlation (3D DIC), for the mechanical analysis of shear deficient prestressed concrete members. Firstly, the experimental program is elaborated. Secondly, the working principle, experimental setup and corresponding accuracy and precision of the considered optical measurement techniques are reported. A novel way to apply synthesised strain sensor patterns for DIC is introduced. Thirdly, the experimental results are reported and an analysis is made of the structural behaviour based on the gathered experimental data. Both techniques yielded useful and complete data in comparison to traditional mechanical measurement techniques and allowed for the assessment of the mechanical behaviour of the reported test specimens. The identified structural behaviour presented in this paper can be used to optimize design procedure for shear-critical structural concrete members.

Keywords: experimental mechanics, prestressed concrete, coordinate measurement machine, stereo-vision digital image correlation.

1. Introduction

Despite the long-established and worldwide research effort, a widely accepted theory for determining the shear capacity of a structural concrete member remains open for discussion (Balázs 2010; Collins 2010). This can be mainly attributed to the complex nature of shear in structural concrete. Indeed, after the occurrence of inclined cracking, various shear transfer mechanisms are activated (Jeong and Kim 2014). The aforementioned mechanisms are interrelated and highly susceptible to various parameters such as the geometry, the amount of prestressing, material properties, the amount and type of shear and longitudinal reinforcement and the loading conditions. As a consequence, numerous analytical modelling approaches can be found in literature. An overview of recent approaches to shear in structural concrete elements can be found in Fédération Internationale du Béton (fib) (2010).

The on-going debate on how to deal with shear in structural concrete members is also reflected by current codes of practice (European Committee for Standardization 2004; Canadian Standards Association 2004; American Concrete Institute 2011) which propose different design provisions resulting in varying design shear capacities and take factors affecting the shear capacity into account in a different way. Due to our incomplete knowledge and the brittle failure modes typically associated with shear, current codes of practice generally propose highly conservative shear provisions, specifically in the case of prestressed concrete elements (Nakamura et al. 2013). For the design of structural concrete members, these design equations lead to excessive material usage and corresponding construction costs. Reversely, using current codes of practice to determine the shear capacity, existing concrete structures are often found to be unable to withstand the applied service loads whereas no structural problems are reported in reality (Valerio et al. 2011; Lantsoght et al. 2013).

Improving analytical modelling approaches for shear in structural concrete members thus remains of utmost importance to optimize the design and analysis of structural concrete elements. A prerequisite for developing suitable models is a clear understanding of the actual mechanical behaviour observed during experimental tests. Traditional measurement techniques, i.e. linear variable differential transformers (LVDTs) or demountable mechanical strain gauges (DEMEC), usually provide limited test data in one or two directions. If the actual mechanical behaviour is to be understood, more elaborate experimental data is required. This paper therefore investigates on the use of advanced

[1]Department of Civil Engineering, Building Materials and Building Technology Section, KU Leuven, 3001 Heverlee, Belgium.
*Corresponding Author;
E-mail: kristof.dewilder@bwk.kuleuven.be

[2]Department of Civil Engineering, Strucutral Mechanics Section, KU Leuven, Kasteelpark Arenberg 40, Box 2448, 3001 Heverlee, Belgium.

optical measurement methods, i.e. 3D coordinate measurement machines (CMMs) and Stereo-vision digital image correlation (3D DIC), for the mechanical analysis of shear-deficient structural concrete elements. The main focus in this paper will be on prestressed concrete beams. Firstly, the experimental program is elaborated. Secondly, the adopted measurement methods are presented and their corresponding setup is described. Thirdly, the experimental results are presented including the results of the precision assessment of the measurements and the full-scale experimental results. Finally, the mechanical behaviour of the reported test beams is investigated based on the acquired measurement data and compared to current modelling approaches found in codes of practice.

2. Experimental Research

2.1 Geometry and Materials

Nine test specimens were constructed and labelled with the descriptive letter B followed by a number ranging from 101 to 109. All specimens were characterized by an I-shaped cross section, were 7000 mm long, 630 mm high. The flange width was equal to 240 mm whereas the web width was 70 mm wide. All specimens were prestressed using 8 or 4 seven-wire strands (nominal diameter 12.5 mm) at the bottom and two seven-wire strands at the top (nominal diameter 9.3 mm). The initial strain given to each prestressing strand was 7.5×10^{-3} mm/mm ($\sigma_{p0} = 1488$ MPa) for specimens B101–B103 and B107–B109 whereas the stress level σ_{p0} in the strands of beams B104–B106 was equal to 750 MPa (initial strain equal to 3.8×10^{-3} mm/mm). Shear reinforcement was provided in six specimens consisting of open single-legged stirrups with a nominal diameter of 6 mm (center-to-center distance equal to 150 mm). To avoid splitting failure at the end of each specimen due to the gradual development of the prestressing force over the member's height, splitting reinforcement was provided in each specimen (nominal diameter 8 mm, center to center distance equal to 50 mm). An overview of the geometry of the reported test specimens is given in Figs. 1a to 1f.

A self-compacting concrete mixture, designed to have a characteristic cylindrical compressive strength f_{ck} equal to 50 MPa, was used to cast each specimen. The concrete mixture consisted of cement CEM I 52.5 R (16.0 wt%), limestone gravel 2/12 (43.8 wt%), sand 0/2 (26.5 wt%), water (6.7 wt%), limestone filler (6.7 wt%) and a high-range water reducer (0.3 wt%). The expected density of this mixture was equal to 2378 kg/m³. Standardized tests were performed to determine the mean cylindrical compressive strength f_{cm} as well as the mean cube compressive strength $f_{cm,cube}$. Moreover, the mean flexural tensile strength $f_{ctm,fl}$ and the secant modulus of elasticity E_{cm} were determined for each specimen. The testing procedures can be found in Bureau for Standardisation NBN (2009a, b, 2014) The results are presented in Table 1. The age of each specimen at the day of testing is also indicated in the aforementioned Table 1.

Standard tensile tests were performed on the reinforcement bars to determine the mean yield stress f_{ym}, the modulus of elasticity E_s, the mean ultimate tensile strength f_{tm} and the corresponding strain at failure ϵ_{su}. The same characteristics were adopted from the technical information provided by the manufacturer of the prestressing strands. The results are shown in Table 2.

2.2 Experimental Setup

All test specimens are subjected to a four-point bending test, schematically shown in Fig. 2. A force-controlled loading scheme was followed using a hydraulic press (Instron, maximum capacity 2.5 MN). The load exerted by the hydraulic press was transferred to two point loads using a steel transfer beam. The total load was applied at a rate equal to 0.250 kN/s (shear force rate $\dot{V} = 0.125$ kN/s). The distance between both support points was equal to 5000 mm. The distance outside the support points was therefore equal to 1000 mm. This allowed the authors to investigate on the mechanical behaviour of shear in the presented test beams outside the length needed for the prestressing force to gradually develop over the member's height. The shear span a, i.e. the distance between the support point and the load point, was equal to 1600 mm (specimens B101, B104 and B107) or 2000 mm (B102–B103, B105–B106, B108–B109). Given the geometry and reinforcement layout, this resulted in shear span-to-effective depth ratios ($\frac{a}{d}$-ratio) between 2.91 and 3.91. An overview of the investigated parameters per specimen is given in Table 3.

2.3 Adopted Measurement Methods
2.3.1 CCD-LED Coordinate Measurement Machine (CMM)

As an optical motion tracking system, coordinate measurement machines [CMMs, also referred to as dynamic measurement machines (DMMs)] have proven to be an ideal tool for a wide range of applications, including biomechanical (Burg et al. 2013), automotive (Deconinck et al. 2004) and civil engineering applications (De Roeck et al. 2004; Sun 2007). The basic principle behind this technique is to determine the three-dimensional location of a discrete amount of points by means of triangulation. Here, two Krypton K600 CMM (Nikon Metrology, formerly known as Metris, Leuven, Belgium) systems were used for the experiment on specimen B101. The points of which the three-dimensional coordinates were to be determined were infra-red (IR) light emitting diodes (LEDs). In total 112 LEDs were used and each LED was placed onto an orthogonal grid covering both zones where a shear force occurred, refer to Fig. 3a, and was glued onto the concrete surface by using a thermoplastic adhesive. Since the LEDs were attached to the material underneath, the displacement of each LED, from which deformations can be derived, is the same as the material under investigation. The IR light emitted by each LED was seen by a camera unit consisting of three 2048 px charge-coupled device (CCD) line-element cameras, as shown in Fig. 3b. All LEDs were connected in series to a camera control unit which also acted as an interface between the cameras and the DAQ laptops.

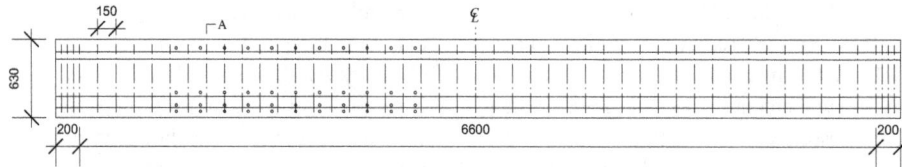

(a) B101-B102, B104-B105, B107-B108

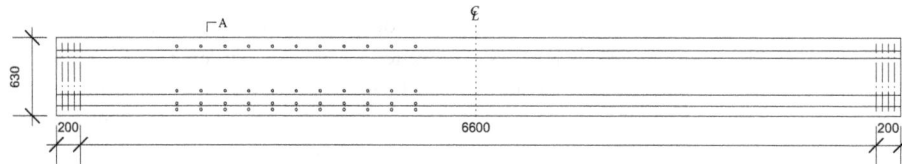

(b) B103, B106, B109

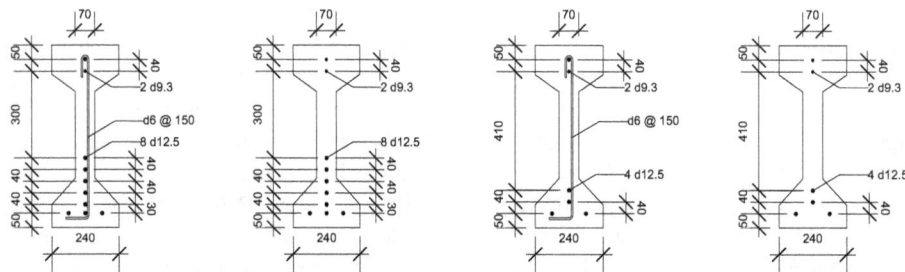

(c) B101-B102, B104- **(d)** B103   **(e)** B107-B108   **(f)** B109
B105

Fig. 1 Geometry and reinforcement layout of the reported test specimens: **a, b** Longitudinal drawings; **c–f** Cross sectional drawings.

Table 1 Concrete material properties for reported specimens.

Specimens	$f_{cm,cube}$ (MPa) (#*, s**)	f_{cm} (MPa) (#, s)	E_{cm} (GPa) (#, s)	$f_{ctm,fl}$ (MPa) (#, s)	Age (days)
B101-B103	87.1 (8, 6.8)	77.5 (5, 10.7)	43.4 (3, 2.3)	5.8 (6, 0.6)	28-233-393
B104-B106	82.8 (9, 10.5)	88.9 (6, 10.2)	43.5 (6, 8.0)	6.5 (6, 1.0)	412-404-407
B107-B109	74.6 (9, 9.6)	89.3 (6, 14.2)	42.2 (6, 4.6)	5.7 (6, 1.1)	428-424-412

* Number of tested specimens, ** Standard deviation.

Table 2 Reinforcement properties.

Reinf. type	Type	d_s d_p* (mm)	E_s E_p (GPa)	f_{ym} $f_{p0.1m}$ (MPa)	f_{tm} f_{pm} (MPa)	ϵ_{su} ϵ_{pu} (%)
Top prestress. reinf.	7-wire	9.3	198.0	1737	1930	5.20
Bot. prestress. reinf.	7-wire	12.5	198.0	1737	1930	5.20
Shear reinf.	Cold worked	6.0	210.0	608	636	2.73
Splitting reinf.	Cold worked	8.0	203.0	542	603	5.97

* Subscript p and s denote properties of prestressing and conventional reinforcement types, respectively.

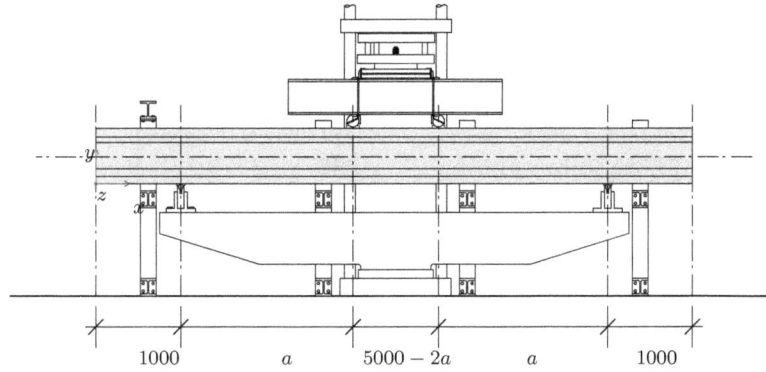

Fig. 2 Schematic representation of the experimental setup (*note* (x, y, z) denote Cartesian coordinates).

Table 3 Overview of the investigated parameters per specimen.

Specimen	Type	d^* (mm)	σ_{p0} (MPa)	a (mm)	$\frac{a}{d}$ (−)	ρ_l^{**} (−)	ρ_w^{***} ($\times 10^{-3}$)
B101	I	511	1488	1600	3.13	0.0208	2.693
B102	I	511	1488	2000	3.91	0.0208	2.693
B103	I	511	1488	2000	3.91	0.0208	0
B104	I	511	750	1600	3.13	0.0208	2.693
B105	I	511	750	2000	3.91	0.0208	2.693
B106	I	511	750	2000	3.91	0.0208	0
B107	I	550	1488	1600	2.91	0.0097	2.693
B108	I	550	1488	2000	3.64	0.0097	2.693
B109	I	550	1488	2000	3.64	0.0097	0

* Effective depth.

** Longitudinal reinforcement ratio $\frac{A_{sl}}{b_w d}$ with A_{sl} area of longitudinal reinforcement and b_w web width.

*** Shear reinforcement ratio $\rho_w = \frac{A_{sw}}{b_w s}$ with $\frac{A_{sw}}{s}$ the area of shear reinforcement per unit length.

To assess the accuracy (closeness of the measurement to the true value, related to systematic errors) and precision (the degree to which repeated measurements under unchanged conditions show the same results, related to random errors) of the CCD-LED system, a reference measurement was performed. During the aforementioned reference measurement, each coordinate of each LED was measured during 50 s at a measurement frequency of 20 Hz. Typical results of the variation of each coordinate around its initial value u_i (with $i = x, y, z$) for one LED are shown in Figs. 4a to 4c. The noise on the measurement data can assumed to be approximately normally distributed. Therefore, Figs. 4d to 4f show the fitted normal probability density functions (PDF) of u_i ($i = x, y, z$). The mean value and corresponding standard deviation are also indicated in the aforementioned figures.

Based on the coordinate measurements obtained during the aforementioned reference measurement, in-plane strain values (both horizontal ϵ_x and vertical ϵ_y) can be derived between two consecutive LEDs. The standard deviation of the in-plane horizontal and vertical strain value, s_{ϵ_x} and s_{ϵ_y} can then be calculated as a function of the surface coordinates. The results are presented in Figs. 5a and 5b. From the aforementioned

Fig. 5a, it can be seen that the value of the standard deviation of the horizontal strain s_{ϵ_x} is well below the typical strain values occurring for concrete in compression (maximum compressive strain $\epsilon_{cu} \simeq 35 \times 10^{-4}$ mm/mm). Figure 5b indicates that less precise measurements of the vertical strain are obtained near the top flange of the beam but good overall results are acquired in the web of the specimens. Indeed, the values of the vertical strain will primarily be used to assess the strain and thus stress levels in the shear reinforcement elements. Yielding of the shear reinforcement occurs at nearly 29×10^{-4} mm/mm, refer to Table 2 whereas rupture occurs at a strain value equal to 273×10^{-4}. Therefore, the expected vertical strains are well outside the obtained noise levels. The variation of the derived strain values as a function of the surface coordinate is primarily due to the different relative positions of each IR LED with respect to the optical centre of the camera control units.

2.3.2 Stereo-Vision Digital Image Correlation (3D-DIC)

As an optical-numerical full-field measurement method, digital image correlation (DIC) has proven to be an ideal tool for a wide range of applications, including the identification of the mechanical material behavior through inverse modeling (Cooreman et al. 2007, 2008), structural health monitoring (Sas et al. 2012) and the study of the deformation characteristics of a wide range of materials (Ivanov et al. 2009; Van Paepegem et al. 2009; Srikar et al. 2016). During the testing of specimens B103–B109, the stereo-vision digital image correlation technique (3D-DIC) has been adopted

Fig. 3 Experimental setup for the adopted coordinate measurement machine (CMM): **a** IR LEDs (indicated with *red rectangle*); **b** Camera units mounted on tripod.

to assess the displacements and deformations during the loading procedure. Stereo-vision implies the use of two cameras to allow for the reconstruction of the three-dimensional geometry and measurement of the three-dimensional displacements opposed to single-camera measurements which yield only two-dimensional data. The basic principle behind this technique is to calculate the displacements on the surface of an object by taking images of a random speckle pattern in undeformed and deformed state. As the speckle pattern is attached to the material underneath, the displacement and deformation of the speckle pattern is the same as the surface material of the object under investigation. To follow the displacement and deformation evolution during the full-scale experiments, a series of images of the object is made and the speckle pattern displacement and deformation in the series of images can be followed. The grey-value images of the speckle pattern are captured with a CCD camera. Similar to the CMM measurements, both zones where a shear force occurs were investigated using two DIC systems, each consisting of two cameras. This setup allowed for stereo-vision measurements. Both cameras of each DIC system take simultaneously a picture of the investigated object, in the following denoted as a frame. Each stereo-vision DIC system consisted of two 8-bit CCD cameras (AVT F201 B; 1628 px × 1236 px resolution) with wide-angle lenses (focal length equal to 12 mm) mounted on a tripod. Each zone under investigation measured approximately 1500 mm × 630 mm. The cameras were located at a perpendicular distance of approximately 2700 mm from the web of the specimen. To ensure good lighting conditions and allow for small exposure times, two 500 W quartz iodine lamps were provided per investigated zone. The image acquisition rate of each camera was equal to 2 Hz. All images were transferred to a desktop computer and synchronized with the analogue data of the hydraulic press (i.e. applied load and corresponding displacement). The experimental setup is depicted in Figs. 6a and 6b.

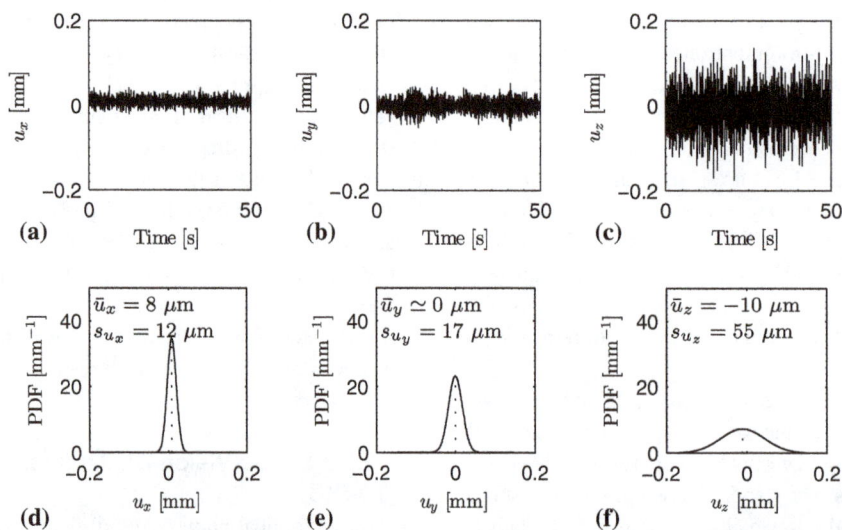

Fig. 4 Accuracy and precision assessment of the Krypton K600 CMM for the *upper left* LED (refer to Fig. 3a): **a** horizontal in-plane displacement u_x; vertical in-plane displacement u_y; out-of-plane displacement u_z; fitted normal PDF for **d** u_x; **e** u_y; **f** u_z.

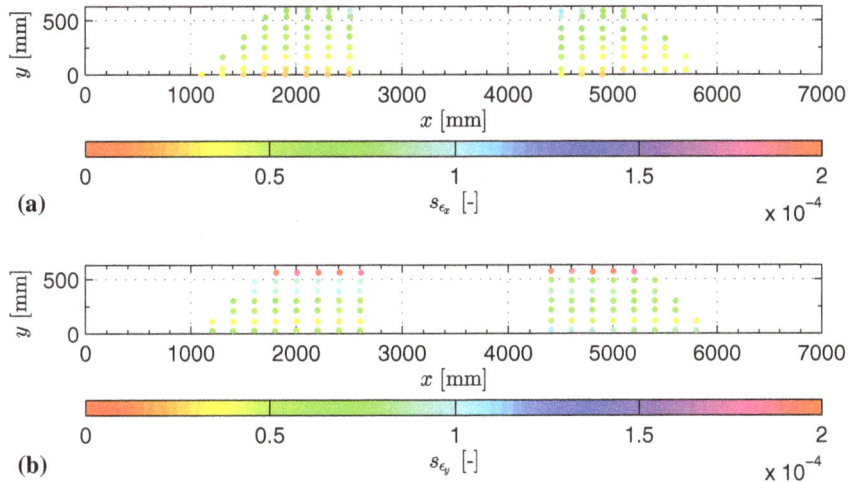

Fig. 5 Precision assessment of the adopted `Krypton K600` CMM for the derived strain measurements: **a** standard deviation of the horizontal strain s_{ϵ_x} as a function of the location on the specimen; **b** standard deviation of the vertical strain s_{ϵ_y} as a function of the location on the specimen.

The analysis of the frames taken during the loading was done with specialized software. In this work, the in-house code `MatchID` (KU Leuven, Campus Ghent) (Lava et al. 2009, 2010, 2011; Wang et al. 2011) was used. To reveal the displacements and deformations of an object during an experiment, typically a square subset of $(2M + 1)$ pixels (px) from the undeformed image is taken and its location in the deformed image is traced. The principle of the stereo-vision DIC algorithm is clearly explained by Lava et al. (2011). Matching of two $(2M+1)$ px subsets in the undeformed image $\mathbf{F}(x_i, y_j)$ and deformed image $\mathbf{G}^t(x_i, y_j)$ at a certain time t (i.e. load step) is performed by adopting an optimization routine for a degree of similarity expressed by a correlation criterion. Here, the Zero Normalized Sum of Squared Differences (ZNSSD) correlation criterion was adopted. This correlation criterion is independent of scale and offset in lighting (Sutton et al. 2009) and is therefore the most suitable correlation criterion to yield accurate results, especially in the zones where the lighting conditions are difficult to control, i.e. transition zones between web and flange of the I-shaped specimens. The mathematical formalism is clearly explained in the reference work by Sutton et al. (2009).

Lecompte (2007) states that *it has already been shown that the subset size is a critical parameter in the correlation process* (Knauss et al. 2003). *On the one hand it should be chosen small enough to allow for a reasonable linear approximation of the displacement field, within the region of the subset. On the other hand the subset size should not be chosen too small, to avoid correlation problems due to the non-uniqueness of the subset information content. This indicates the importance of an adequate speckle pattern for digital image correlation.* From a solely black or white pattern, no valuable displacement information can be gathered so that the subset size should be larger than the speckle size. Destrycker (2012) states that *to isolate the effect of the speckle pattern, there should be a way of applying the speckle pattern in a controlled way, e.g. controlling the*

Fig. 6 Experimental setup for DIC measurements: **a** numerically generated speckle pattern applied onto specimen B104; **b** 8-bit CCD cameras mounted on tripod located at a perpendicular distance of 2700 mm from the web of the test specimens.

speckle size, the speckle size distribution, the grey value distribution and the actual colour of the black and white paint. Therefore, to be able to generate suitable DIC speckle patterns, a numerical technique recently proposed by Bossuyt (2012) was adopted. In his work, Bossuyt (2012) firstly defines two concepts which are used to assess the suitability of a speckle pattern for DIC measurements. Firstly, the autocorrelation peak sharpness radius of a pre-processed image of the considered pattern is proposed to quantitatively evaluate how a particular strain sensor pattern influences the sensitivity of a DIC measurement. Secondly, the autocorrelation margin is proposed to evaluate how that pattern influences the robustness of the DIC measurement. The former is related to the measurement precision whereas the latter is correlated to the measurement accuracy. Ideal patterns for DIC would combine a sharp autocorrelation peak with a well-defined autocorrelation margin. For simple patterns, these characteristics vary in direct proportion to each other. However, Bossuyt (2012) proposes a method based on morphological image processing and Fourier transform to

synthesize a DIC pattern with wide autocorrelation margins even though the autocorrelation peaks are sharp. Such patterns are exceptionally well-suited for DIC measurements. A detail of the numerically generated speckle pattern is shown in Fig. 7a. The generated pattern is then applied onto each specimen where the DIC technique was used by adopting a heat-sensitive stencil printing technique which consists of three layers: a vinyl base layer, the inverse of the speckle pattern and a top protective heat-sensitive polypropylene layer. The printed speckles have a precalculated oversampling of at least five pixels in order to avoid aliasing effects in the obtained results due to the expected small magnitude of the displacement and deformation field. Given the camera sensor properties and the dimensions of the field of view, speckles with a diameter of nearly 5 mm were required. Obtaining

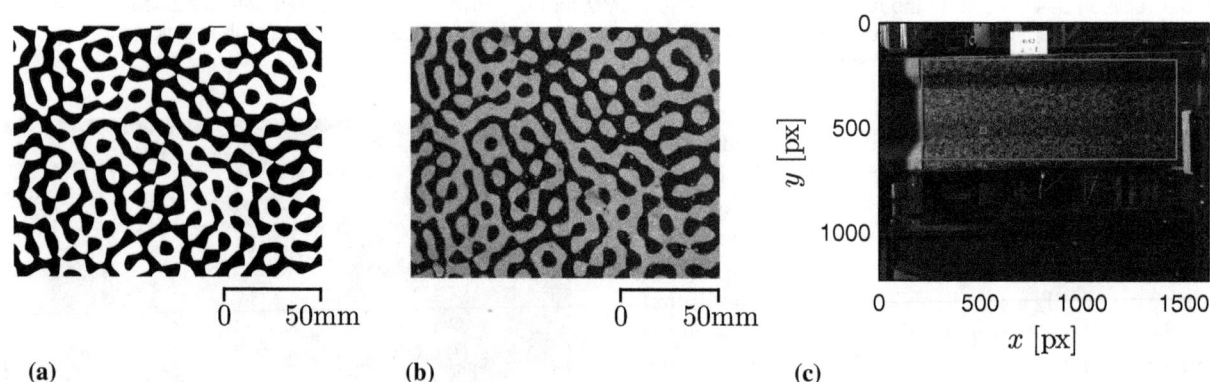

(a)   **(b)**   **(c)**

Fig. 7 a Detail of numerically generated speckle pattern; **b** Detail of the speckle pattern shown in (**a**) applied onto specimen B105; **c** indication of area of interest (AOI) and subset size in *red* for the left measurement field of specimen B103.

Table 4 Measurement settings for the DIC tests.

	Unit	B103	B104	B105	B106	B107	B108	B109
Subset	(px)	27	27	27	27	27	27	27
Step	(px)	3	3	3	3	3	3	3
Measurement points	(−)	68327	67482	67819	70290	68843	68868	75864
Temporal resolution	(fps)	2	2	2	2	2	2	2
Camera distance	(mm)	2700	2700	2700	2700	2700	2700	2700
Interpolation	(−)	b.c.s.[†]	b.c.s.	b.c.s.	b.c.s.	b.c.s.	b.c.s.	b.c.s.
Displacement								
Spatial resolution[a]	(mm)	31.2	30.9	29.9	29.5	29.9	30.8	28.4
In-plane resolution[†]	(mm)	0.018	0.019	0.034	0.027	0.020	0.021	0.023
Out-of-plane resolution[‡]	(mm)	0.125	0.120	0.165	0.167	0.147	0.149	0.152

[†] Bicubic spline.

[‡] Maximum mean value of the standard deviation of the in-plane horizontal and vertical displacement; mean value of the standard deviation of the out-of-plane displacement.

[a] Physical dimension of subset.

(a)

(b)

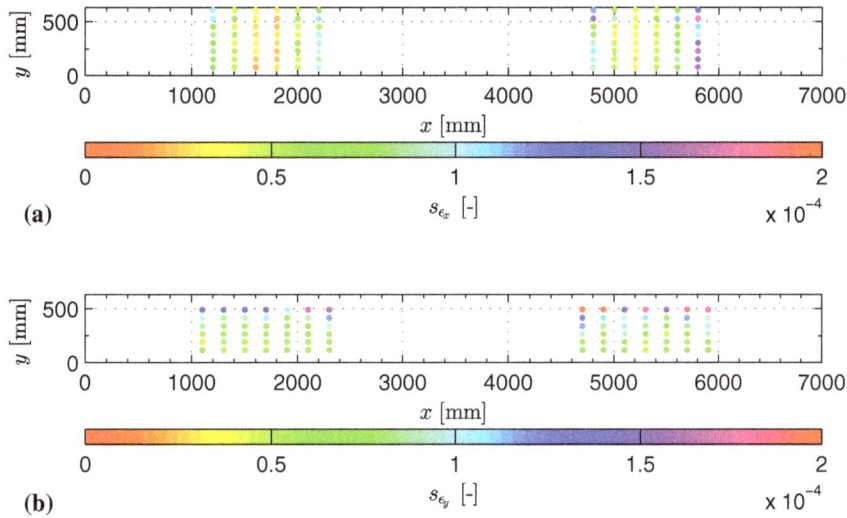

Fig. 8 Precision assessment of the adopted 3D DIC system for the derived strain measurements for specimen B103: **a** standard deviation of the horizontal strain s_{ϵ_x} as a function of the location on the specimen; **b** standard deviation of the vertical strain s_{ϵ_y} as a function of the location on the specimen.

large speckles is nearly impossible with traditional speckle techniques (i.e. spray painting). However, the adopted numerical technique allows for the generation of a speckle pattern tailored to the needs of the experiment. Figure 7b shows the same detail as presented in Fig. 7a applied onto beam B105. Figure 7c shows the speckle pattern, area of interest and adopted subset size. Since full-field displacement data is readily available, Green-Lagrange strains can be easily derived from the aforementioned displacement data. Therefore, the displacement data is smoothed over a certain zone to damp out the effect of noise and local uncertainties. A bilinear plane can be fitted through the displacement values in the points around the center of the strain window.

To assess the precision of the 3D-DIC setup for the presented experiments, a number ($N = 50$) of reference images were taken prior to each test under a zero-loading condition. The subset size was taken equal to 27 px whereas the stepsize was chosen equal to 3 px. Given the experimental setup, the physical dimension of 1 pixel approximated 1 mm. The subset was allowed to undergo an affine transformation thus taking into account translation, rotation, shear and normal straining. As the displacements from one frame to another may be smaller than one pixel, the subset in the image of the deformed state is not likely to fit on the pixel grid and an interpolation method between the pixels is needed. Therefore, a bicubic interpolation scheme has been adopted in the MatchID software during the analyses. Additionally, Gaussian prefiltering (5×5 px kernel size) of the subset information was adopted as a low pass-filter to attenuate high-frequency signals and allow for proper interpolation. Based on the measured displacements u_i (with $i = x, y, z$), Green-Lagrange strains can be determined by means of bilinear-quadrilateral smoothing of the displacement field. This method was adopted for the analyses of the frames during the loading procedure. For the purpose of assessing the precision, strains were derived directly from the displacement data with similar initial base lengths L_0 as the CMM and without

smoothing of the displacement field to allow for a reasonable comparison between both adopted systems. The base length to determine the horizontal strains L_0^- was chosen approximately equal to 200 mm whereas the base length to determine vertical strains $L_0^|$ was taken roughly equal to 75 mm. An overview of the adopted DIC settings is presented in Table 4. Similar to the CMM system, Figs. 8a and 8b present the standard deviation of the horizontal and vertical strain s_{ϵ_i} (with $i = x, y$) based on the N pairs of reference images obtained from both DIC systems for specimen B103. Comparable results were found for the remaining specimens where the DIC technique has been adopted.

From the aforementioned Figs. 8a and 8b, it can be firstly seen that similar values for the standard deviation of both the horizontal and vertical strain are found if the data of the 3D DIC system is compared to the data obtained from the Krypton K600 system, refer to Figs. 5a and 5b. Secondly, it can be seen that less precise measurements of the horizontal strain ϵ_x occur near the edge of the area of interest, refer to Fig. 8a. This can be attributed to optical aberration effects near the edge of the field of view. However, the value of the expected strains during the loading procedure well exceed the presented values of the standard deviation of the horizontal strain. Thirdly, due to the shaded area near the transition between the top flange and the web, refer to Fig. 7c, less precise measurements of the vertical displacement are obtained resulting in less accurate vertical strain measurements near the top flange. However, vertical strains will primarily be investigated in the web of the specimens to assess the strain and stress levels in the shear reinforcement elements.

3. Results and Discussion

Figure 9a–c depict the experimentally observed load-displacement response curves of the presented test specimens. The onset of bending (○) and diagonal (△) cracking is also

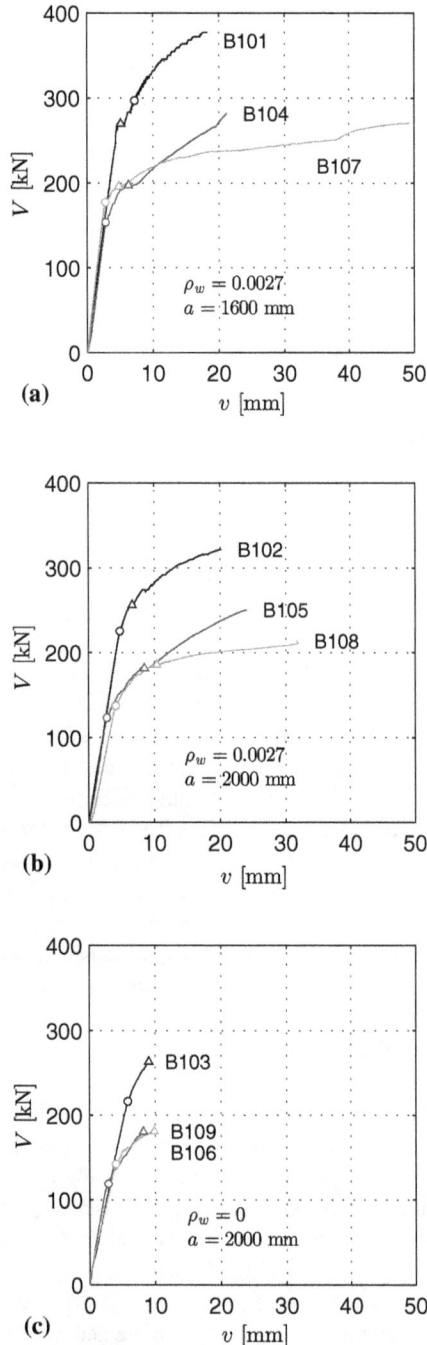

Fig. 9 Experimentally determined load-displacement response curves measured at 1200 mm from the support point: **a** specimens B101, B104 and B107; **b** specimens B102, B105 and B108; **c** specimens B103, B106 and B109 (*note* onset of bending and diagonal cracking indicated with (*circle*) respectively (*triangle*).

indicated in the aforementioned Figs. 9a to 9c. All I-shaped test specimens but beams B107 and B108 failed due to shear in a very brittle manner. The aforementioned specimens exhibited severe diagonal web cracking leading to excessive yielding and sudden rupture of the shear reinforcement bars. Crushing of the diagonal compressive struts was not

observed. Beams B107 and B108 exhibited a ductile bending failure mode leading to rupture of the prestressing strands. Typical experimentally observed failure modes of the presented test beams are shown in Figs. 10a to 10d. A number of observations can be made based on Figs. 9a to 9c and 10a to 10d:

1. Prior to the onset of cracking, the stiffness in the elastic regime is comparable for all specimens with the same shear span and overall height. Indeed, prior to the occurrence of cracking, the response of the test specimens to the applied load is governed by the bending stiffness EI. Due to the comparable secant modulus of elasticity, refer to Table 1, and the negligible influence of the area of longitudinal reinforcement on the second moment of inertia, it can be concluded that the bending stiffness is similar for the reported test specimens.

2. The occurrence of cracks determines the transition between linear and nonlinear behaviour. All specimens exhibited both bending and web cracks. The load at which web cracks occurs, is function of the amount of prestressing and the concrete tensile strength respectively.

3. Specimens where shear reinforcement was provided and which failed in shear (B101–B102, B104–B105) exhibited a significant post-cracking stiffness and post-cracking bearing capacity resulting in a brittle shear failure mode due to diagonal tension. Specimens B107 and B108 which failed in bending, show a highly ductile behaviour with a limited post-cracking bearing capacity. Finally, beams without shear reinforcement (B103, B106 and B109) failed immediately after the occurrence of the first inclined web crack.

The failure load and failure mode of each specimen is summarized and presented in Table 5. The experimentally determined failure load and failure mode can be compared to analytical predictions of the bearing capacity according to the shear design procedures outlined in Eurocode 2 (EC 2) (European Committee for Standardization 2004; Bureau for Standardisation NBN 2010). For the calculation of the shear capacity of a structural concrete member, a distinction is to be made between members with and without shear reinforcement. Structural concrete members with shear reinforcement are to be designed according to EC 2 using the variable angle truss model (VATM). This approach assumes that the behaviour of a structural concrete member near failure can be idealized by means of a parallel chord truss. The bottom and top flanges resist the applied bending moment whereas the combination of a compressive stress field with constant inclination and vertical tension bars resist the applied shear force. The design shear capacity V_{Rd} of a member with shear reinforcement is the minimum of the shear force required to obtain yielding of the shear reinforcement $V_{Rd,s}$ and the shear force which causes crushing of the compressive struts $V_{Rd,max}$, refer to Eq. (1).

$$V_{Rd} = \min \left\{ V_{Rd,s}, V_{Rd,max} \right\} \tag{1}$$

The expressions for $V_{Rd,s}$ and $V_{Rd,max}$ can be derived from equilibrium conditions of the adopted truss model where the

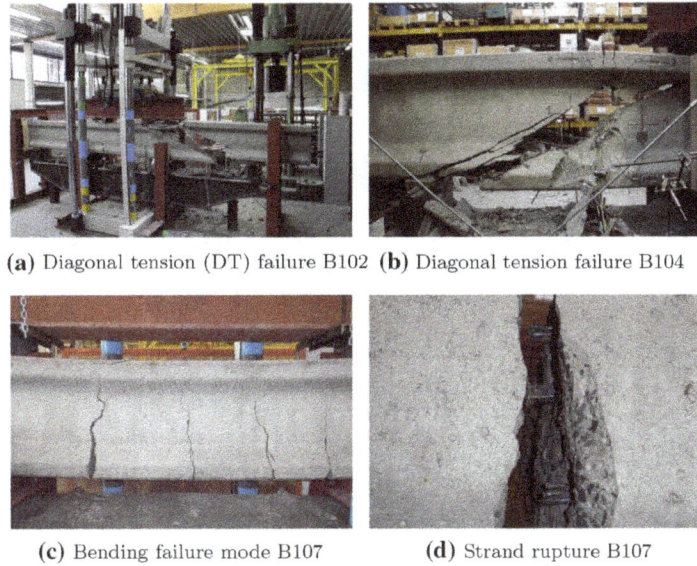

(a) Diagonal tension (DT) failure B102 (b) Diagonal tension failure B104

(c) Bending failure mode B107   (d) Strand rupture B107

Fig. 10 Typical experimentally observed failure modes for the reported test specimens.

Table 5 Experimentally observed and analytically predicted failure load and failure mode.

Specimen	Experiment		Eurocode 2 (European Committee for Standardization 2004, Bureau for Standardisation NBN 2010)			
	$V_{u,exp}$ (kN)	Failure mode	$V_{u,pred}$ (kN)	$V_{u,bend}$ (kN)	Failure mode	$\frac{V_{u,exp}}{V_{u,pred}}$ (−)
B101	377.7	S-DT†	158.1	412.2	S	2.39
B102	321.6	S-DT	158.1	329.6	S	2.03
B103	262.8	S-DT	243.3‡	329.6	S	1.08
B104	281.8	S-DT	135.2	406.6	S	2.08
B105	251.2	S-DT	135.2	325.3	S	1.86
B106	179.7	S-DT	206.9‡	325.3	S	0.87
B107	271.3	B	147.9	236.5	S	1.83
B108	213.8	B	147.9	189.2	S	1.45
B109	181.0	S-DT	197.0*	189.2	B	0.92
					$\frac{V_{u,exp}}{V_{u,pred}}$	1.61
					COV[a]	55.4 %

† Shear failure mode due to diagonal tension.
‡ Obtained using Eq. (12).
[a] Coefficient of Variation.

angle between the horizontal and the inclined compressive stresses is denoted by θ, refer to Eqs. (2)–(3).

$$V_{Rd,s} = \frac{A_{sw}}{s} z f_{ywd} \cot \theta \qquad (2)$$

$$V_{Rd,max} = \frac{\alpha_{cw} v_1 b_w z f_{cd}}{\cot \theta + \tan \theta} \qquad (3)$$

In Eqs. (2)–(3), A_{sw}/s denotes the area of shear reinforcement per unit length, z is the internal lever arm equal to $0.9d$ whereas f_{ywd} is the design value of the yield

strength of the shear reinforcement bars. Factors α_{cw} and v_1 take into account the stress distribution of the compressive chord respectively the effect of lateral tensile straining on the ultimate compressive strength. The width of the web is denoted by b_w. The inclination angle θ can be chosen freely between certain limits as presented in Eq. (4).

$$1 \leq \cot \theta \leq \cot \theta_{max} \qquad (4)$$

The maximum value of $\cot \theta$, thus the minimum allowable value for the angle θ can be determined using Eq. (5) (Bureau for Standardisation NBN 2010).

$$\cot\theta_{\max} = 2 + \frac{0.15\sigma_{cp}b_w d}{\frac{A_{sw}}{s}zf_{ywd}} \leq 3 \tag{5}$$

Equation (5) clearly indicates that for highly prestressed members (average normal stress due to prestressing σ_{cp}), a lower angle of inclination is allowed with a minimul value of $\theta = 18.4°$.

In the case of structural concrete members without shear reinforcement, a distinction has to be made between members cracked respectively uncracked due to the acting bending moment, refer to Eqs. (6)–(7).

$$V_{Rd} = V_{Rd,c1}$$
$$= \left[\frac{0.18}{\gamma_c}\left(1+\sqrt{\frac{200}{d}}\right)\left(100\frac{A_{sl}}{b_w d}f_{ck}\right)^{\frac{1}{3}}+0.15\sigma_{cp}\right]b_w d \tag{6}$$

$$V_{Rd} = V_{Rd,c2} = \frac{Ib_w}{S}\sqrt{f_{ctd}^2 + \alpha_l\sigma_{cp}f_{ctd}} \tag{7}$$

In the previous equations, γ_c is a partial safety factor (equal to 1.5), A_{sl} is the area of longitudinal tensile reinforcement, I is the second moment of area whereas S is the first moment of area. Finally, f_{ctd} is the design value of the characteristic uni-axial concrete tensile strength. If Eqs. (1)–(7) are used to estimate the actual failure load, partial safety factors should be omitted and average material properties are to be used rather than characteristic or design values. Eqs. (2)–(3) and Eqs. (5)–(7) are then transformed to Eqs. (8)–(12).

$$V_{R,s} = \frac{A_{sw}}{s}zf_{ym}\cot\theta_{\max} \tag{8}$$

$$V_{r,max} = \frac{\alpha_{cw}\nu_1 b_w zf_{cm}}{\cot\theta_{\max} + (\cot\theta_{\max})^{-1}} \tag{9}$$

$$\cot\theta_{\max} = 2 + \frac{0.15\sigma_{cp}b_w d}{\frac{A_{sw}}{s}zf_{ym}} \leq 3 \tag{10}$$

$$V_{R,c1} = \left[0.18\left(1+\sqrt{\frac{200}{d}}\right)\left(100\frac{A_{sl}}{b_w d}f_{cm}\right)^{\frac{1}{3}}+0.15\sigma_{cp}\right]b_w d \tag{11}$$

$$V_{R,c2} = \frac{Ib_w}{S}\sqrt{(0.67f_{ctm,fl})^2 + \alpha_l\sigma_{cp}(0.67f_{ctm,fl})} \tag{12}$$

In Eq. (12), the mean value of the uni-axial tensile strength is written as a function of the experimentally determined mean flexural tensile strength. The adopted relation is given in Müller et al. (2013). The results of the aforementioned calculation using Eqs. (8)–(12) are also indicated in Table 5. The theoretical failure load required to obtain a bending failure mode, derived from a general plane section analysis $V_{u,bend}$, is also indicated in the aforementioned Table 5.

Based on the results presented in Table 5, the following observations and preliminary conclusions can be made:

1. In general, a poor correlation is found between the experimental results and analytical calculations according to EC 2 for all specimens apart for beams B103, B106 and B109 without shear reinforcement. Even if all partial safety factors are omitted and average material strength properties are used rather than characteristic or design values, an average experimental-to-predicted failure load ratio of 1.61 is found with a coefficient of variation (COV) equal to 55.4 %.

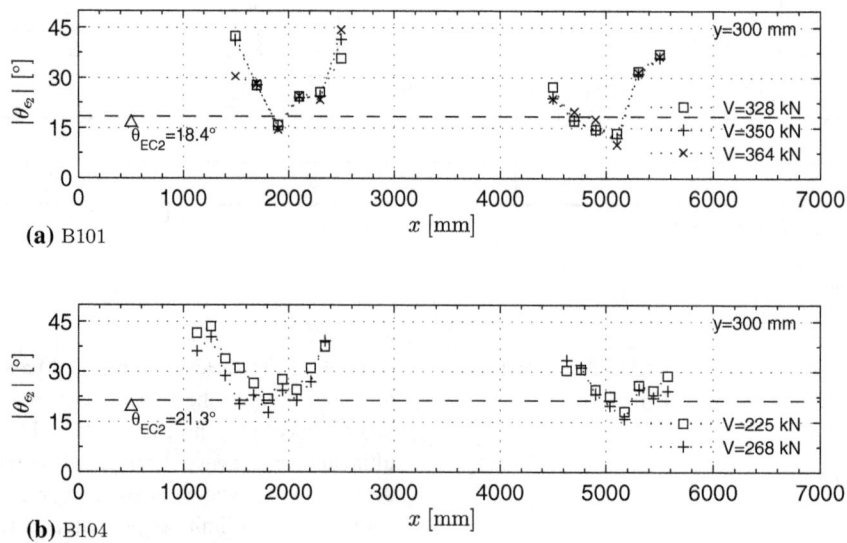

Fig. 11 Experimentally measured angle $|\theta_{\epsilon_2}|$ between the direction of the principal compressive strain ϵ_2 as a function of the location along the x-axis and applied load for specimens **a** B101 (derived from Krypton K600 CMMs) and **b** B104 (derived from 3D-DIC).

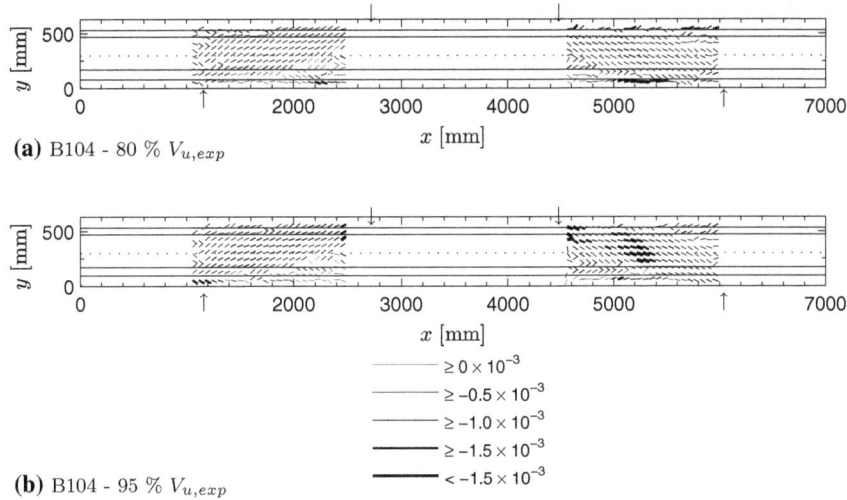

(a) B104 - 80 % $V_{u,exp}$

(b) B104 - 95 % $V_{u,exp}$

$$\geq 0 \times 10^{-3}$$
$$\geq -0.5 \times 10^{-3}$$
$$\geq -1.0 \times 10^{-3}$$
$$\geq -1.5 \times 10^{-3}$$
$$< -1.5 \times 10^{-3}$$

Fig. 12 Experimentally determined direction and magnitude of the principal compressive strain field ϵ_2 of specimen B104 as a function of the surface coordinates and the applied load level (*Note* line where the angle $|\theta_{\epsilon_2}|$ was determined in Fig. 11b indicated with a *dotted line*).

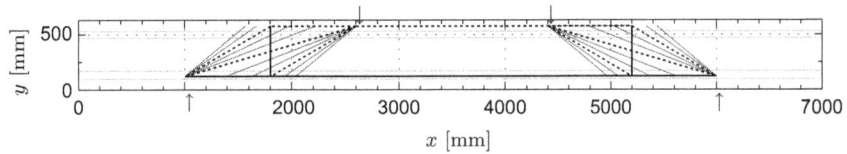

Fig. 13 Schematic representation of the identified mechanical behaviour of the presented test beams (Members in compression indicated with a *dashed line*, member in tension indicated with a *solid line*).

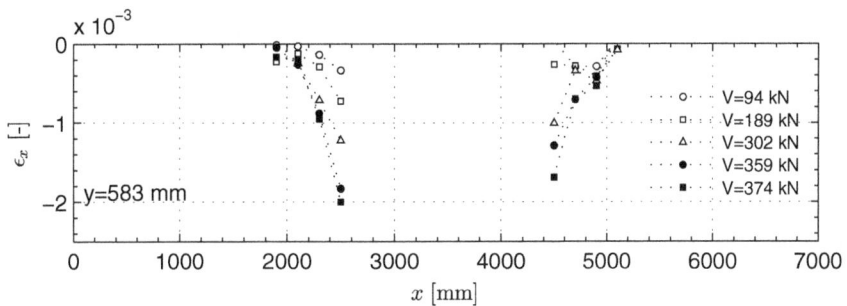

Fig. 14 Typical experimentally observed horizontal strain profile at the flange of specimen B101 as a function of the applied load and location along the longitudinal x-axis.

2. The failure mode is correctly predicted for all specimens apart for beams B107–B109. Specimens B107 and B108 failed due to bending despite having a lower shear capacity in comparison to the corresponding bending strength. Indeed, the experimental failure load correlates well with the theoretical load required to obtain the bending capacity for the aforementioned specimens. Moreover, the wrongly predicted failure mode for specimen B109 can be attributed to the small difference in the analytically calculated shear and bending capacity.

3. As expected, increasing the shear reinforcement ratio, increases the shear capacity (B102–B103, B105–B106, B108–B109).

4. Increasing the shear span-to-effective depth ratio while keeping all other investigated parameters constant,

consistently decreases the experimentally observed shear failure load (B101–B102, B104–B105).

5. Increasing the prestressing force while keeping the longitudinal reinforcement ratio approximately constant, increases the shear capacity of specimens with (B101–B104, B102–B105) and without shear reinforcement (B103–B106).

6. Decreasing the longitudinal reinforcement ratio while keeping the prestressing force constant does not significantly influence the failure load (B104–B07, B105–B108 and B106–B109). However, the failure mode shifts from a brittle shear induced failure mode towards a more ductile bending induced failure mode for specimens with shear reinforcement.

The found discrepancy between the experimentally observed and analytically calculated failure load in the case of prestressed concrete members with shear reinforcement is certainly to be further investigated. Therefore, the experimentally observed mechanical behaviour is assessed based on the acquired data of the optical(-numerical) measurement methods. Figure 11a, b present the experimentally determined angle $|\theta_{\epsilon_2}|$ between the principal compressive strain ϵ_2 with respect to the horizontal at mid-depth determined from the Krypton K600 CMMs respectively 3D-DIC as a function of the location along the x-axis and the applied load. The measured value for $|\theta_{\epsilon_2}|$ is compared to the adopted value for θ used for the presented strength calculations, refer to Table 5. Similar results were obtained for the remaining test specimens.

Figures 11a and 11b indicate that the value of the experimentally determined angle $|\theta_{\epsilon_2}|$ in the middle of the shear span ($x = 1800$ and 5200 mm) correlates well with the adopted value of the angle θ in the presented shear strength calculations, refer to Table 5. However, the value of $|\theta_{\epsilon_2}|$ is not constant along the longitudinal x-axis of the specimen contrary to the assumption made by the VATM approach. Instead, a parabolic course of the angle $|\theta_{\epsilon_2}|$ is observed which does not tend to change significantly if the loading is furthermore increased. This observation if clarified by Figs. 12a and 12b which presents the full-field magnitude and direction of the principal compressive strain field ϵ_2 for specimen B104 as a function of the surface coordinates and the applied load. Similar results were observed for the remaining specimens where the DIC technique has been adopted.

Based on the direction and magnitude of the principal compressive strain field ϵ_2 presented in Figs. 12a and 12b, it can be observed that a direct compression strut is developed between the support point and the load application point which carries a significant amount of the applied load. The remaining part of the applied load fans out towards the bottom of the specimen. A similar observation can be made for the support point load. The force in both aforementioned fan regions is equilibrated by the vertical force in the shear reinforcement elements. A schematic representation of the identified structural behaviour as presented in Figs. 12a and 12b is depicted in Fig. 13. The identified structural behaviour as described above corresponds with the observed parabolic course of the direction of the principal compressive strain $|\theta_{\epsilon_2}|$. In view of the mechanical behaviour of the presented test specimens presented in Figs. 12a, 12b and 13, the experimentally observed failure mode should be interpreted as a splitting failure mode due to the forces in the fan regions.

Direct strut action is generally considered to be an important bearing mechanism for beams with a shear span-to-effective depth ratio less than or equal to 2.5 (Ramirez et al. 1998). However, the presented test beams were characterized by a shear span-to-effective depth ratio varying between 2.91 and 3.91, refer to Table 3. The identified structural behaviour, as presented in Figs. 12a, 12b and 13, provides a plausible explanation for the following experimentally observed phenomena.

1. The possibility of carrying the applied shear force by means of direct strut action significantly increases the shear carrying capacity in comparison to the shear capacity obtained using the variable angle truss model as proposed by Eurocode 2. This provides a possible explanation why the current sectional shear design provisions found in EC2 performed poorly in predicting the shear capacity of the presented prestressed concrete beams.

2. Figure 14 shows the typically observed profile of the horizontal strain ϵ_x at the top flange of the presented test beams. Figure 14 clearly shows that the horizontal strain at the top of the presented beams rapidly decreases to relatively low strains, and thus relatively low stresses, away from the loading point. Due to the inclined strut action, it is indeed expected that low strain values occur at the top of the specimen near the support point.

4. Conclusions

This paper aims to investigate on the use of advanced optical(-numerical) measurement methods for the mechanical analysis of shear-critical prestressed concrete beams. Therefore, an experimental program consisting of nine full-scale prestressed I-shaped beams was drafted. The main investigated parameters were the amount of prestressing, the amount of longitudinal reinforcement and shear reinforcement and the shear span-to-effective depth ratio respectively. All specimens were subjected to a load-controlled four-point bending test until failure. During the experimental research, the use of two advanced optical(-numerical) measurement methods, i.e. *3D coordinate measurement machines* (CMM) and *stereo-vision digital image correlation* (3D-DIC), was explored. Firstly, the experimental setup was elaborated in detail. Specifically in the case of the DIC technique, a novel technique to apply numerically synthesised strain sensor patterns, i.e. speckle patterns, in a controlled way was presented. The presented technique allows for the application of tailor-made strain sensor patterns to virtually any given object's surface. A reference measurement was performed in unloaded state to asses the measurement precision. Both techniques were found to be comparable in terms of displacement and strain resolution. The maximum standard deviation of the in-plane displacements was equal to approximately 20×10^{-3} mm for the CMMs whereas a value of approximately 30×10^{-3} mm was found for the DIC technique. Moreover, the expected value of the strains occurring during the experiments well exceeded the observed noise levels on the in-plane horizontal and vertical strains. It can thus be concluded that both techniques are well suited for assessing the structural behaviour of the reported test specimens. However, due to the brittle and highly energy releasing failure modes observed during the tests on specimens failing in shear, the DIC technique is preferred over the CMM technique since the latter technique requires relatively expensive IR LED sensors to be glued onto the concrete side surface which can sustain damage at the moment of failure.

All specimens were designed to fail in shear. However, seven specimens failed in a brittle manner due to shear (diagonal tension failure mode) whereas two specimens failed due to bending in a ductile manner. The experimental results were compared to analytical calculations according to the current design procedures found in Eurocode 2 (EC 2). Based on the work presented in this paper, it can be concluded that EC 2, adopting the variable angle truss model, in general significantly underestimates the experimentally determined failure load. Omitting all partial safety factors and using average material strength properties rather than characteristic values resulted in an average experimental-to-predicted failure load ratio equal to 1.61 (coefficient of variation equal to 55.4 %). Based on the extensive amount of experimental displacement and deformation data, it was found that the applied load was primarily carried by means of a direct compression strut in combination with fan regions contrary to the model adopted by EC 2 which assumes that a compression field with constant inclination along the member's axis is developed. The identified structural behaviour for the reported test specimens can be used to optimize current shear design provisions as proposed by codes of practice.

References

American Concrete Institute. (2011). Aci 318–11 building code requirements for structural concrete and commentary.

Balázs, G. L. (2010). A historical review of shear. In V. Sigrist, F. Minelli, G. Plizzari, & S. Foster (Eds.), *Bulletin 57: Shear and punching shear in RC and FRC elements* (pp. 1–13). Lausanne, Switzerland: Fédération Internationale du Béton (fib).

Bossuyt, S. (2012). Optimized patterns for digital image correlation. In Proceedings of the SEM International Conference and Exposition on Experimental and Applied Mechanics, volume 3: Imaging Methods for Novel Materials and Challenging Applications, Costa Mesa, CA, USA, Society for Experimental Mechanics.

Bureau for Standardisation NBN. (2009a). Nbn en 12390–3.

Bureau for Standardisation NBN. (2009b). Nbn en 12390–5.

Bureau for Standardisation NBN. (2010). Nbn en 1992-1-1 anb.

Bureau for Standardisation NBN. (2014). Nbn en 12390–13.

Burg, J., Peeters, K., Natsakis, T., Dereymaeker, G., Vander Sloten, J., & Jonkers, I. (2013). In vitro analysis of muscle activity illustrates mediolateral decoupling of hind and mid foot bone motion. *Gait and Posture*, *38*(1), 56–61.

Canadian Standards Association. (2004). Csa a23.3 design of concrete structures (csa a23.3-04), 2004.

Collins, M. P. (2010). Improving analytical models for shear design and evaluation of reinforced concrete structures. In V. Sigrist, F. Minelli, G. Plizzari, & S. Foster (Eds.), *Bulletin 57: Shear and punching shear in RC and FRC elements* (pp. 77–92). Lausanne, Switzerland: Fédération Internationale du Béton (fib).

Cooreman, S., Debruyne, D., Coppieters, S., Lecompte, D., & Sol, H. (2008). *Identification of the mechanical material parameters through inverse modelling* (pp. 337–342). Non-Destructive Testing: Emerging Technologies in.

Cooreman, S., Lecompte, D., Sol, H., Vantomme, J., & Debruyne, D. (2007). Elasto-plastic material parameter identification by inverse methods: Calculation of the sensitivity matrix. *International Journal of Solids and Structures*, *44*(13), 4329–4341.

De Roeck, G., Wens, L., & Jacobs, S. (2004). 3d static and dynamic displacement measurements by a system of three linear ccd-units. In P. Sas & M. De Munck (Eds.), *International Conference on Noise and Vibration Engineering (ISMA)* (pp. 2289–2300), Leuven, Belgium: KU Leuven.

Deconinck, F., Desmet, W., & Sas, P. (2004). Increasing the accuracy of mdof road reproduction experiments: Calibration, tuning and a modified twr approach. In P. Sas & M. De Munck (Eds.), *International Conference on Noise and Vibration Engineering (ISMA)* (pp. 709–722), Leuven, Belgium: KU Leuven.

Destrycker, M. (2012). Experimental validation of residual stress simulations in welded steel tubes with digital image correlation. PhD thesis, KU Leuven, Department of Civil Engineering.

European Committee for Standardization (CEN). (2004). Eurocode 2: Design of concrete structures—part 1–1: General rules and rules for buildings.

Fédération Internationale du Béton (fib). (2010). Shear and punching shear in RC and FRC elements—Workshop 15–16 October 2010, Salò (Italy), volume 57 of Bulletin, Lausanne (Switzerland).

Ivanov, D., Ivanov, S., Lomov, S., & Verpoest, I. (2009). Strain mapping analysis of textile composites. *Optics and Lasers in Engineering*, *47*(3–4), 360–370.

Jeong, J.-P., & Kim, W. (2014). Shear resistant mechanism into base components: Beam action and arch action in shear-critical rc members. *International Journal of Concrete Structures and Materials*, *8*(1), 1–14.

Knauss, W. G., Chasiotis, I., & Huang, Y. (2003). Mechanical measurements at the micron and nanometer scales. *Mechanics of Materials*, *35*(3–6), 217–231.

Lantsoght, E., van der Veen, C., Walraven, J., & de Boer, A. (2013). Shear assessment of reinforced concrete slab bridges. *Structural Engineering International*, *4*, 418–426.

Lava, P., Cooreman, S., Coppieters, S., De Strycker, M., & Debruyne, D. (2009). Assessment of measuring errors in dic using deformation fields generated by plastic fea. *Optics and Lasers in Engineering*, *47*(7–8), 747–753.

Lava, P., Cooreman, S., & Debruyne, D. (2010). Study of systematic errors in strain fields obtained via dic using heterogeneous deformation generated by plastic fea. *Optics and Lasers in Engineering, 48*(4), 457–468.

Lava, P., Coppieters, S., Wang, Y., Van Houtte, P., & Debruyne, D. (2011). Error estimation in measuring strain fields with dic on planar sheet metal specimens with a non-perpendicular camera alignment. *Optics and Lasers in Engineering, 49*(1), 57–65.

Lecompte, D. (2007). Elasto-Plastic material parameter identification by inverse modeling of static tests using digital image correlation. PhD thesis, Vrije Universiteit Brussel, Department of Mechanics of Materials and Constructions.

Müller, H., Breiner, R., Anders, I., Mechterine, V., Curbach, M., Speck, K., et al. (2013). *Code-type models for concrete bahviour* (Vol. 70)., Bulletin Lausanne, Switzerland: Fédération Internationale du Béton.

Nakamura, E., Avendano, A. R., & Bayrak, O. (2013). Shear database for prestressed concrete members. *ACI Structural Journal, 110*(6), 909–918.

Ramirez, J. A., et al. (1998). Recent approaches to shear design of structural concrete. *Journal of Structural Engineering-Asce, 124*(12), 1375–1417.

Sas, G., Blanksvard, T., Enochsson, O., Taljsten, B., & Elfgren, L. (2012). Photographic strain monitoring during full-scale failure testing of ornskoldsvik bridge. *Structural Health Monitoring-an International Journal, 11*(4), 489–498.

Srikar, G., Anand, G., & Suriya Prakash, S. (2016). A study on residual compression behavior of structural fiber reinforced concrete exposed to moderate temperature using digital image correlation. *International Journal of Concrete Structures and Materials,* pp. 1–11.

Sun, S. (2007). Shear behavior and capacity of large-scale prestressed high-strength concrete bulb-tee girders. PhD thesis, University of Illinois at Urbana-Champaign, Department of Civil and Environmental Engineering.

Sutton, M. A., Orteu, J. J., & Schreier, H. W. (2009). *Image correlation for shape, motion and deformation measurements: basic concepts, theory and applications.* New York, NY: Springer.

Valerio, P., Ibell, T., & Darby, A. P. (2011). Shear assessment of prestressed concrete bridges. *Proceedings of the Institution of Civil Engineers-Bridge Engineering, 164*(4), 195–210.

Van Paepegem, W., Shulev, A. A., Roussev, I. R., De Pauw, S., Degrieck, J., & Sainov, V. C. (2009). Study of the deformation characteristics of window security film by digital image correlation techniques. *Optics and Lasers in Engineering, 47*(3–4), 390–397.

Wang, Y., Lava, P., Debruyne, D., & Van Houtte, P. (2011). Error estimation of dic for heterogeneous strain states. *Advances in Experimental Mechanics VIII, 70,* 177–182.

9

Seismic Behavior Factors of RC Staggered Wall Buildings

Jinkoo Kim*, Yong Jun, and Hyunkoo Kang

Abstract: In this study seismic performance of reinforced concrete staggered wall system structures were investigated and their behavior factors such as overstrength factors, ductility factors, and the response modification factors were evaluated from the overstrength and ductility factors. To this end, 5, 9, 15, and 25-story staggered wall system (SWS) structures were designed and were analyzed by nonlinear static and dynamic analyses to obtain their nonlinear force–displacement relationships. The response modification factors were computed based on the overstrength and the ductility capacities obtained from capacity envelopes. The analysis results showed that the 5- and 9-story SWS structures failed due to yielding of columns and walls located in the lower stories, whereas in the 15- and 25-story structures plastic hinges were more widely distributed throughout the stories. The computed response modification factors increased as the number of stories decreased, and the mean value turned out to be larger than the value specified in the design code.

Keywords: staggered wall systems, seismic design, overstrength factors, ductility factors, response factors.

1. Introduction

Reinforce concrete (RC) shear walls are key elements to resist both gravity and lateral loads in building structures, and the seismic performance and analysis modeling of RC shear wall structures have been widely investigated by many researchers (e.g., Wallace 2012). Recently an alternative building structure system, a staggered wall system, has drawn attention due mainly to its capability to provide wider open space. The staggered-wall system consists of a series of storey-high RC walls spanning the total width between two rows of exterior columns and arranged in a staggered pattern on adjacent column lines. With the columns only on the exterior of the building, a full width of column-free area can be created. Compared with traditional shear wall structures, the structures with vertical walls placed at alternate levels have advantage for their enhanced spatial flexibility. Currently Korean government provides various incentives for apartment buildings designed with increased spatial flexibility. In this regard the apartment buildings with vertical walls placed at alternate levels have advantage for their enhanced spatial flexibility. Such a structural system has already been widely applied in steel residential buildings, which is typically called a staggered truss system.

The system was first proposed by Fintel (1968), who found out that the staggered wall systems are very competitive with the conventional form of construction and are more economical. Mee et al. (1975) carried out shaking table tests of 1/15 scaled models for the staggered wall systems and found that the consistent mass analysis gave reasonable estimation of dynamic behavior of the system. Kim and Jun (2011) evaluated the seismic performance of partially staggered wall apartment buildings using non-linear static and dynamic analysis, and compared the results with those of conventional shear wall apartment buildings. Lee and Kim (2013) investigated the seismic performance of six and 12-story staggered wall structures with a middle corridor based on the FEMA P695 procedure. It was found that the collapse margin ratios of the model structures obtained from incremental dynamic analyses turned out to be larger than the limit states specified in the FEMA P695. Kim and Han (2013) investigated the sensitivity of design variables to the seismic response of staggered wall structures. It was observed that when the earthquake intensity is relatively small, the yield stress of rebars and the concrete strength in the link beams are important factors as well as inherent damping ratio. As the intensity of seismic load increased, the strength of columns became another important factor. Lee and Kim (2013) derived empirical formulas for fundamental natural period of reinforced concrete staggered wall structures. They found that the natural periods of the staggered wall structures are similar to those of the shear wall structures having the same overall configuration.

The staggered wall systems, however, have not been considered as one of the basic seismic-force-resisting systems in most design codes due mainly to the vertical discontinuity of the main structural elements. ASCE 7 (2010) requires that lateral systems that are not listed as the basic seismic-force-resisting systems shall be permitted if analytical and test data are submitted to demonstrate the lateral force resistance and energy dissipation capacity. The American Institute of Steel

Department of Civil and Architectural Engineering,
Sungkyunkwan University, Suwon, Korea.
*Corresponding Author; E-mail: jkim12@skku.edu

Construction (AISC) Design Guide 14 (AISC 2002) recommends the response modification factor of 3.0 for seismic design of staggered truss system buildings; however none is specified for reinforced concrete staggered wall systems.

Seismic behavior factors including the response modification factors are essential for seismic design of structures. The factors are provided for typical structure systems in most design codes. However for non-typical structures the determination of the behavior factors is an important issue (e.g. Tomažević and Weiss 2010, Skalomenos et al. 2015). In this study the behavior factors such as overstrength factors, ductility factors, and the response modification factors of reinforced concrete staggered wall system (SWS) structures were evaluated following the procedure recommended in the ATC 19 (1995). To this end, 5, 9, 15, and 25-story SWS structural models were designed and were analyzed by nonlinear static and dynamic analyses to obtain their force–displacement relationship up to failure. The response modification factors were computed based on the overstrength and the ductility capacities obtained from the capacity envelopes.

2. Analysis Model Structures

2.1 Configuration of Staggered Wall System Analysis Models

To evaluate seismic performance and behavior factors of reinforced concrete staggered wall structures, the rectangular

plan staggered wall systems with 5, 9, 15, and 25 stories and wall length of 6 and 9 m were designed. Figure 1 shows the structural plan and the three-dimensional view of the 5-story model structure with 9 m parallel staggered walls. In the model structures, the story-high RC walls that span the width of the building are located in a staggered pattern. The staggered arrangement of the floor-deep walls placed at alternate levels on adjacent column lines allows an interior floor space of twice the column spacing to be available for freedom of floor arrangements. The floor system spans from the top of one staggered wall to the bottom of the adjacent wall serving as a diaphragm. The horizontal load is transferred to staggered walls below through diaphragm action of floor slabs. In this study the staggered walls were designed as story-high deep beams. The combined system of floor diaphragm and staggered wall acts like H-shaped deep beam which resists the applied load efficiently. With RC walls located at alternate floors, flexibility in spatial planning can be achieved compared with conventional wall-type structures with vertically continuous shear walls.

Columns and beams are located along the longitudinal perimeter of the structures providing a full width of column-free area within the structure. Along the longitudinal direction, the column-beam combination resists lateral load as a moment resisting frame. Along the transverse direction, the columns are expected to have minimum bending moments because of the cantilever action of the double-frame deformation configuration as illustrated in Fig. 2. The naming

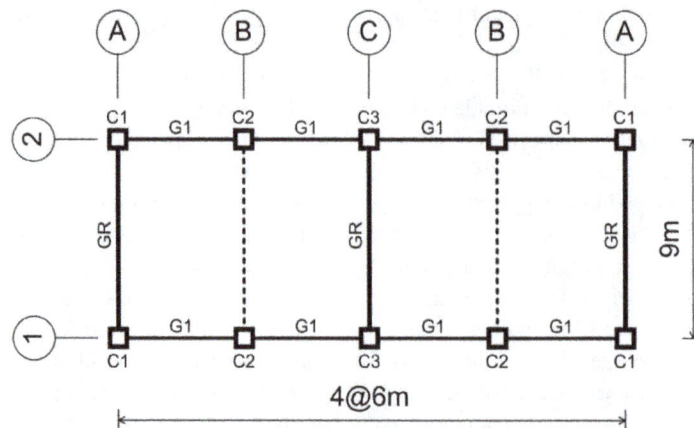

(a)

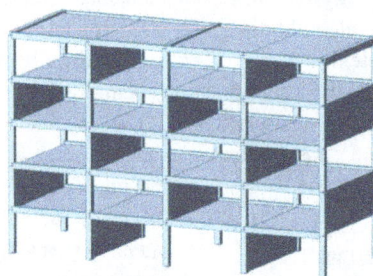

(b)

Fig. 1 Staggered wall analysis model structure. a Structural plan. b Three-dimensional view.

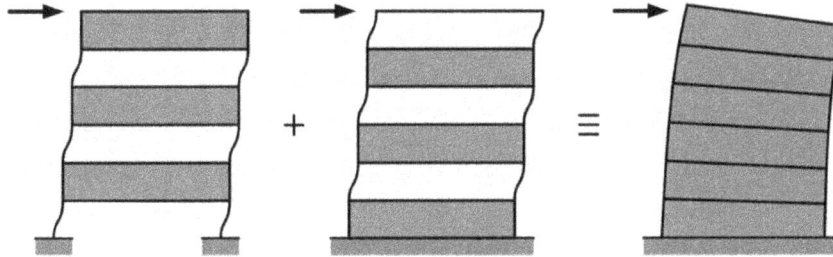

Fig. 2 Behavior of a staggered wall system subjected to lateral loads.

Table 1 Naming plan for analysis model structures.

Length of wall (m)	Seismic region	Number of story	Name
6	Low seismic region	5	5F_6 m low
		9	9F_6 m low
		15	15F_6 m low
		25	25F_6 m low
	Medium seismic region	5	5F_6 m medium
		9	9F_6 m medium
		15	15F_6 m medium
		25	25F_6 m medium
9	Low seismic region	5	5F_9 m low
		9	9F_9 m low
		15	15F_9 m low
		25	25F_9 m low
	Medium seismic region	5	5F_9 m medium
		9	9F_9 m medium
		15	15F_9 m medium
		25	25F_9 m medium

plan for the analysis model structures depending on the design variables such as length of staggered walls, design seismic load level, and number of story are presented in Table 1.

2.2 Structural Design of Analysis Model Structures

The model structures were designed per the ACI 318-14 (ACI 2014) using the design loads specified in the ASCE 7-13 (2010). The dead load was estimated to be 4.71 kN/m^2 including the weight of the structure itself and immovable fixtures, and live loads of 1.92 kN/m^2 was used assuming that the structure was used as residential buildings. The staggered wall structures, as well as the staggered truss structures, have not been included in seismic load resisting systems due mainly to the fact that the lateral load resisting system, the staggered walls, is not vertically continuous. In addition, as the staggered walls act like story-high deep beams, the structures are similar to typical weak column-strong beam systems. Therefore in this study response modification factor of 3.0 was used in the structural design of the staggered wall

systems, which is generally used for the structures to be designed without consideration of seismic detailing. Table 2 shows the seismic coefficients used for evaluation of design seismic load following the ASCE 7-13 specification, where the parameters S_s and S_1 represent the maximum considered earthquake (MCE) spectral response acceleration parameters at short period and at 1 s period, respectively, and the parameters S_{DS} and S_{D1} represent the design spectral response acceleration at short and at a period of 1 s, respectively. To consider the effect of design seismic load levels, the model structures were designed with two different levels of seismic loads corresponding to low and medium seismic regions. The design seismic loads for the structures located in the low and the medium seismic regions were determined based on the assumption that the structures were located in the class B site (rock) and C site (very dense soil or soft rock), respectively. The design spectrum for low seismic region was constructed using the design spectral response acceleration parameters, S_{DS} and S_{D1}, of 0.31 and 0.13, respectively. The design spectrum for medium seismic region was constructed using

Table 2 Seismic coefficients used for evaluation of design seismic load.

Seismic load level	Low	Medium
Maximum considered earthquake		
S_S	0.46	0.80
S_1	0.19	0.30
Site class	B	C
Design earthquake		
S_{DS}	0.31	0.57
S_{D1}	0.13	0.20
$S_a (T = 0)$	0.12	0.23
R-factor	3	

the acceleration parameters of 0.57 and 0.20, respectively. At zero natural period (T = 0), the spectral response accelerations for the Low and the Medium earthquakes are 0.12 and 0.23, respectively. The ultimate strength of concrete is 24 MPa and the tensile yield stress of re-bars is 400 MPa. The thickness of the staggered walls is 20 cm throughout the stories. The thickness of the floor slabs is 21 cm which is the minimum thickness required for wall-type apartment buildings in Korea to prevent transmission of excessive noise and vibration through the floors. The thickness of the staggered walls is 20 cm throughout the stories. The rebar placements in the columns and staggered walls are presented in Table 3. The reinforcements of the columns followed the seismic detailing of ordinary moment resisting frames specified in the ACI 318 (2014). The staggered walls were designed as deep beams, for which only minimum reinforcement of D10@400 was needed both horizontal and vertical directions due to their large depth. Table 4 shows the fundamental natural periods of the analysis model structures along the transverse direction where the staggered walls are located, where it can be observed that the natural periods of the structures designed for medium seismic load are slightly shorter than those of the structures designed for low seismic load.

2.3 Modeling for Analysis

The seismic performances of the model structures were evaluated using the nonlinear analysis program CANNY (Li 2004) which utilizes fiber model for modeling elements. Fiber model has been proven to be effective in nonlinear analysis of structures by various researchers (e.g. Sfakianakis 2002; Calabrese et al. 2010; Li and Hatzigeorgiou 2012; Li et al. 2013). In this study both the material and the geometric nonlinearities were considered in the analysis. The geometric nonlinearity was considered by member level and the frame level p-delta effect. The material nonlinear behavior of the concrete was modeled using the work of Kent and Park (1971) as shown in Fig. 3 where the ultimate strength $f'_c = 24$ MPa and ε_{50u} = the strains corresponding to the stress equal to 50 % of the maximum concrete strength for unconfined concrete. The reinforcing steel was modeled by bi-linear lines with yield stress of 400 MPa and 2 % of post-yield stiffness.

E is the elastic modulus of steel rebars. The expected ultimate strengths of the concrete and steel were taken to be 1.5 and 1.25 times the nominal strengths based on the recommendation of the FEMA-356 (FEMA 2000). As the model structures were designed without considering seismic detailing, the confinement effect of concrete was neglected in the stress–strain relationship. The columns and walls were modeled by the multi-axial spring model with fiber elements as shown in Fig. 4. The axial/bending deformation was simulated by elongation or contraction of each fiber element. The in-plane shear force is resisted by the spring W and the out-of-plane shear is resisted by the springs $C1$ and $C2$, respectively. The symbols I_1 and A_{s1} denote the moment of inertia and rebar cross-sectional area of the element $C1$, and I_2 and A_{s2} denote the moment of inertia and rebar cross-sectional area of the element $C2$, respectively. The symbols I and A_s denote the moment of inertia and rebar cross-sectional area of the element W. The symbols d and θ denote the displacement and the rotation at a joint, respectively. The hysteretic behavior of the shear springs was idealized by the origin-oriented model based on the tri-linear hysteresis curve as described in Fig. 5, which can consider the decrease in gradient of loading as the loading cycle and the deformation increase. It is assumed that the cross section of the shear walls remains plane when an in-plane wall deformation occurs. Following the plane section remain plane assumption, strain of the fiber element in the cross section is proportional to the distance from the neutral axis. The stress of each slice is calculated using the stress–strain relation from the strain of each fiber slice, and the bending moment is calculated by summing the moments to the center of the cross section. Figure 6 shows the modified Clough model used to simulate the bending deformation of the elements (Clough and Johnston 1966). The model is composed of bi-linear lines and may represent the degradation of stiffness after yielding. Even though the simplified origin-oriented hysteresis model may not be quite accurate for predicting shear response of the wall element, especially under high shear stresses, it was employed in this study for the following reasons: (i) the staggered walls act more like deep beams rather than shear walls, and (ii) for design level earthquakes, inelastic deformations are concentrated mostly

Table 3 Rebar details in the analysis model structure.

(a) Beams

			Low seismic region				Medium seismic region			
	Type		Dimension (mm)		Rebar		Dimension (mm)		Rebar	
			H	B	Upper	Lower	H	B	Upper	Lower
5F	G1	1–3F	350	220	4-D22	2-D22	400	250	5-D25	3-D25
		4–5 F	350	220	4-D19	2-D19	350	220	4-D22	2-D22
	GR	5 F	400	250	3-D19	3-D19	350	220	4-D19	3-D19
9F	G1	1–3 F	350	220	4-D22	2-D22	400	250	5-D25	4-D25
		4–6 F	350	220	4-D22	2-D22	400	250	5-D25	4-D25
		7–9 F	350	220	4-D19	2-D19	400	250	5-D22	3-D22
	GR	9 F	400	250	2-D22	3-D22	350	220	4-D19	3-D19
15F	G1	1-3 F	350	220	4-D25	2-D25	450	300	6-D25	5-D25
		4–6 F	350	220	4-D25	2-D25	450	300	6-D25	5-D25
		7–9 F	350	220	5-D22	2-D22	450	300	5-D25	4-D25
		10–12 F	350	220	5-D19	2-D19	400	250	5-D25	3-D25
		13–15 F	350	220	4-D19	2-D19	400	250	5-D22	3-D22
	GR	15 F	400	250	2-D22	3-D22	350	220	4-D19	3-D19
25F	G1	1–3 F	400	250	4-D25	2-D25	500	320	6-D25	5-D25
		4–6 F	400	250	4-D25	2-D25	500	320	6-D25	5-D25
		7–9 F	400	250	4-D25	2-D25	500	320	7-D25	5-D25
		10–12 F	400	250	4-D25	2-D25	450	300	6-D25	5-D25
		13–15 F	400	250	4-D25	2-D25	450	300	6-D25	5-D25
		16–18 F	350	220	4-D22	2-D22	400	250	6-D25	4-D25
		19–21 F	350	220	5-D19	2-D19	400	250	5-D25	3-D25
		22–25 F	350	220	4-D19	2-D19	350	220	4-D25	2-D25
	GR	25 F	400	250	3-D19	3-D19	350	220	4-D19	3-D19

(b) Columns

Bar-placement detail (column)

		Low seismic region						Medium seismic region					
	Type	C1		C2		C3		C1		C2		C3	
		BH	Rebar	BH	Rebar	BH	Rebar	BH	Rebar	BH	Rebar	BH	Rebar
5F	1–3 F	400	8-D22	600	16-D22	400	12-D22	600	16-D22	400	8-D22	600	12-D22
	4–5 F	400	8-D22	550	8-D22	400	12-D22	550	8-D22	400	6-D22	550	16-D22
9F	1–3 F	400	12-D22	650	16-D22	500	16-D22	650	12-D25	500	12-D22	650	16-D22
	4–6 F	400	6-D19	600	14-D19	450	8-D19	600	10-D22	450	6-D22	600	10-D22
	7–9 F	400	8-D22	550	6-D25	400	12-D22	550	6-D25	400	6-D19	550	8-D25
15F	1–3 F	550	6-D25	750	12-D25	600	8-D25	650	12-D25	600	12-D25	750	16-D25
	4–6 F	500	6-D25	700	10-D25	550	6-D25	600	10-D22	550	16-D19	700	10-D25
	7–9 F	450	4-D25	650	10-D25	550	6-D25	550	6-D25	500	16-D16	650	10-D25
	10–12 F	420	4-D25	600	10-D22	500	6-D25	500	6-D25	500	16-D16	600	8-D25
	13–15 F	420	8-D25	550	6-D25	500	8-D25	500	6-D25	450	4-D25	550	6-D25

Table 3 continued

(b) Columns

Bar-placement detail (column)

	Type	Low seismic region						Medium seismic region					
		C1		C2		C3		C1		C2		C3	
		BH	Rebar	BH	Rebar	BH	Rebar	BH	Rebar	BH	Rebar	BH	Rebar
25F	1–3 F	700	10-D25	750	20-D29	800	14-D25	600	20-D25	800	16-D25	750	20-D29
	4–6 F	650	10-D25	700	20-D25	750	12-D25	550	12-D25	750	16-D25	700	18-D29
	7–9 F	600	8-D25	650	20-D25	700	10-D25	500	10-D25	700	10-D25	650	16-D29
	10–12 F	550	6-D25	600	16-D25	650	10-D25	450	10-D25	650	10-D25	600	16-D29
	13–15 F	500	6-D25	550	10-D25	600	10-D25	400	8-D25	600	10-D25	550	12-D29
	16–18 F	450	4-D25	500	10-D25	550	6-D25	400	8-D25	550	6-D25	500	8-D29
	19–21 F	450	4-D25	450	12-D25	500	6-D25	400	10-D25	500	6-D25	450	12-D29
	22–25 F	400	8-D25	400	10-D25	450	10-D25	400	8-D25	450	4-D25	400	8-D29

Table 4 Fundamental natural periods of model structures along the transverse direction.

Wall length (m)	Seismic load level	Story	Period (s)
6	Low	5	0.15
		9	0.43
		15	1.09
		25	2.37
	Medium	5	0.14
		9	0.36
		15	0.88
		25	2.04
9	Low	5	0.13
		9	0.32
		15	0.69
		25	1.47
	Medium	5	0.11
		9	0.26
		15	0.59
		25	1.36

in columns and most staggered walls remain elastic. At the first yield point, f_e, the post-yield stiffness was set to be 16 % of the initial stiffness, and after the final yield point, f_y, the stiffness was reduced to 0.1 % of the initial stiffness. The slabs were considered as rigid diaphragm.

3. Seismic Performance Evaluation

To evaluate the nonlinear behavior of the model structures subjected to seismic load, pushover analyses were carried out along the transverse direction by applying incremental lateral load proportional to the fundamental mode of vibration. To define the failure limit state of the model structures, the following two approaches were followed: first, the structure was assumed to have reached a limit state when the inter-story drift reached 1.5 % of the story height as recommended by most seismic design codes such as the ASCE 7 (2010). Second, a structural failure was defined when formation of plastic hinges leaded to failure mechanism. The model structures were assumed to have failed when either of the two limit states occurred.

The base shear versus roof displacement relationship for each model structure is depicted in Fig. 7 where it can be

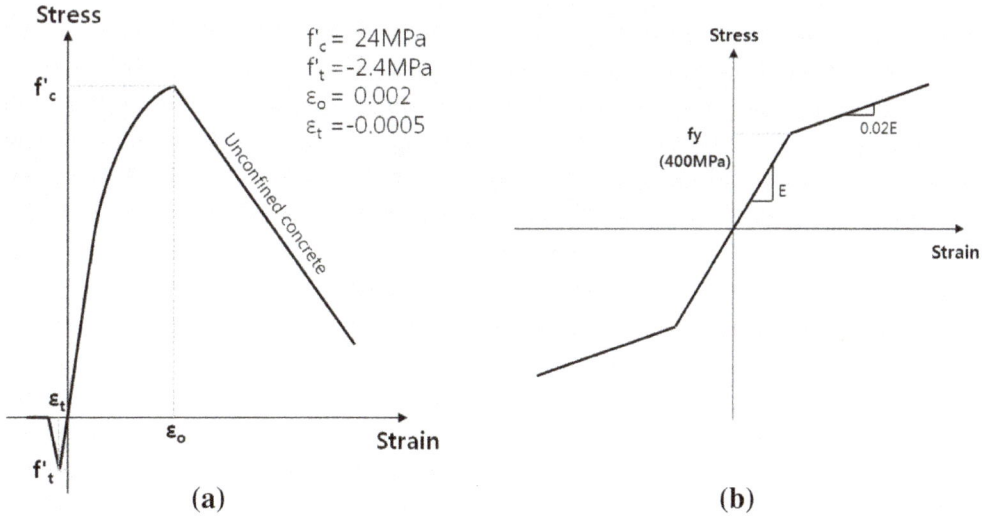

Fig. 3 Stress–strain relationships of structural materials. **a** Concrete. **b** Re-bars.

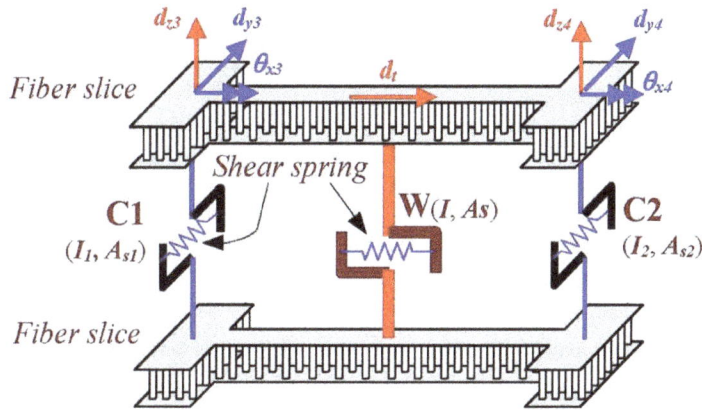

Fig. 4 Fiber model for staggered walls.

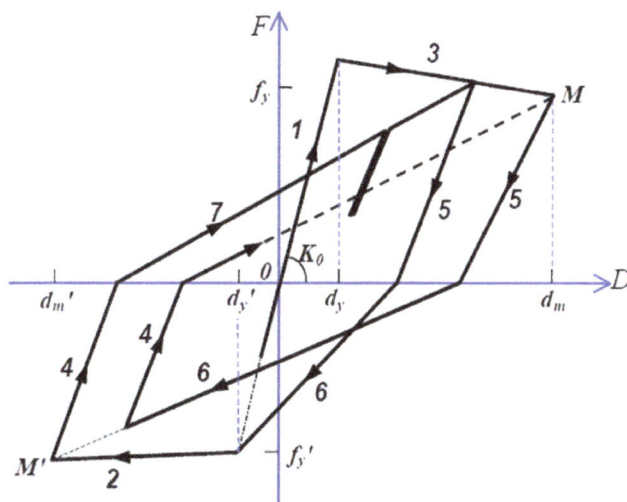

Fig. 5 Modified Clough model for flexural deformation.

observed that the five-story structure with 9 m-long wall designed for medium-level seismic load have highest strength and the 20 five-story structure with 6 m wall designed for low-level seismic load have lowest strength. The strength decreased as the number of story increased. The structures with 9 m-long staggered walls showed higher strength than those with 6 m staggered walls. However the maximum displacements were generally larger in the structures with 6 m walls. The model structures designed with medium level seismic load showed higher strength than the

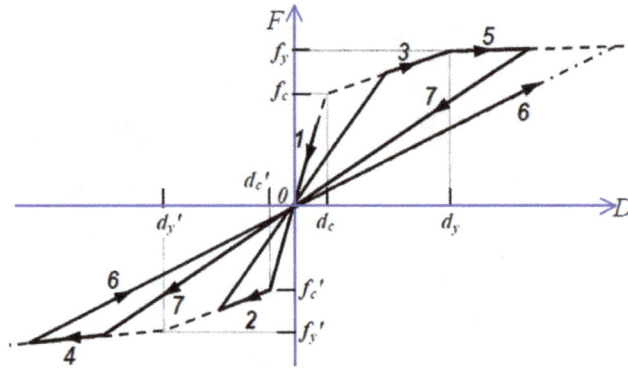

Fig. 6 Origin-oriented model for shear deformation.

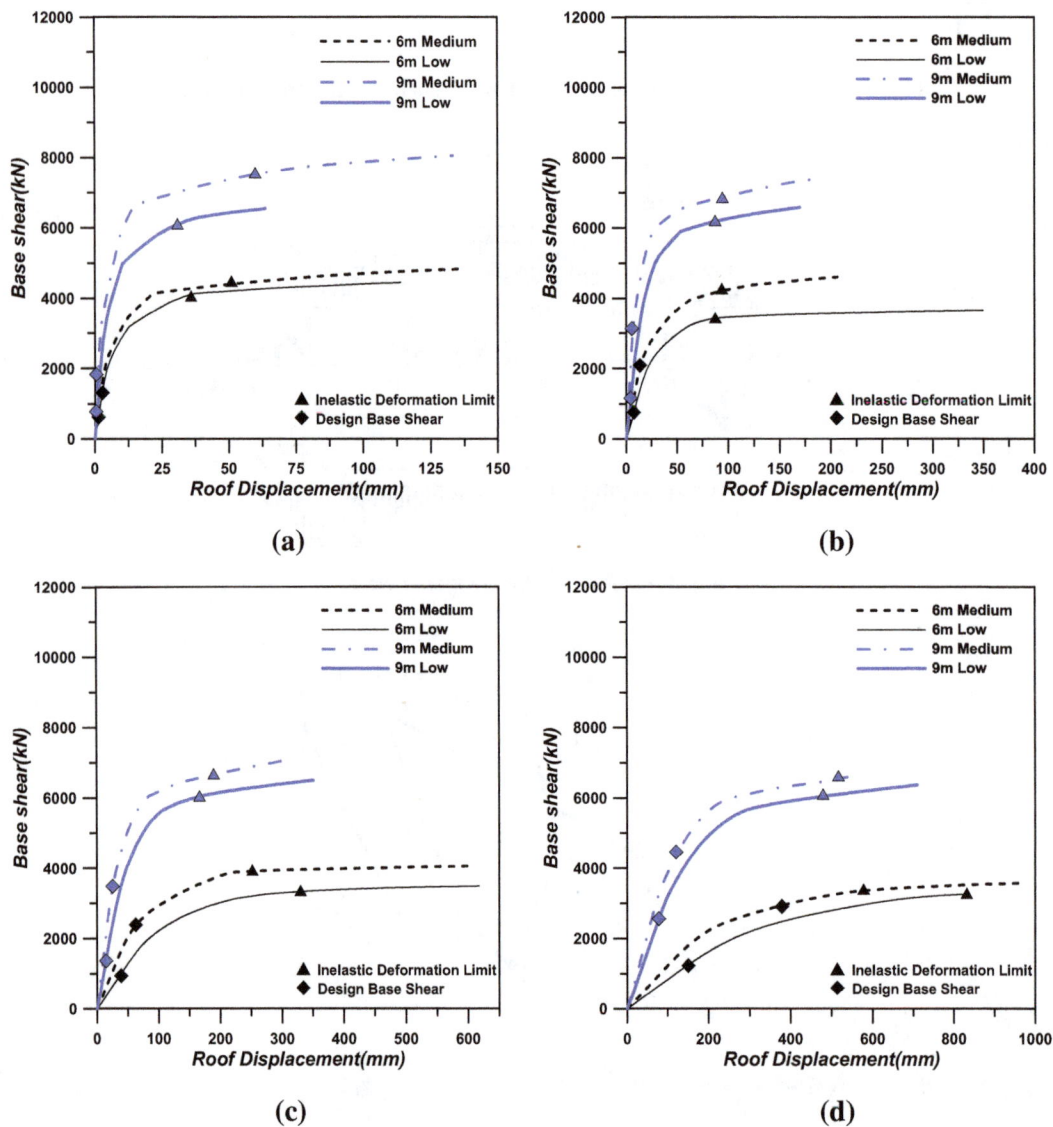

Fig. 7 Pushover curves of analysis model structures. **a** 5-story, **b** 9-story, **c** 15-story, **d** 25-story.

structures designed with low level seismic load. It also can be observed that, even though the design base shear increases as the number of story increases, the maximum strength does not increase proportionally to the design base shear. This is due to the fact that as the number of story increases the damage is concentrated in the lower few stories, as can be observed in the plastic hinge formation presented in Fig. 13. This implies that the SWS in its standard form may not be efficient in the medium to highrise buildings.

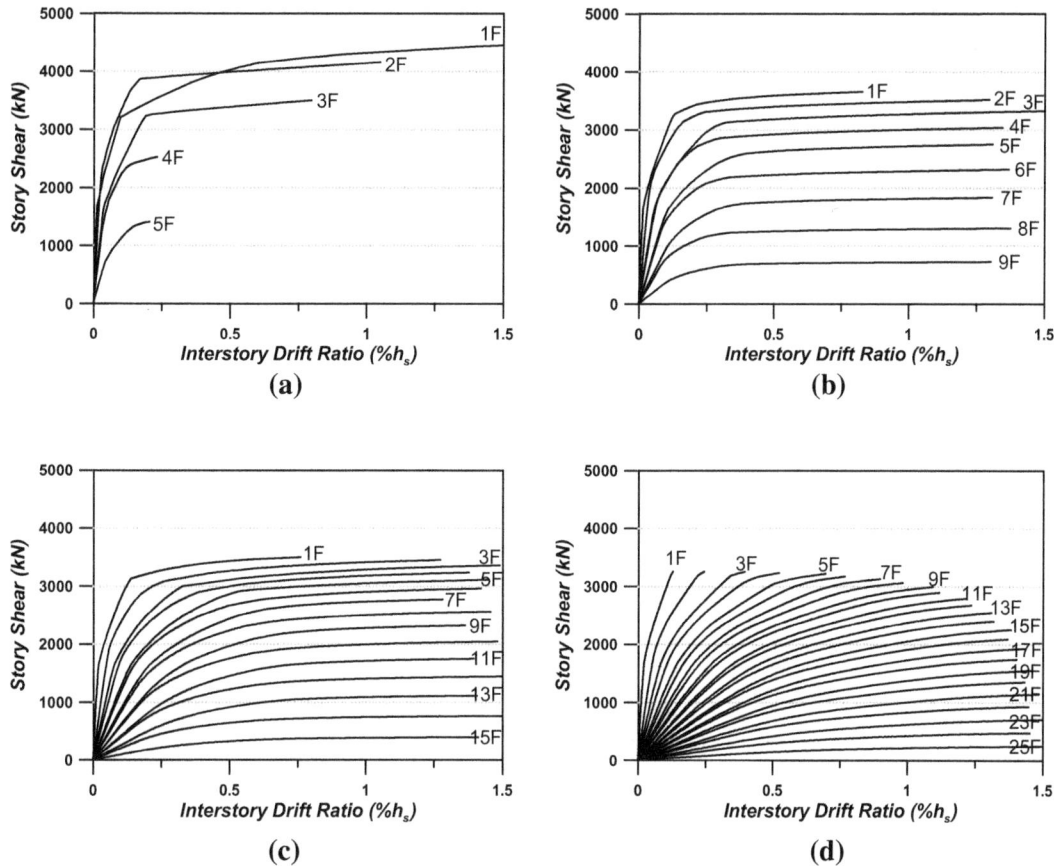

Fig. 8 Story shear versus inter-story drift curves of the model structures with 6 m-long staggered walls designed for low-level seismic load. **a** 5-story, **b** 9-story, **c** 15-story, **d** 25-story.

Figures 8, 9, 10 and 11 show the story shear versus inter-story drift curves of model structures. It can be noticed that both story stiffness and story strength are larger in lower stories. Compared with the story shear versus inter-story drift curves of higher stories, the curves of lower stories generally show distinct yield points. In the low-rise structures, large deformation is concentrated in the lower few stories. This trend is more noticeable in the structures with 9 m staggered walls designed for low seismic load. As the building height increases and the wall length decreases, deformation is more uniformly distributed throughout the stories.

Figure 12 shows the inter-story drifts of the model structures obtained from pushover analysis. The pushover analyses were performed until the maximum inter-story drift reached 1.5 % of the story height. It can be observed that the inter-story drifts of the structures with 6 m wall length are relatively uniformly distributed along the height. It also can be noticed that in the five- and nine-story structures with 6 m span, large inter-story drifts occurred in the lower stories, whereas the inter-story drifts are more uniformly distributed in the fifteen- and 20 five-story structures. In the structure with 9 m span length, large inter-story drifts occurred in the lower-stories in all structures. This implies that the staggered wall structures with 9 m span length behave more like moment frames, whereas the structures with 6 m span behave more like shear wall structures. It was observed in all

model structures that the maximum inter-story drift of 1.5 % of the story height was reached prior to the story collapse mechanism. Figure 13 shows the plastic hinge formation of the model structures with 6 m walls designed for medium seismic load when the maximum inter-story drift reached 1.5 % of the story height. It was observed that in the low-rise model structures the first story walls yielded first followed by yielding of the first-story columns. In the higher structures plastic hinges formed first at the top and bottom ends of the columns located in the higher stories due to the higher mode effects. However at the ultimate state most plastic hinges were concentrated in the lower story columns as can be observed in the figure. Similar results were also observed in Kim and Han (2013).

For seismic performance evaluation of the model structures with 6 m staggered walls designed for medium seismic load, incremental dynamic analyses (IDA) were carried out using the El Centro and the Taft earthquakes. Figure 14 shows the response spectra of the Taft and El Centro earthquakes scaled to the design spectra for the low and the medium seismic loads. IDA were carried out with the following procedure: (1) Scale the earthquake records so that the pseudo acceleration S_a at the fundamental period of the structure becomes 0.1 g; (2) Carry out nonlinear dynamic analysis and estimate the maximum inter-story drift and base shear of the structure; (3) Increase S_a by 0.1 g and carry out the same analysis procedure. Figure 15 compares the base

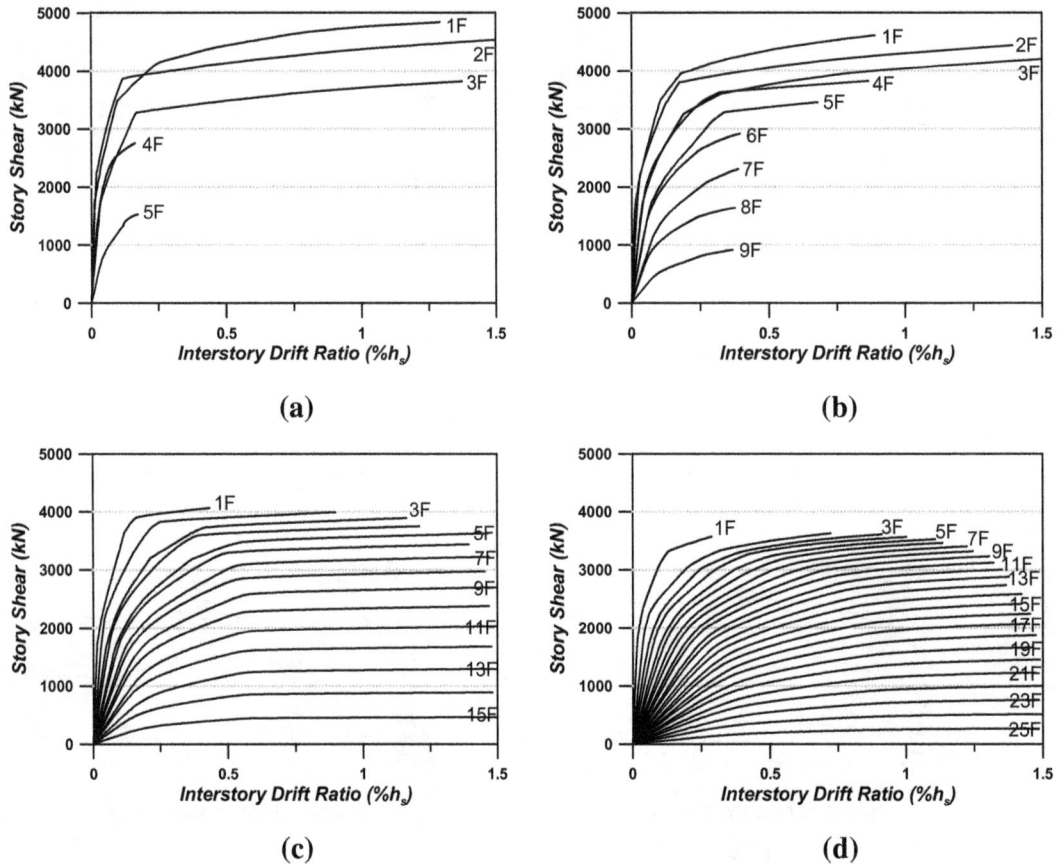

Fig. 9 Story shear versus inter-story drift curves of the model structures with 6 m-long staggered walls designed for medium-level seismic load. **a** 5-story, **b** 9-story, **c** 15-story, **d** 25-story.

shear-roof displacement relationships of the model structures obtained by incremental dynamic analyses and nonlinear static pushover analyses. Except for the slight discrepancy in the results of the 15- and 25-story model structures, the base shear-roof displacement curves obtained from IDA and pushover analyses generally coincide well with each other. In the linear elastic deformation stage the two results are almost identical. After yielding slight difference is observed between the two results; however the difference is not significant.

4. Behavior Factors of the Model Structures

The ATC-19 (1995) proposed simplified procedure to estimate the response modification factors by the product of the three parameters that profoundly influence the seismic response of structures:

$$R = R_o R_\mu R_\gamma \qquad (1)$$

where R_o is the overstrength factor to account for the observation that the maximum lateral strength of a structure generally exceeds its design strength. Similar procedure was applied to evaluate the seismic design factors for reinforced concrete moment frames (AlHamaydeh et al. 2011), reinforced masonry structures (Shedid et al. 2011), and steel moment resisting frames with buckling restrained braces

(Abdollahzadeh et al. 2012). The FEMA (2000) specified three components of overstrength factors in Table C5.2.7-1: design overstrength, material overstrength, and system overstrength. R_μ is a ductility factor which is a measure of the global nonlinear response of a structure, and R_γ is a redundancy factor to quantify the improved reliability of seismic framing systems constructed with multiple lines of strength. In this study the redundancy factor was assumed to be 1.0 based on the fact that there are more than four seismic load-resisting frames along the transverse direction. Then the response modification factor is determined as the product of the overstrength factor and the ductility factor. From the base-shear versus roof displacement relationships, the overstrength factor and the ductility factor are obtained as follows (ATC-19 1995):

$$R_o = \frac{V_y}{V_d} \qquad (2a)$$

$$R_\mu = \frac{V_e}{V_y} \qquad (2b)$$

where V_d is the design base shear, V_e is the maximum seismic demand for elastic response, and V_y is the base shear corresponding to the yield point, which can be obtained from the capacity curves. To find out the yield point, straight lines are drawn on the pushover curve as depicted in Fig. 16 in such a way that the area under the original curve is equal to

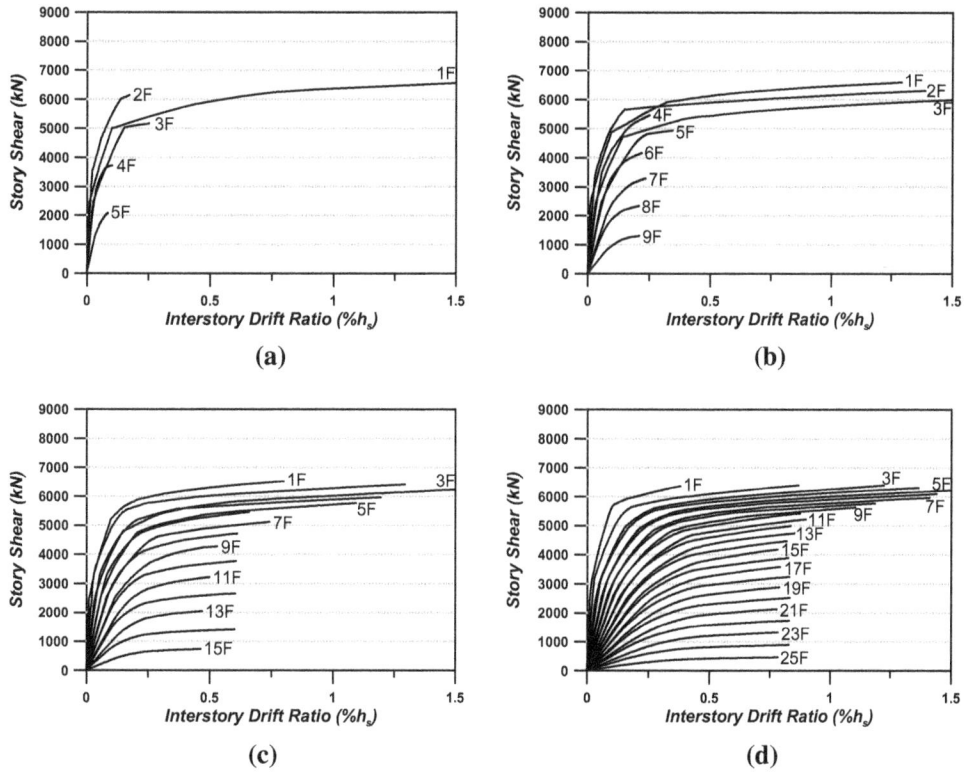

Fig. 10 Story shear versus inter-story drift curves of the model structures with 9 m-long staggered walls designed for low-level seismic load. **a** 5-story, **b** 9-story, **c** 15-story, **d** 25-story.

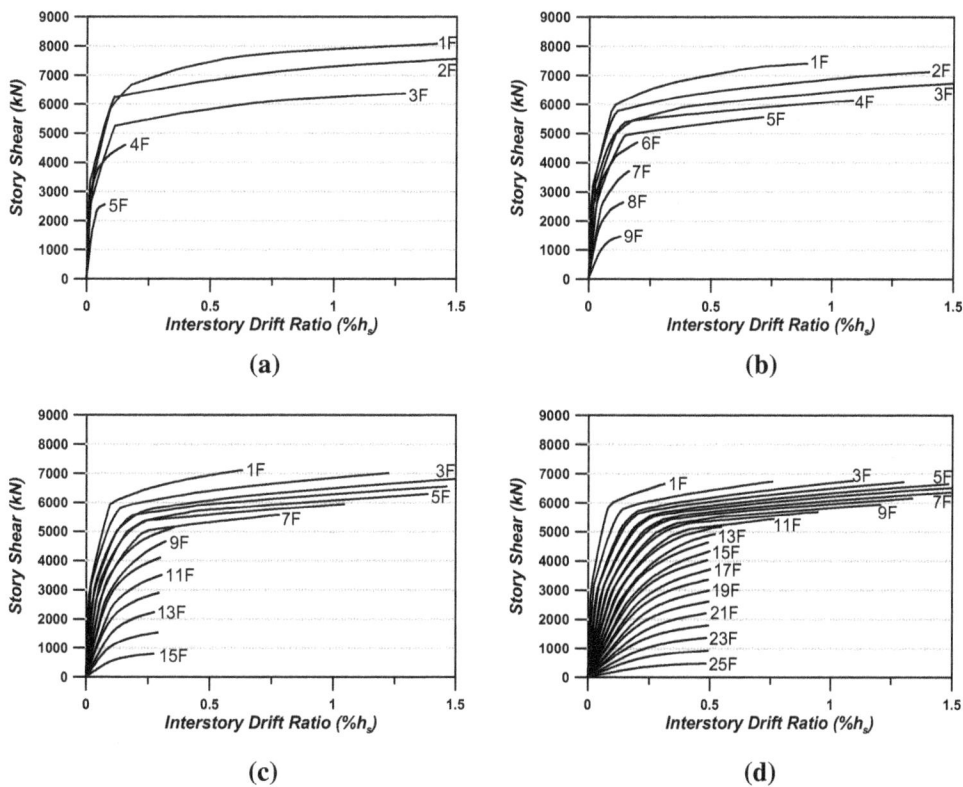

Fig. 11 Story shear versus inter-story drift curves of the model structures with 9 m-long staggered walls designed for medium-level seismic load. **a** 5-story, **b** 9-story, **c** 15-story, **d** 25-story.

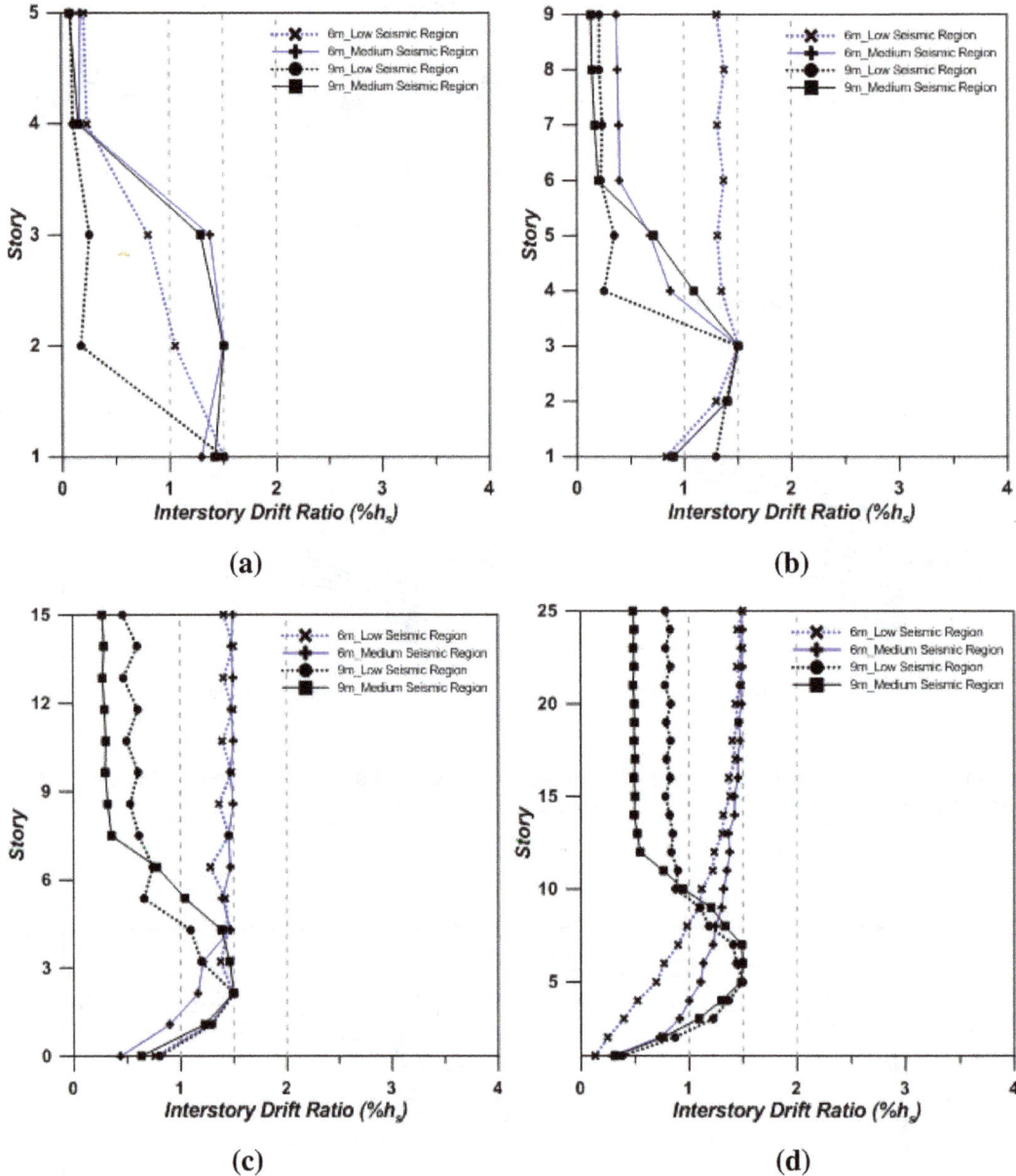

Fig. 12 Inter-story drifts of model structures. **a** 5-story, **b** 9-story, **c** 15-story, **d** 25-story.

that of the idealized one as recommended in the FEMA-356 (2000). In this study the ductility factor was obtained using the system ductility ratio μ as proposed by Newmark and Hall (1982)

$$\begin{aligned} R_\mu &= 1.0 && (T < 0.003 \text{ sec}) \\ R_\mu &= \sqrt{2\mu - 1} && (0.12 < T < 0.5 \text{ sec}) \\ R_\mu &= \mu && (T > 1.0 \text{ sec}) \end{aligned} \quad (3)$$

where T is the fundamental natural period of the structure and the ductility ratio was obtained by dividing the roof displacement at failure with the displacement at yield. Equation (3) is plotted in Fig. 17.

The overstrength factors of the model structures were computed using Eq. (2a) based on the capacity curves presented in Figs. 5 and 6, and are plotted in Fig. 18. It can be

observed that as the height of the structure increases the overstrength factors decrease. The overstrength factors of the structures designed with medium-level seismic load turned out to be smaller than those of the structures designed with low-level seismic load. The structures with 9 m-long staggered walls showed higher overstrength than the structures with 6 m-long walls.

Figure 19 plots the ductility factors of the model structures. No distinct pattern was observed in the distribution of ductility factors depending on the building height and the seismic load levels, and they were relatively uniform regardless of the height of the model structures and the length of the staggered walls with average value of 2.34. This implies that the ratios of the ultimate and the yield displacements are similar in most model structures. It can be noticed that the five-story

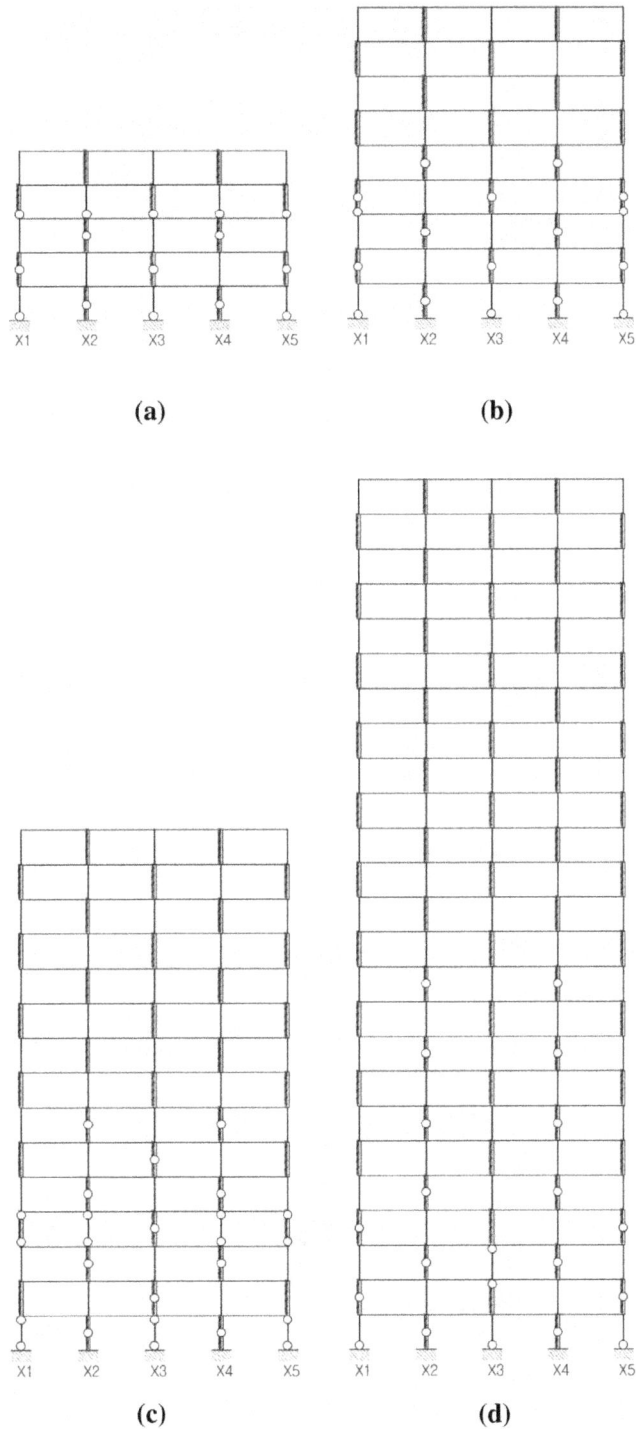

Fig. 13 Plastic hinge formation of model structures with 6 m-long staggered walls designed for medium-level seismic load. **a** 5-story, **b** 9-story, **c** 15-story, **d** 25-story.

structures designed for medium level earthquake load have slightly larger ductility than the structures designed for low-level seismic load. It was observed from the pushover analysis that the collapse was initiated by yielding of the columns located at the end of the staggered walls. Therefore to reinforce the columns, especially those located in the lower stories, would help to increase the

ductility of the SWS structures by preventing or delaying the formation of a story collapse mechanism.

The response modification factors are presented in Fig. 20, which are computed by multiplying the overstrength and the ductility factors. It can be observed that the response modification factors decrease as the height of the structure increases, which conforms to the results of the previous

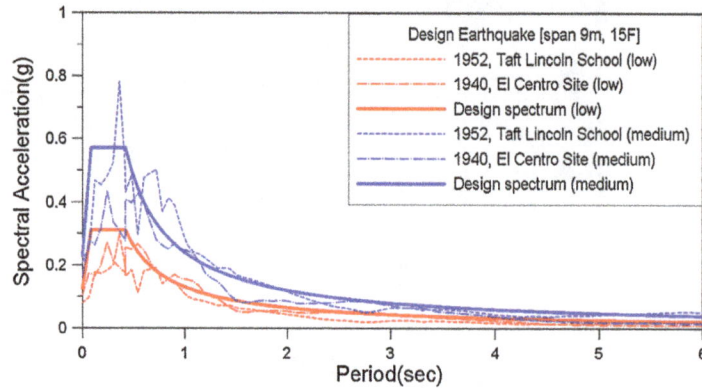

Fig. 14 Response spectra of the Taft and El Centro earthquakes scaled to the design spectra for the low and the medium seismic loads.

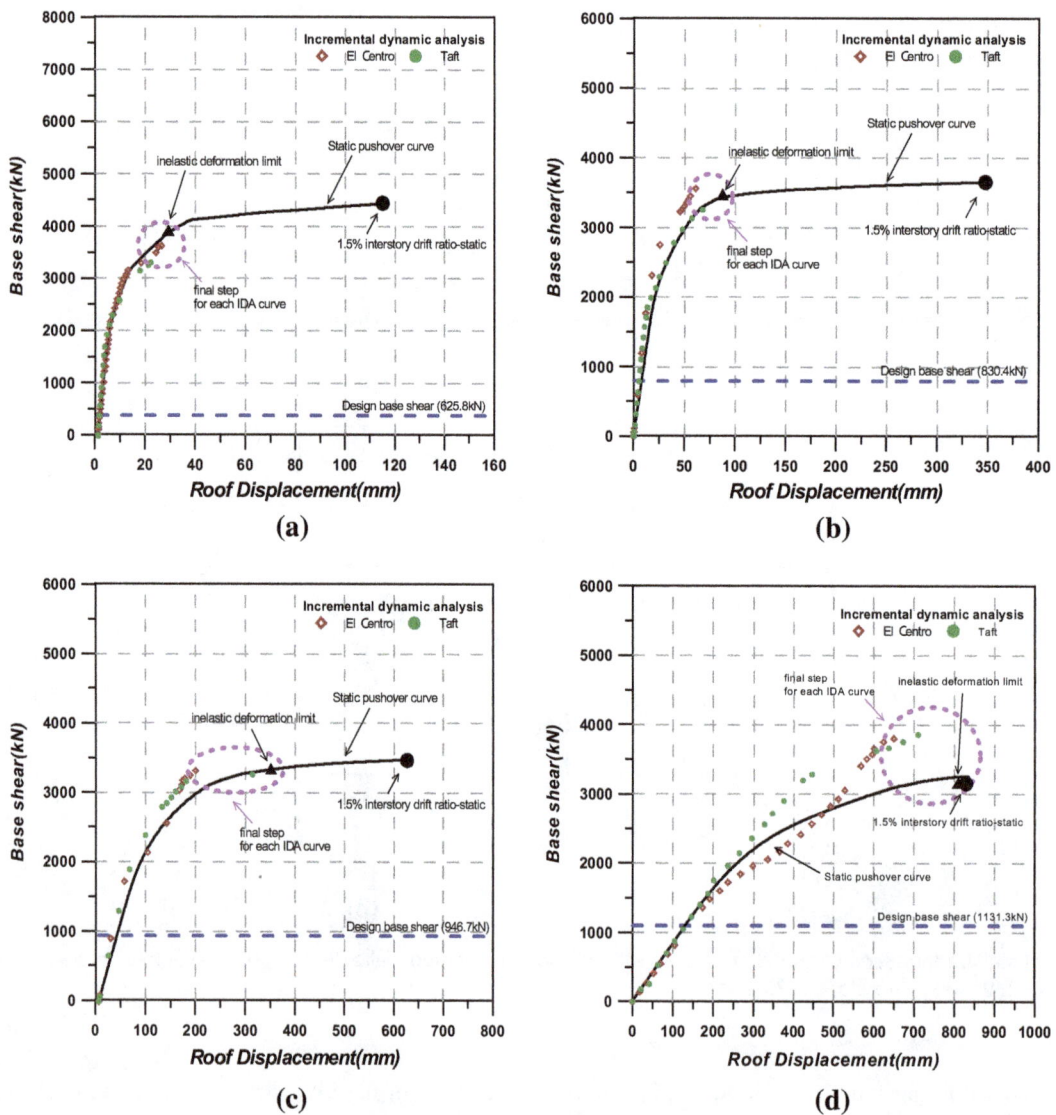

Fig. 15 Comparison of base shear–roof displacement relationships obtained by pushover analyses and incremental dynamic analyses. **a** 5-story, **b** 9-story, **c** 15-story, **d** 25-story.

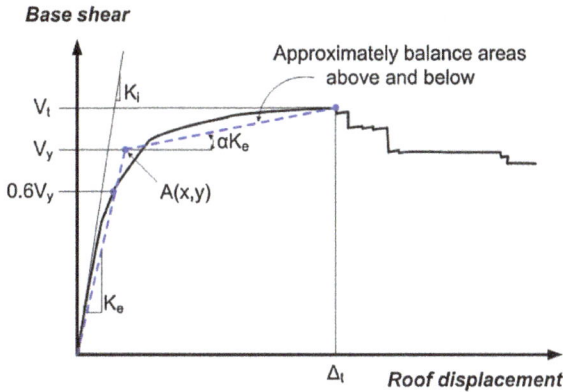

Fig. 16 Bi-linear idealization of a pushover curve.

researches on the structures with staggered trusses (Kim et al. 2007). The mean value for the response modification factors is much larger than 3.0, the code-recommended value for the structures not specified as one of the seismic-load resisting systems. In the 5-story structures, the computed response modification factors are larger than 7.0; in the 20-story structures the factors become as low as 2.0 in the structure with 6 m-long staggered walls designed for medium-level seismic load. However in the most model structures used in this study the computed response modification factors turned out to be larger than 3.0. This implies that the reinforced concrete structures with staggered walls may have enough resistance against design level seismic load. As in the case of the overstrength factors, the structures designed for medium-level earthquake had smaller response modification factors than those of the structures designed with low-level seismic load. The structures with 9 m-long staggered walls showed higher response modification factors than the structures with 6 m-long walls except for the 15-story structure designed for low-level seismic load. It can be noticed that the variation of the response modification factors is mainly contributed from the variation of the overstrength factors.

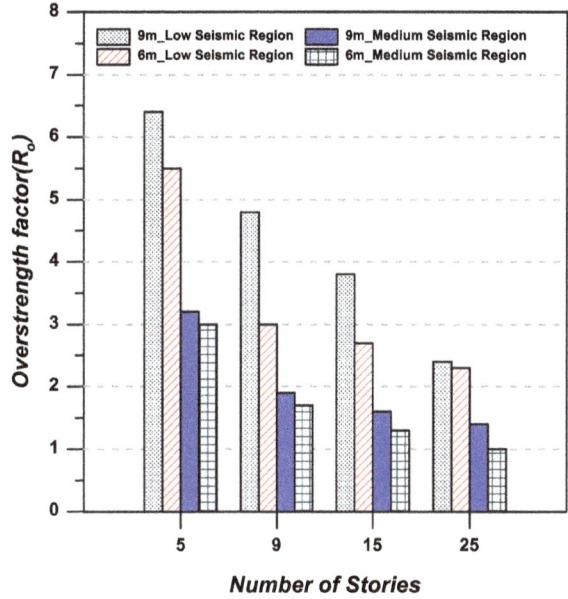

Fig. 18 Overstrength factors of model structures.

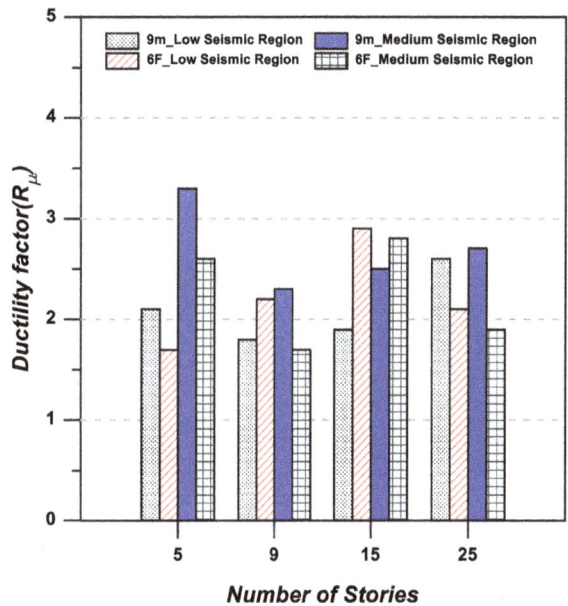

Fig. 19 Ductility factors of model structures.

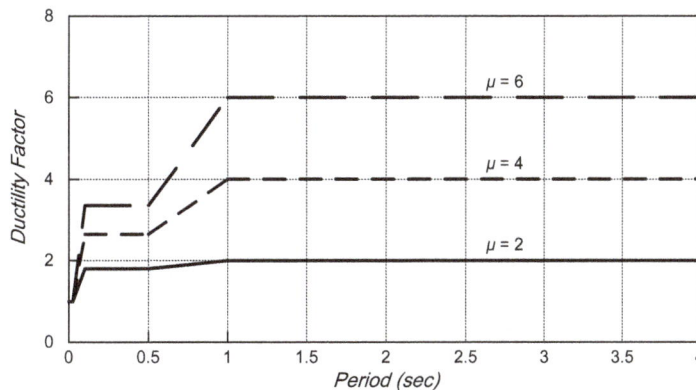

Fig. 17 Relationship between ductility factor and natural period proposed by Newmark and Hall (1982).

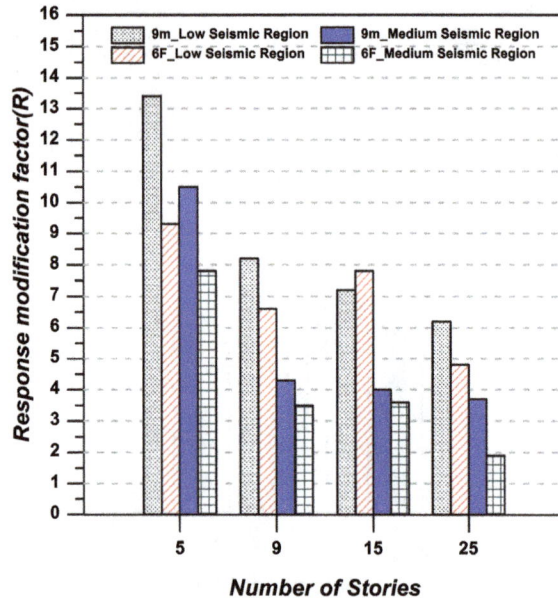

Fig. 20 Response modification factors of model structures.

5. Conclusions

One of the main obstacles to be overcome for application of staggered wall systems is to ensure the seismic safety of the systems and to provide valid seismic design coefficients. In this study seismic performance and the behavior factors such as overstrength factors, ductility factors, and the response modification factors of reinforced concrete SWS structures were evaluated. The analysis results showed that the behavior factors obtained by pushover analysis and incremental dynamic analysis turned out to be similar to each other. The overstrength factors of the structures designed with medium-level seismic load turned out to be smaller than those of the structures designed with low-level seismic load. This is due mainly to the fact that the participation of gravity load is more significant in the design of the latter system. The structures with 9 m-long staggered walls showed higher overstrength than the structures with 6 m-long walls. The ductility factors were relatively uniform regardless of the height of the model structures and the length of the staggered walls with average value of 2.34. The response modification factors obtained by multiplying overstrength factor and ductility factor decreased as the number of stories increased. Except for the structure with 6 m-long staggered walls designed for medium-level seismic load, the response modification factors turned out to be higher than 3.0 which was used for evaluation of design seismic load. The magnitude of the response modification factors were contributed mainly from large overstrength rather than from large deformability. The response modification factors of the structures designed for low-level seismic load were higher than those of the structures designed for higher seismic load. Based on the analysis results, it is concluded that the RC SWS structures generally have adequate strength and ductility capacities to resist design seismic load. As the response modification factor of the model

structures analyzed in this study ranged from 3.5 to 8, the current response modification factor of 3.0 seems to be in the conservative side and a little higher value of 3.5 or 4.0 may be more appropriate value for seismic design of staggered wall structures.

It was also observed that the maximum strength of the model structures did not increase proportionally to the design base shear, even though the design base shear increased as the number of story increased. This is due to the fact that as the number of story increased the damage was concentrated in the lower few stories. Even though no story failure mechanism was observed until maximum inter-story drift of 1.5 % was reached in all model structures, it would be necessary to delay the occurrence of story failure mechanism by reinforcing lower story columns to increase seismic-load resisting capacity of the structures. Also the adoption of seismic joint details specified in the ACI code will help increase the ductility of the system.

Finally it needs to be stated that, as the seismic performance of the staggered wall structures has not been validated by proper tests, further experimental research is still required for accurate evaluation of the seismic load resisting capacity of the staggered wall structures. Also the use of more accurate nonlinear concrete model will help enhance the validity of this study

Acknowledgments

This research was supported by a Grant (13AUDP-B066083-01) from Architecture & Urban Development Research Program funded by Ministry of Land, Infrastructure and Transport of Korean government.

References

Abdollahzadeh, G. H., Banihashemi, M. R., Elkaee, S. & Esmaeelnia A. M. (2012). Response modification factor of dual moment-resistant frame with buckling restrained brace (BRB). *15th World Congress of Earthquake Engineering, 2012.* Portugal: Lisbon.

ACI 318. (2005). *Building code requirements for structural concrete and commentary (ACI 318 M-05).* Farmington Hills, Michigan: American Concrete Institute.

AlHamaydeh, M., Abdullah, S., Hamid, A., & Mustapha, A. (2011). Seismic design factors for RC special moment resisting frames in Dubai, UAE. *Earthquake Engineering and Engineering Vibration, 10*(4), 495–506.

ASCE 7. (2010). *Minimum design loads for buildings and other structures.* Reston, VA: American Society of Civil Engineers.

ATC-19. (1995). *Structural response modification factors* (pp. 5–32). Redwood City, CA: Applied Technology Council.

ATC. (1995). *A critical review of current approaches to earthquake-resistant design, ATC-34* (p. 31-6). Redwood City, CA: Applied Technology Council.

Calabrese, A., Almeida, J. P., & Pinho, R. (2010). Numerical issues in distributed inelasticity modeling of RC frame elements for seismic analysis. *Journal of Earthquake Engineering, 14*(S1), 38–68.

Clough R. W. & Johnston S. B. (1966). Effect of stiffness degradation on earthquake ductility requirements. *In Proceedings of the Japan Earthquake Engineering Symposium.*

FEMA. (2000). *Prestandard and commentary for the seismic rehabilitation of buildings, FEMA-356.* Washington, D.C.: Federal Emergency Management Agency.

Fintel, A. M. (1968). *Staggered transverse wall-beams for multistory concrete buildings—a detailed study.* Skokie, IL: Portland Cement Association.

ICC (International Code Council). (2009). *International building code.* Falls Church, VA: International Code Council.

Kent, D. C., & Park, R. (1971). Flexural members with confined concrete. *ASCE, 97*(7), 1969–1990.

Kim, J., & Han, S. (2013). Sensitivity analysis for seismic response of reinforced concrete staggered wall structures. *Magazine of Concrete Research, 65*(22), 1048–1059.

Kim, J., & Jun, Y. (2011). Seismic performance evaluation of partially staggered wall apartment buildings. *Magazine of Concrete Research, 63*(12), 927–939.

Kim, J., & Lee, M. (2013). Fundamental period formulas for RC staggered wall buildings. *Magazine of Concrete Research, 66*, 325–338. Accepted for publication.

Kim, J., Lee, J., & Kim, Y. (2007). Inelastic behavior of staggered truss systems. *The Structural Design of Tall and Special Buildings, 16*(1), 85–105.

Lee, J., & Kim, J. (2013). Seismic performance evaluation of staggered wall structures using FEMA P695 procedure. *Magazine of Concrete Research, 65*(17), 1023–1033.

Li, K-N. (2004). *CANNY:3-dimensional nonlinear static/dynamic structural analysis computer program-user manual.* Vancouver, Canada: CANNY Structural Analysis.

Li, Z., & Hatzigeorgiou, G. D. (2012). Seismic damage analysis of RC structures using fiber beam-column elements. *Soil Dynamics and Earthquake Engineering, 32*(1), 103–110.

Li, N., Li, Z., & Xie, L. (2013). A fiber-section model based Timoshenko beam element using shear-bending interdependent shape function. *Earthquake Engineering and Engineering Vibration, 12*(3), 421–432.

Mee, A. L. (1975). Wall-beam frames under static lateral load. *ASCE, 101*(2), 377–395.

Newmark, N. M., & Hall, W. J. (1982). *Earthquake spectra and design, EERI, monograph series.* Oakland, CA: Earthquake Engineering Research Institute.

Orakcal, K., Wallace, J. W., & Conte, J. P. (2004). Nonlinear modeling and analysis of slender reinforced concrete walls. *ACI Structural Journal, 101*(5), 688–698.

Sfakianakis, M. G. (2002). Biaxial bending with axial force of reinforced, composite and repaired concrete sections of arbitrary shape by fiber model and computer graphics. *Advances in Engineering Software, 33*(4), 227–242.

Shedid, M., El-Dakhakhni, W., & Drysdale, R. (2011). Seismic response modification factors for reinforced masonry structural walls. *Journal of Performance of Constructed Facilities, 25*(2), 74–86.

Skalomenos, K. A., Hatzigeorgiou, G. D., & Beskos, D. E. (2015). Seismic behavior of composite steel/concrete MRFs: Deformation assessment and behavior factors. *Bulletin of Earthquake Engineering, 13*(12), 3871–3896.

Tomaževič, M., & Weiss, P. (2010). Displacement capacity of masonry buildings as a basis for the assessment of behavior factor: An experimental study. *Bulletin of Earthquake Engineering, 8*(6), 1267–1294.

Vamvatsikos, D., & Cornell, C. A. (2001). Incremental dynamic analysis. *Earthquake Engineering and Structural Dynamics, 31*, 491–514.

Wallace, J. W. (2012). Behavior, design, and modeling of structural walls and coupling beams—lessons from recent laboratory tests and earthquakes. *International Journal of Concrete Structures and Materials, 6*(1), 3–18.

Wu, J., & Hanson, R. D. (1989). Study of inelastic spectra with high damping. *Journal of the Structural Engineering, ASCE, 115*(6), 1412–1431.

Zou, X., He, Y., & Xu, L. (2009). Experimental study and numerical analyses on seismic behaviors of staggered truss system under low cyclic loads. *Thin-walled Structures, 47*(11), 1343–1353.

10

Computing the Refined Compression Field Theory

A. M. Hernández-Díaz[1],*, and M. D. García-Román[2]

Abstract: In recent years, some modifications were introduced in the stress–strain relationship of the steel in order to develop a more efficient shear model for reinforced concrete members. The last contribution in this sense corresponding to the Refined Compression Field Theory (RCFT, 2009); this theory proposed a steel constitutive model that has account the tension stiffening area prescribed by technical codes, what simplifies all the design process. However, under certain design conditions supported by such codes, the RCFT model does not provide a real (non-complex) solution for the steel yield strain when the prescribed tension stiffening area is considered; then the load-strain response cannot be computed. In this technical note, the tension stiffening area is fixed in order to guarantee the application of the embedded steel constitutive model for all the standard design range.

Keywords: reinforced concrete, compression field theories, steel constitutive model, tension stiffening area, solvability.

1. Introduction

The design and analysis of reinforced concrete members subjected to shear may be performed taking into consideration different strategies reported in the literature, among several others (ASCE-ACE Committee 445 on Shear and Torsion 1998; Hernández-Díaz and Gil-Martín 2012; Jeong and Kim 2014; Mofidi and Chaallal 2014). One of the most widely known is the so-called Modified Compression Field Theory (MCFT) (Vecchio and Collins 1986). In the MCFT, the stress–strain relationship for the steel reinforcement is assumed to be elastic-perfectly plastic, being the Young's modulus constant up to the yield strength (f_y) and then zero upon yielding at the crack location. To allow new increments of shear force, MCFT introduces the notion of local shear stress, and as a consequence, requires the check of equilibrium conditions for local shear stresses at the crack location in order to ensure that the steel stress does not exceed the steel yield strength.

A few years ago, Gil-Martín et al. (2009) proposed a new steel constitutive model leading to the Refined Compression Field Theory (RCFT). In the line of a few other shear theories, such as the Rotating Angle-Softened Truss Model (RA-STM, Belarbi and Hsu 1994), the RCFT proposes a stress–strain relationship for the reinforcing bars stiffened by

[1] Department of Civil Engineering, Universidad Católica de Murcia (UCAM), Campus de Los Jerónimos, 30107 Murcia, Spain.
*Corresponding Author; E-mail: amhd83@gmail.com
[2] Department of Civil Engineering, University of La Laguna, Campus de Anchieta, 38271 Santa Cruz de Tenerife, Spain.

concrete ("embedded bar model"); the novelty is that the embedded bar stress–strain relationship is obtained imposing equilibrium on the tension stiffening effect; so new formulation for the steel model would no longer be needed (compared with RA-STM) and the crack check can be avoided (compared with MCFT). According to Gil-Martín et al. (2009), from equilibrium of forces between a cracked section and a generis section (see Fig. 1a), the RCFT predicts the average stress (along the bar between cracks) of an embedded bar as a function of the average strain (i.e., measured on certain length including several cracks):

$$\sigma_{s,av} = \begin{cases} f_y - \dfrac{A_c}{A_s} \dfrac{f_{ct}}{1+\sqrt{3.6M\varepsilon_s}} & \text{if } \varepsilon_s \geq \varepsilon_{max} \\ E_s\varepsilon_s & \text{if } \varepsilon_s < \varepsilon_{max} \end{cases}$$

with

$$\varepsilon_{max} = \frac{f_y}{E_s} - \frac{\frac{f_{ct}}{1+\sqrt{3.6M\,\varepsilon_{max}}}}{E_sA_s}A_c \qquad (1)$$

$$M = \frac{A_c}{\sum \phi\pi}$$

where $\sigma_{s,av}$ is the average tensile stress in steel (for longitudinal or transverse reinforcement), A_s is the cross section of steel bar (longitudinal or transverse), ε_s is the average tensile strain in steel and concrete, f_{ct} is the tensile strength of concrete, E_s is the elastic modulus of reinforcement, ε_{max} is the apparent yield strain (or average tensile strain when first yielding occurs at the crack location, Fig. 1) and A_c is the area of concrete bonded to the bar. Technical codes (e.g., EHE 2008) usually define A_c as a value equal to the rectangular area (tributary to and surrounding the bar) over a distance not exceeding 7.5Ø from the center of the bar (Fig. 1b), and Ø is the diameter of the bar. Hereafter, we refer to A_c as the prescribed tension stiffening area. In Eq. (1) the embedded bar stress–strain relationship is established for the concrete tension stiffening model proposed by Bentz (2005).

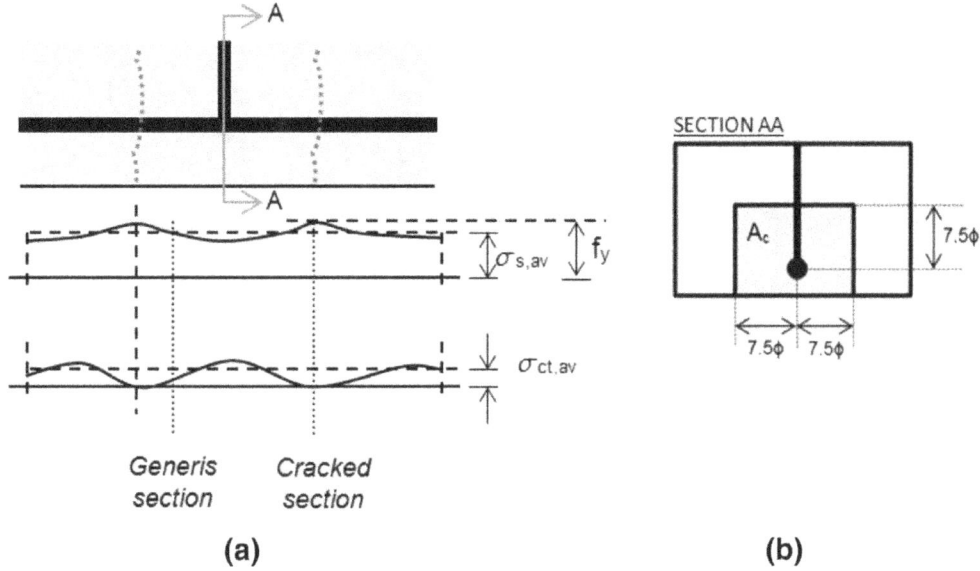

Fig. 1 **a** Steel embedded bar in tension (adapted from Gil-Martín et al. 2009); **b** Area A_c of concrete bonded to the bar (according to EHE-08).

Numerical results obtained from RCFT for different tested specimens (Gil-Martín et al. 2009; Palermo et al. 2013) show a better fitting of the experimental results, in particular near the peak point in the shear response curve, where the MCFT significantly deviates from the experimental evidences. Nevertheless, it can be proved that when the prescribed tension stiffening area (A_c) is adopted, for certain specimens (specifically those with high values of the ratio f_{ct}/ρ, being ρ the reinforcement ratio) it is not possible to obtain a positive real solution for the apparent yield strain (ε_{max}) defined in Eq. (1). If all terms in the expression of the apparent yield strain are moved to the right hand side, the following function G is obtained:

$$\varepsilon_{max} = \frac{f_y}{E_s} - \frac{\frac{f_{ct}}{1+\sqrt{3.6M\,\varepsilon_{max}}} A_c'}{E_s A_s} \rightarrow G[f_{ct}, \varepsilon_{max}]$$

$$= (\varepsilon_y - \varepsilon_{max}) - \frac{\frac{A_c' f_{ct}}{1+\sqrt{3.6M\,\varepsilon_{max}}}}{E_s A_s} \quad (2)$$

where ε_y is the strain corresponding to f_y (i.e., $\varepsilon_y = f_y/E_s$). A new variable, A_c', has been introduced in Eq. (2) in order to discuss the solvability of this equation in terms of the ratio A_c'/A_c. To illustrate the effect of the tension stiffening area in the solvability of the RCFT model, the top longitudinal reinforcement of a beam (specimen H75/2) tested in shear by Cladera in 2002 has been considered. The top longitudinal bar diameter is 8 mm, the side cover is 25 mm, the prescribed tension stiffening area (according to EHE) is 9025 mm², the tensile strength of concrete (f_{ct}) is 4.5 MPa and the yield stress of steel (f_y) is 530 MPa. The function G [f_{ct}, ε_{max}] has been represented in Fig. 2 for different values of A_c', resulting in a set of curves. For this specimen, the apparent yield strain (ε_{max}) corresponds to the intersection points of these curves with the abscissa $f_{ct} = 4.50$ MPa.

The interval adopted for the strength f_{ct} in Fig. 2 coincides with the range established for this parameter by EC-2 (2002).

It can be seen that, for high values of A_c' (like $A_c' = 0.7A_c$ and $A_c' = 0.9A_c$), the curve G [f_{ct}, ε_{max}] = 0 presents a knee that breaks the bijection between f_{ct} and ε_{max} (*problem of uniqueness*), or even, no solutions exists for ε_{max} (*problem of existence*), as it occurs by taking $A_c' = A_c$ in this specimen. In relation to the problem of uniqueness, by continuity and taking into account that $\varepsilon_{max} = \varepsilon_y$ when $A_c' = 0$ (cf. Gil-Martín et al. 2009), the actual value of ε_{max} is that of the solution closest to the yield strain ε_y, that is, the greatest one of the two positive real solutions of Eq. (2). However, the absence of solution in the steel constitutive model proposed by RCFT indicates that the equilibrium of internal forces along the cracked member (see Fig. 1) is not verified, and therefore, the stress–strain relationship for the steel must be corrected.

2. Fixing the Tension Stiffening Area

Hereafter, assume f_{ct} is a parameter, and abusing the notation, denote also by G [ε_{max}, A_c'] the above mentioned bivariate function G [f_{ct}, ε_{max}]. In Fig. 3a the function G [ε_{max}, A_c'] has been represented for three values of A_c'; this figure shows that the equation G [ε_{max}, A_c'] = 0 ceases to have positive real solutions when the value of A_c' is greater than the one that makes the graphic of G tangent to the positive part of the abscise axis. This turns out to happen when both functions $G[\varepsilon_{max}, A_c']$ and G' [ε_{max}, A_c'] vanish simultaneously, where:

$$G'[\varepsilon_{max}, A_c'] = \frac{\sqrt{3.6M}\, A_c' f_{ct}}{2A_s E_s \left(1 + \sqrt{3.6M\varepsilon_{max}}\right)^2 \sqrt{\varepsilon_{max}}}$$
$$- 1, \, \varepsilon_{max} > 0 \quad (3)$$

is the derivative of $G[\varepsilon_{max}, A_c']$. Solving the system of equations $\{G[\varepsilon_{max}, A_c'] = 0, \, G'$ [ε_{max}, A_c'] = 0$\}$ in the unknowns ε_{max} and A_c', the positive real solution for the apparent yield strain is

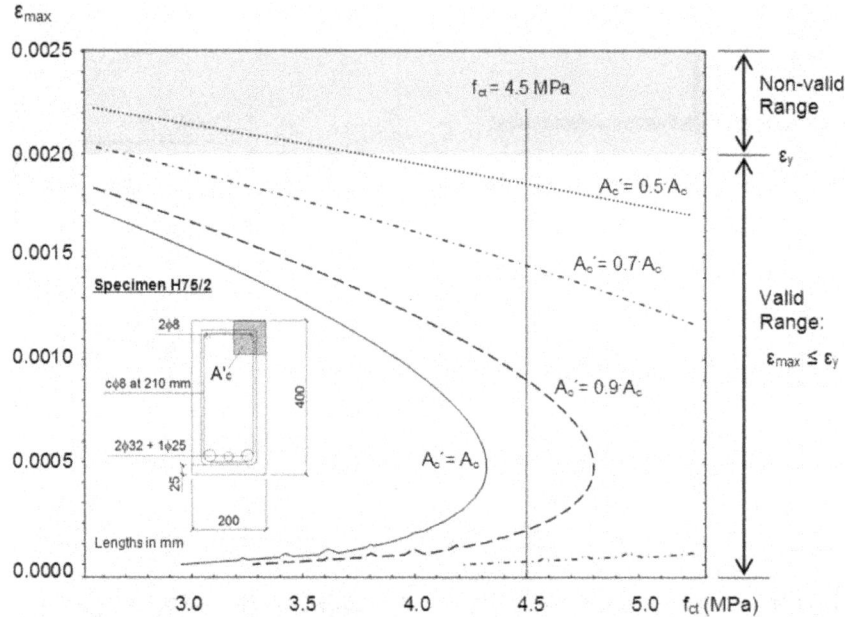

Fig. 2 Specimen H 75/2 (Cladera Bohigas 2002): solvability graphical analysis for apparent yield strain (ε_{max}) in the top longitudinal reinforcement.

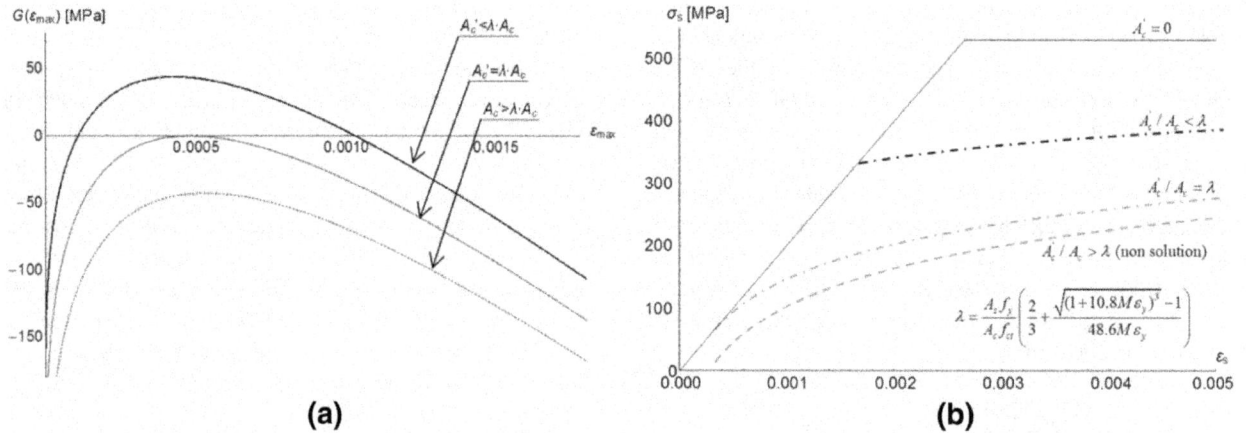

(a)   (b)

Fig. 3 Parameters of RC section: $f_y = 530$ MPa, $f_{ct} = 4.50$ MPa, $A_s = 50.27$ mm^2 (1 Ø 8), $A_c = 9025$ mm^2: **a** Function $G[\varepsilon_{max}]$ for different values of A_c'; the equation $G[\varepsilon_{max}] = 0$ has no positive real solutions when A_c'/A_c is greater than the limit value λ. **b** Effect of tension stiffening area over the embedded steel behavior model.

$$\varepsilon_{max} = \frac{2 + 10.8 M \varepsilon_y - 2\sqrt{1 + 10.8 M \varepsilon_y}}{32.4M},$$

and it is solvable only when

$$A_c' = \frac{A_s f_y}{f_{ct}} \left(\frac{2}{3} + \frac{\sqrt{(1 + 10.8 M \varepsilon_y)^3} - 1}{48.6 M \varepsilon_y} \right)$$

This value of A_c' is the greatest one for which $G\,[\varepsilon_{max}, A_c'] = 0$ has, at least, a positive real solution. Let us denote λ the factor:

$$\lambda = \frac{A_s f_y}{A_c f_{ct}} \left(\frac{2}{3} + \frac{\sqrt{(1 + 10.8 M \varepsilon_y)^3} - 1}{48.6 M \varepsilon_y} \right). \tag{4}$$

In a general sense, the factor λ represents the greatest portion of the tension stiffening area which may be taken in order to

preserve the internal equilibrium of forces, in such a way that as concrete participation increases, the steel stress diminishes. According to this consideration, the effect of tension stiffening area over the embedded steel behavior model is illustrated in Fig. 3b for the case of specimen H75/2.

Several studies (Gil-Martín et al. 2009; Palermo et al. 2013; Hernández-Díaz 2012; Palermo et al. 2014) show the convenience of correcting the prescribed tension stiffening area in order to adjustment the shear response of reinforced concrete members, particularly for high shear strains, where the technical codes underestimate the concrete tension stiffening (Gil-Martín et al. 2009; Hernández-Díaz 2012). This strategy only can be performed for those values of A_c' such that $0 \leq A_c'/A_c \leq \lambda$. In order to display the usefulness of the coefficient λ, a widely validated shear test (Abersman and Conte 1973), *apud* (Collins and Mitchell 1991;

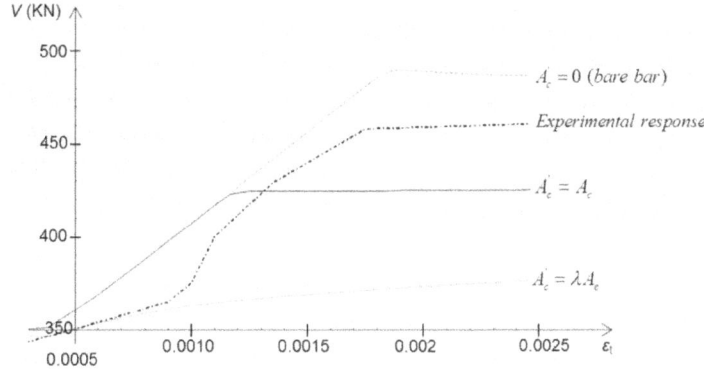

Fig. 4 Prestressed concrete beam CF1 (Abersman and Conte 1973): feasible search domain for the experimental adjustment of the RCFT model. (V shear force; ε_t transverse strain).

Hernández Montes and Gil-Martín 2014), has been considered; to this aim, the load-strain curve of the tested specimen has been predicted using the RCFT under two assumptions: (1) neglecting the contribution of the concrete tension stiffening area surrounding the reinforcement bars (i.e., $A'_c \approx 0$), and (2) considering the limit value of the tension stiffening area established by the coefficient λ. These two curves have been illustrated in Fig. 4 together the experimental shear response obtained in (Abersman and Conte 1973). As shown, the experimental results lie about halfway the shear curve corresponding to the assumption (1) and the limit curve corresponding to the coefficient λ; in particular, for a low-intermediate range of the shear strain, the limit tension stiffening area proposed in this work coincides approximately with the experimental response. In this sense, the factor λ establishes a feasible (solvable) search domain for the experimental adjustment of RCFT model using meta-heuristic methods (cf. Hernández-Díaz 2012), what improves the computational effectiveness of this process.

In the preceding results, the tension-stiffening curve proposed by Bentz has been assumed; however, the formulation of coefficient λ may be adapted to other tension stiffening models proposed in the literature (cf. Vecchio and Collins 1986; Palermo et al. 2013; Stramandinoli et al 2008; Hernández-Montes et al. 2013; Wu and Gilbert 2008) Recently, Hernández-Montes et al. (2013) proposed a tension stiffening curve, based on EC-2 formulation, which takes into account both the reinforcement ratio and the mechanic characteristics of the involved materials. Such expression is only valid until the reinforcement reaches the yield strain (ε_y) at any crack location. Once more, the apparent yield strain (ε_{max}) may be obtained from internal equilibrium; in this case, the above-defined function G adopts the following expression (see Eq. (13) at Carbonell-Márquez et al. 2014):

$$G[\varepsilon_{max}, \rho_{eff}] = (\varepsilon_y - \varepsilon_{max})$$

$$- \frac{\sqrt{\left(\frac{\rho_{eff}}{2}E_s\varepsilon_{max}\right)^2 + f_{ct}^2\left(1 + n\rho_{eff}\right)} - \frac{\rho_{eff}}{2}E_s\varepsilon_{max}}{E_s\rho_{eff}}$$

$$(5)$$

where $n = E_s/E_c$ and ρ_{eff} is the effective reinforcement ratio ($\rho_{eff} = A_s/A'_c$). Equation (5) represents a monotonically

decreasing function over the whole strain domain; therefore, in this case the value of A'_c must be only constrained in order to avoid values of ε_{max} lower than the average strain ε_{ct} corresponding to the concrete tensile strength (f_{ct}), then the expression of coefficient λ is given by:

$$\lambda = \frac{\rho_{eff}}{2}\left(\frac{\sqrt{f_{ct}^2 n^2 - 4f_y\left(f_{ct}n - f_y\right)}}{f_{ct}} - n\right) \qquad (6)$$

3. Conclusions

A practical relationship is obtained between the reinforcement area (A_s), the tensile concrete strength (f_{ct}) and the yield stress (f_y) for a given value of the tension stiffening area (A_c). These four parameters are not independent, and three of them constrain the fourth one in order to preserve the internal equilibrium of a cracked member. Therefore, such relation must be satisfied in order to make operative the embedded steel constitutive model for every reinforced concrete section. Finally, this result is extensive to every structural system involving cracked embedded reinforcing bars.

References

Abersman, B., Conte, D. F (1973) *The design and testing to failure of a prestressed concrete beam loaded in flexure and shear*, BASc thesis, Department of Civil Engineering, University of Toronto, Canada.

ASCE-ACE Committee 445 on Shear and Torsion. (1998). Recent approaches to shear design of structural concrete. *Journal of Structural Engineering, 124*(12), 1375–1416.

Belarbi, A., & Hsu, T. T. C. (1994). Constitutive laws of concrete in tension and reinforcing bars stiffened by concrete. *ACI Structural Journal, 91*(4), 465–474.

Bentz, E. C. (2005). Explaining the riddle of tension stiffening models for shear panel experiments. *ASCE Journal of Structural Engineering, 131*(9), 1422–1425.

Carbonell-Márquez, Juan F., Gil-Martín, Luisa M., Fernández-Ruíz, M. Alejandro, & Hernández-Montes, E. (2014). Effective area in tension stiffening of reinforced concrete piles subjected to flexure according to Eurocode 2. *Engineering Structures, 76*, 62–74.

Cladera Bohigas, A. (2002) *Shear Design of Reinforced High-Strength Concrete Beams*, Ph.D. Dissertation, Technical University of Catalonia, Barcelona, Spain.

Collins, M. P., & Mitchell, D. (1991). *Prestressed concrete structures*. Upper Saddle River, NJ: Prentice Hall.

EC-2 (2002), *Eurocode. Design of concrete structures—part 1*, Brussels, Belgium.

EHE. (2008). *Instrucción de Hormigón Estructural*. Spain: Ministerio de Fomento, Gobierno de España.

Gil-Martín, L. M., Hernández-Montes, E., Aschheim, M., & Pantazopoulou, S. (2009). Refinements to compression field theory, with application to wall-type structures, *ACI Special publication, 265*, 123–142.

Hernández Montes, E., & Gil-Martín, L. M. (2014). *Hormigón armado y pretensado. Concreto reforzado y preesforzado*. Madrid, Spain: Colegio de Ingenieros de Caminos, Canales y Puertos.

Hernández-Díaz, A. M. (2012) *Revisión de las Teorías de Campo de Compresiones en Hormigón Estructural*, Ph.D. Dissertation, University of Granada, Granada, Spain.

Hernández-Díaz, A. M., & Gil-Martín, L. M. (2012). Analysis of the equal principal angles assumption in the shear design

of reinforced concrete members. *Engineering Structures, 42*, 95–105.

Hernández-Montes, E., Cesetti, A., & Gil-Martín, L. M. (2013). Discussion of "An efficient tension-stiffening model for nonlinear analysis of reinforced concrete members", by Renata S.B. Stramandinoli, Henriette L. La Rovere. *Engineering Structures, 48*, 763–764.

Jeong, J.-P., & Kim, W. (2014). Shear resistant mechanism into base components: Beam action and arch action in shear-critical RC members. *International Journal of Concrete Structures and Materials, 8*(1), 1–14.

Mofidi, A., & Chaallal, O. (2014). Tests and design provisions for reinforced-concrete beams strengthened in shear using FRP sheets and strips. *International Journal of Concrete Structures and Materials, 8*(2), 117–128.

Palermo, M., Gil-Martín, L. M., Hernández-Montes, E., & Aschheim, M. (2014). Refined compression field theory for plastered straw bale walls. *Construction and Building Materials, 58*, 101–110.

Palermo, M., Gil-Martín, L. M., Trombetti, T., & Hernández-Montes, E. (2013). In-plane shear behaviour of thin low reinforced concrete panels for earthquake reconstruction. *Materials and Structures, 46*(5), 841–856.

Stramandinoli, Renata S. B., Rovere, La, & Henriette, L. (2008). An efficient tension-stiffening model for nonlinear analysis of reinforced concrete members. *Engineering Structures, 30*(7), 2069–2080.

Vecchio, F. J., & Collins, M. P. (1986). The modified compression field theory for reinforced concrete elements subjected to shear. *ACI Journal, 83*(2), 219–231.

Wu, H. Q., Gilbert, R. I. (2008) *An experimental study of tension stiffening in reinforced concrete members under short-term and long-term loads*. UNICIV REPORT No. R-449, University of New South Wales, Australia.

11

Simplified Design Procedure for Reinforced Concrete Columns Based on Equivalent Column Concept

Hamdy M. Afefy[1],*, and El-Tony M. El-Tony[2]

Abstract: Axially loaded reinforced concrete columns are hardly exist in practice due to the development of some bending moments. These moments could be produced by gravity loads or the lateral loads. First, the current paper presents a detailed analysis on the overall structural behavior of 15 eccentrically loaded columns as well as one concentrically loaded control one. Columns bent in either single curvature or double curvature modes are tested experimentally up to failure under the effect of different end eccentricities combinations. Three end eccentricities ratio were studied, namely, $0.1b$, $0.3b$ and $0.5b$, where b is the column width. Second, an expression correlated the decay in the normalized axial capacity of the column and the acting end eccentricities was developed based on the experimental results and then verified against the available formula. Third, based on the equivalent column concept, the equivalent pin-ended columns were obtained for columns bent in either single or double curvature modes. And then, the effect of end eccentricity ratio was correlated to the equivalent column length. Finally, a simplified design procedure was proposed for eccentrically loaded braced column by transferring it to an equivalent axially loaded pin-ended slender column. The results of the proposed design procedure showed comparable results against the results of the ACI 318-14 code.

Keywords: columns, double curvature mode, eccentric loading, equivalent column concept, single curvature mode, reinforced concrete.

1. Introduction

Eccentrically loaded reinforced concrete columns are commonly exist in practice due to the existence of some bending moments. The eccentricity of the supported beams as well as the unavoidable imperfections of construction are the main sources of the developed bending moments in the columns under gravity loads. In addition, lateral loads due to wind or earthquake loading are another source of the developed bending moments on the columns. Therefore, the strength of the columns is controlled by the compressive strength of concrete, the tensile strength of the longitudinal reinforcements and the geometry of the column' cross-section (Park and Paulay 1975; Nilson 2004; McCormac 1998; Yalcin and Saatcioglu 2000; MacGregor and Wight 2009). Contrasting to reinforced concrete beams, the compression failure cannot be avoided for eccentrically loaded columns since the type of failure is mainly dependent on the axial load level (Park and Paulay 1975).

Reinforced concrete columns are classified as short columns while the slenderness effect can be neglected or slender columns where the slenderness effect has to be included in the design. In order to distinguish between these two types, there are two important limits for slenderness ratio/index which are the lower and the upper slenderness limits. Most of the limit expressions provided by codes were derived assuming a certain loss of the column ultimate capacity due to the second order effect. Lower slenderness limits may be defined as the slenderness producing a certain reduction, usually 5–10 %, in the column ultimate capacity compared to that of a non-slender column (Mari and Hellesland 2005). Inspite that the lower slenderness limit of short column is mostly dependent on the adopted design standards. Figure 1 shows comparison among the limiting slenderness indices stipulated by the American Concrete Institute Code, ACI 318-14 (2014), the Canadian Standard Code, CSA A23.3-04 (R 2010) and the Egyptian Code of Practice, ECP 203-2007 (2007), where H_e is the effective length of the column and i is the radius of gyration of the column cross-section. It can be noted that the limit stipulated by the ACI 318-14 depends on the relative end moments, while the limit adopted by the CSA A23.3-04 depends on both the end moments ratio and the axial load level. On the other hand, the ECP 203-07 adopts a fixed limit for the upper slenderness limit for the short column regardless of the end moments, the axial load level and the concrete strength, in order to distinguish between the short and the slender column.

[1]Department of Structural Engineering, Faculty of Engineering, Tanta University, Tanta 31511, Egypt.
*Corresponding Author; E-mail:
hamdyafefy@hotmail.com
[2]Department of Structural Engineering, Faculty of Engineering, Alexandria University, Alexandria, Egypt.

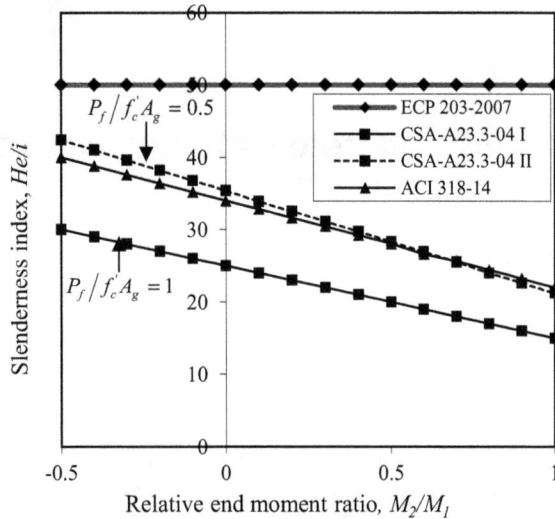

Fig. 1 Limitations of the upper slenderness limits for short column stipulated in different standards.

As for the upper slenderness limit, there is no explicit definition for that limit at most of the design standards (American Concrete Institute 2014; CAN/CSA-A23.3-04 (R2010) 2010; ECP 203-2007). In addition, the amount of reduction in the column capacity corresponding to that limit is not well defined. Although the upper slenderness limit can be considered as the limit required to avoid instability failure of the column (Ivanov 2004; Barrera et al. 2011). Despite this common basis, and even though most relevant factors governing the behavior of slender columns are well identified, a lack of uniformity can be observed in the conceptual treatment of the lower/upper slenderness limits in different codes. Not surprisingly, large differences may be obtained when applying the above code provisions. Also, there are

different values of the lower/upper slenderness limits for columns based on the bracing conditions.

In this paper the proposed design approach is aimed to consider any imperfection on the original column as well as the acting end moments when designing the column. That can be done be transforming the original column considering any initial bending moments to an equivalent pin-ended axially loaded column. And then, the additional bending moment including the end eccentricities as well as slenderness effect can be calculated. Therefore, the lower slenderness ratio could be bypassed. In addition, in order to verify the instability failure of the column, the acting axial load on the equivalent column is compared with the critical buckling load of the column.

Hinged-ended columns braced against side-sway may be bent in either single or double curvature mode with loading depending on the direction of acting end moments as depicted in Fig. 2 (Park and Paulay 1975; Cranston 1972). For both curvature modes, the bending deformations cause additional bending moments that can affect the primary end moments. If the additional moments are large, the maximum moments may move from ends to within the height of the columns. Since the lateral deformation for the case of single curvature mode is greater than that of the double curvature mode, the maximum bending moment in the single curvature case is higher than that in the double curvature one (Park and Paulay 1975). Therefore, the greatest reduction in the ultimate load capacity will occur for the case of equal end eccentricities for columns bent in single curvature mode, while the smallest reduction will occur for the case of equal end eccentricities for columns bent in double curvature mode (MacGregor et al. 1970; Milner et al. 2001).

It is accepted that the deflected axis of any column may be represented by a portion of the column deflected shape of

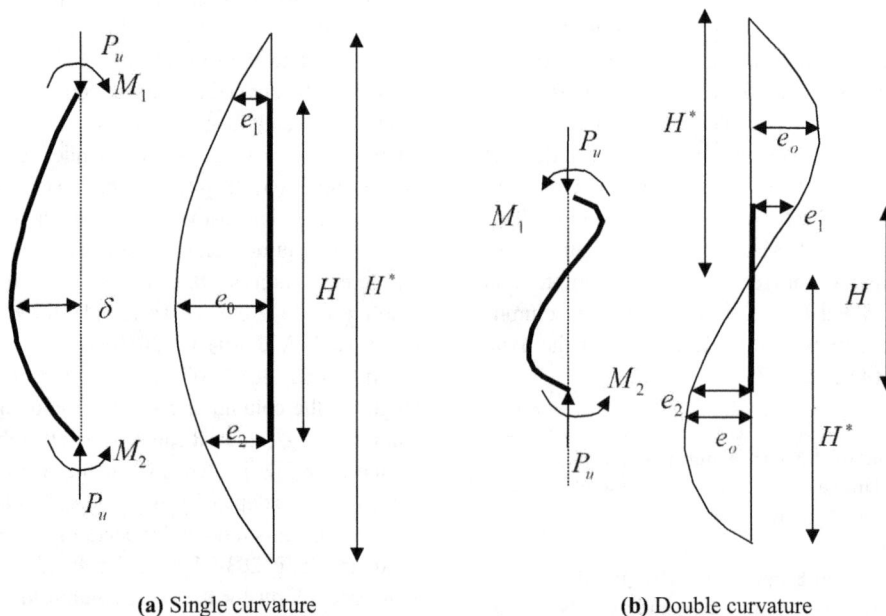

(a) Single curvature

(b) Double curvature

Fig. 2 Curvature modes of RC columns under end eccentric loading.

axially loaded pin-ended column (Chen and Lui 1987). Therefore, for a given column subjected to end moments, an equivalent column exists. Making use of Fig. 2, the column deflected shape of the equivalent pin-ended column can be represented by sinusoidal curve as illustrated in Eq. (1).

$$e = e_o \sin\frac{\pi x}{H^*} \tag{1}$$

where e is the lateral deflection of the column at a distance x from one end of the column, H^* is the length of the equivalent pin-ended column and e_o is the maximum deflection at the mid-height of the equivalent column that can be calculated using Eq. (2).

$$e_o = \phi_m \frac{H^{*2}}{\pi^2} \tag{2}$$

where ϕ_m is the curvature of the column based on the column's mode of failure.

This concept is adopted in order to reduce uni-axially loaded column to an axially loaded equivalent pin-ended column with greater length (El-Metwally 1994; Afefy et al. 2009; Afefy 2012).

In the current paper, the behavior of eccentrically loaded column bent in both single and double curvature modes is studied experimentally. In addition, based on the experimental test results, an expression was derived in order to predict the capacity lost due to column end eccentricities. And then, the equivalent column concept is employed in order to switch eccentrically loaded columns bent in either single or double curvature mode to axially loaded pin-ended equivalent columns. The end eccentricity ratio is correlated to the equivalent column length. Finally, a simplified design procedure for eccentrically loaded braced columns is proposed and compared against the design procedure stipulate in the ACI 318-14 Code.

2. Experimental Work Program

2.1 Test Columns

The experimental work program included 15 reduced-scale columns (1/3 scale model) divided into 4 groups as well as a control axially loaded column. The first two groups represented columns bent in single curvature modes, while the remaining two groups represented columns bent in double curvature modes. For both curvature modes, equal and unequal end eccentricities combinations about minor axis were studied.

The nominal axial capacity of the column cross-section was about 600 kN based on Eq. (3) as recommended by ACI 318-14.

$$P_o = 0.85f'_c(A_c - A_s) + A_sf_y \tag{3}$$

where P_o is the nominal axial capacity of the column cross-section, f'_c is the concrete compressive cylinder strength, f_y is the yield strength of the longitudinal steel bars, A_c is the

cross-sectional area of concrete section, and A_s is the cross-sectional area of the longitudinal steel bars.

It was noted that the usual end eccentricity value, e/b, for columns in reinforced concrete buildings is varying from 0.1 to 0.65 (Mirza and MacGregor 1982). In addition, recent researches showed that exposing the reinforced concrete column to an end eccentricity ratio more than half the column side exhibited drastic reduction in the ultimate capacity of the column (MacGregor et al. 1970; Milner et al. 2001; Chuang and Kong 1997; Afefy 2007). Hence, the studied end eccentricity ratios were chosen to be 0.1, 0.3 and 0.5.

Table 1 summarizes the details of the tested columns. The column cross-section was 100 mm width by 150 mm length and the overall height of 1200 mm. The column longitudinal reinforcement was four deformed bars of 10 mm diameter corresponding to reinforcement ratio of 2.09 %. The stirrups were made from mild smooth bars of 6 mm diameter spaced every 100 mm, while both ends were provided by additional stirrups as depicted in Fig. 3a.

All specimens were cast horizontally in wooden forms. Two days after casting, the standard cubes and the sides of the specimens were stripped from the molds and covered with plastic sheets until the seventh day, when the plastic sheets were removed and the specimens allowed air-drying until testing day.

2.2 Material Properties

The used concrete was normal strength concrete of 40 MPa target cube strength, which was the average of three standard cubes of 150 mm side length. The cement used was normal Portland cement (Type I) with 4.75 kN/m³ cement content and the water to cement ratio was kept as 0.38. The concrete mix contained type I crushed pink limestone as the coarse aggregates whose maximum aggregate size was 10 mm. The sand was supplied from a local plant around the site and its fineness modulus was 2.7 %. The volumes of limestone and sand in one cubic meter were 0.73 and 0.37, respectively. The average concrete strength at the testing day of columns was 42.89 MPa, while the test of all columns had been carried out in two consecutive days.

In order to determine the mechanical properties of the longitudinal deformed steel bars of 10 mm diameter as well as the transverse smooth bars of 6 mm diameter, tensile tests were performed on three specimens for each bar size. For the 10 mm deformed bars, the mean value of tensile yield strength, ultimate strength and Young's modulus were 418 MPa, 580 MPa and 202 GPa, respectively, while the relevant values for the 6 mm mild steel bars were 250 MPa, 364 MPa and 205 GPa, respectively. The used steel to form the pile caps in order to facilitate the application of eccentric loading at both ends of columns was mild steel of 12 mm thickness and yield strength of 280 MPa.

2.3 Test Setup and Instrumentation

Steel rig had been fabricated and assembled at the Reinforced Concrete Laboratory of the Faculty of Engineering, Alexandria University, Alexandria, Egypt, in order to facilitate the execution of the experimental work program. Steel

Table 1 Test matrix.

Group no.	Column	Eccentricity at upper end		Eccentricity at lower end		Curvature mode
		mm	e/b	mm	e/b	
Control	C-0-0	0.0	0..0	0.0	0.0	
1	S-1-1	10	0.1	10	0.1	Single
	S-3-3	30	0.3	30	0.3	
	S-5-5	50	0.5	50	0.5	
2	S-0-1	10	0.1	0.0	0.0	
	S-1-3	30	0.3	10	0.1	
	S-1-5	50	0.1	10	0.5	
	S-0-3	30	0.3	0.0	0.0	
	S-3-5	50	0.5	30	0.3	
	S-5-0	50	0.5	0.0	0.0	
3	D-1-1	10	0.1	10	0.1	Double
	D-3-3	30	0.3	30	0.3	
	D-5-5	50	0.5	50	0.5	
4	D-1-3	30	0.3	10	0.1	
	D-1-5	50	0.5	10	0.1	
	D-3-5	50	0.5	30	0.3	

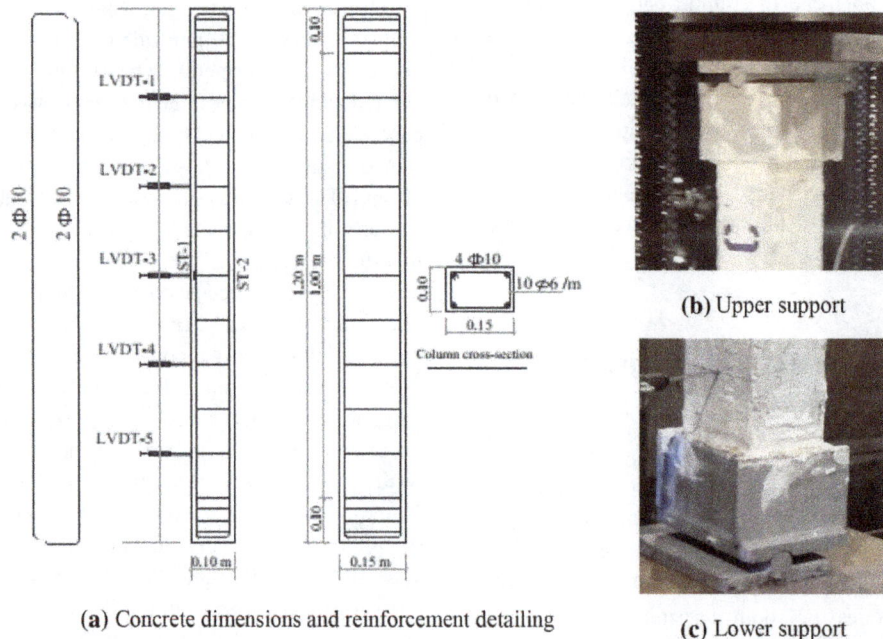

(a) Concrete dimensions and reinforcement detailing

(b) Upper support

(c) Lower support

Fig. 3 Concrete dimensions, reinforcement detailing and supports details for typical column.

caps were provided at both ends of the column in order to distribute the column compression load at both ends as well as to facilitate the application of eccentric loading. Figures 3b and 3c show both supports, while the column was loaded using compression test machine of 3000 kN capacity. Five 100 mm LVDTs were used in order to measure the lateral deformation about the minor axis as depicted in Fig. 3. Hence, the final deformed shape can be obtained. In addition, 2 strain gauges of 6 mm gauge length were mounted on the mid-height of the column longitudinal bars in order to measure the developed normal strain at the mid-height section. Load was applied in a force-control protocol to the column through moving lower head of the testing machine incrementally every 5 kN. The loading was continued until the specimen cannot sustain further loading. After each loading step, the acting loads on the column, the strain gauges readings and the LVDTs readings were recorded. An automatic data logger unit had been used in order to record and store data during the test for load cells, strain gauges and the LVDTs.

2.4 Specimen Nomenclature

Table 1 presents test parameters and the associated specimen descriptions. The specimen nomenclature consists of 3 symbols separated by hyphens. The first symbol indicates the curvature mode (S = single curvature, D = double curvature). The second nomenclature stands for amount of lower end eccentricity (0 = concentric loading, 1 = 10 mm end eccentricity, 3 = 30 mm end eccentricity, 5 = 50 mm end eccentricity). The third number indicates the same end eccentricity as presented in the second nomenclature but for the upper end. For instance, S-1-5 can be interpreted as follows: S = single curvature; 1 = the eccentricity at lower end = 10 mm; 5 = the eccentricity at the upper end = 50 mm.

3. Results and Discussion

The test results of the concentrically loaded column as well as eccentrically loaded columns under the effect of different end eccentricities combinations are presented and discussed in detailed. In general, all eccentrically loaded columns sustained ultimate loads lower than that sustained by the concentrically loaded column. In addition, the ultimate load reduction for columns bent in single curvature modes were higher than those of the columns had the same end eccentricities but bent in double curvature modes. A summary of the test results is given in Table 2 and further discussed is presented including modes of failure, deformed shapes, ultimate capacity and developed normal strain on the longitudinal bars at the mid-height section.

3.1 Modes of Failure

The failure of axially loaded column C-0-0 was sudden compressive failure since after yielding of the longitudinal steel bars in compression, the concrete had been crushed at the upper half of the column. The application of equal end eccentricities as for columns S-1-1, S-3-3 and S-5-5 resulted in employing constant moment along the entire height of the column. For columns S-1-1 and S-3-3, cracks began to appear very close to the ultimate load near the mid-height section. On the other hand, increasing the end eccentricity to be 0.5b resulted in regular flexural failure. For column S-5-5, cracks began to appear at the tensile side at acting load of about 62 % of the failure load. With further loading, cracks spread on the tensile side till the concrete crushed at the compression side near the mid-height section. Figure 4 shows the failed columns of group No. 1.

For the case of unequal end eccentricities, failure was either regular tension failure or sudden flexural failure (compression failure). Cracks began to appear near the end support of the higher end eccentricity, and then failure was triggered by concrete crushing at such support. For all cases of end eccentricity of 0.5b, cracks appeared at the tension side near the end support at acting load of about 82 % of the failure load, while for other end eccentricities (0.1b and 0.3b) cracks appeared at a vertical load very close to the failure load. Figure 5 depicts the failure shapes for all columns of group No. 2.

For all columns bent in double curvature mode, failures were similar to the case of single curvature modes with unequal end eccentricities where all columns failed near the end support of the higher end eccentricity in flexural mode of failure. Figures 6 and 7 show the failure shapes for all columns of groups No. 3 and No. 4. It can be noted that column bent in double curvature mode sustained higher load than the opponent column bent in single curvature mode. For instance, columns D-1-3, D-1-5, and D-3-5 sustained ultimate loads of 480, 300, and 379 kN, respectively, while columns S-1-3, S-1-5 and S-3-5 sustained ultimate loads of 395, 245, and 220 kN, respectively. That can be attributed to that the section of the maximum lateral deformation due to axial compression is around the mid-height point, while this location has minimal effect of bending moment for column bent in double curvature mode. On the other hand, for column bent in single curvature mode, this location, mid-height section, has considerable bending moment, which magnifies the primary moment on the column leading to lower sustained load.

3.2 Deformed Shapes

The measured deformed shapes about minor axis for all columns near failure are depicted in Fig. 8. Figures 8a, b show the deformed shapes for columns bent in single curvature modes. It can be noted that inspite the column C-0-0 was consider as short column it exhibited slight lateral deformation of about 0.03b. This value is within the limits stipulated by the Egyptian Code of Practice, ECP 203-2007. This limit states that the upper limit for short column in order to neglect the slenderness effect is 0.05b. Increasing the equal end eccentricities to be 10 mm (S-1-1) resulted in increased the measured lateral deformation by about 0.05b compared to that of the axially loaded column (C-0-0). Increasing the end eccentricities to be 30 mm (S-3-3) resulted in increased lateral deformation by about 0.06b. Increasing the end eccentricities further to 50 mm (S-5-5) resulted in increased lateral deformation by about 0.12b. The measured lateral deformations of all columns having equal end eccentricities and bent in single curvature mode were approximately symmetrical about the mid-height point as depicted in Fig. 8a. As for the case of unequal end eccentricities, the maximum value for the measured lateral deformation was bias to the end having the higher end eccentricity as depicted in Fig. 8b. For the case of columns bent in single curvature mode, the upper bound was

Table 2 Experimental results.

Group no.	Column	Cracking load, kN	Failure load, kN	Maximum measured steel strain at mid-height, micro-strain		Maximum lateral deflection, mm	Dominant mode of failure
				Compressive	Tensile		
Control	C-0-0	NA	675	2247	NA	3.06	Sudden compressive
1	S-1-1	NA	450	2326	NA	7.72	Compression failure
	S-3-3	270	300	2063	3453	8.91	Tension failure
	S-5-5	105	170	980	2362	14.77	Tension failure
2	S-0-1	NA	530	1982	NA	2.72	Compression failure
	S-1-3	NA	395	1809	NA	8.47	Compression failure
	S-1-5	210	245	1103	474	7.60	Tension failure
	S-0-3	NA	410	1612	NA	4.04	Compression failure
	S-3-5	180	220	1157	958	11.37	Tension failure
	S-5-0	220	265	856	446	5.65	Tension failure
3	D-1-1	NA	580	1460	NA	2.82	Compression failure
	D-3-3	300	473	1004	NA	5.40	Tension failure
	D-5-5	250	416	528	NA	4.90	Tension failure
4	D-1-3	NA	480	1321	NA	2.56	Compression failure
	D-1-5	180	300	905	NA	5.77	Tension failure
	D-3-5	240	379	569	NA	4.57	Tension failure

NA not applicable.

exhibited by column S-5-5, while the lower bound was manifested by axially loaded column C-0-0.

For columns bent in double curvature mode, it can be noted that the columns showed un-symmetric deformed shape compared to initial center line of the column. However, when consider the final deformed shape due to axial load as exhibited by column C-0-0, the final deformed shapes showed symmetric configuration with respect to the deformed shape of column C-0-0, for the case of equal end eccentricities as depicted in Fig. 8c. As for unequal end eccentricities, the maximum lateral deformations were shifted to the end having the higher end eccentricity as shown in Fig. 8d.

Figure 9a shows the relationships between the vertical load and the developed lateral deflection at the mid-height section for all columns of Group No. 1. It can be noted that increasing the end eccentricity ratio resulted in decreasing the ultimate load carrying capacity and increasing the corresponding lateral defection. The column S-5-5 showed the highest reduction in the ultimate capacity as well as the highest lateral deflection among all columns subjected to different end eccentricity combinations and bent in either single or double curvature modes as depicted in Figs. 9b, c.

For columns having unequal end eccentricity combinations bent in single curvature modes and the columns bent in double curvature modes the maximum lateral deflections were noticed to be developed at the upper half of the columns as shown in Fig. 8. Therefore, the lateral deflections for those columns were presented at a distance 0.67 of the column height as depicted in Figs. 9b and 9c. It can be observed that the columns bent in double curvature modes showed higher ultimate capacity and lower lateral defections than those of columns bent in single curvature modes and having the same end eccentricities combinations.

3.3 Ultimate Capacity

Table 2 summarizes the ultimate sustained loads for all columns. It can be noted that the highest ultimate capacity exhibited by the concentrically loaded column C-0-0, while the lowest ultimate capacity was achieved by column S-5-5 having single curvature mode and equal end eccentricities of $0.5b$, as expected. The column S-5-5 sustained only 25 % of the relevant capacity of concentrically loaded column C-0-0. This means that with further end eccentricity the column will drop its normal capacity significantly.

(a) S-1-1   **(b)** S-3-3   **(c)** S-5-5

Fig. 4 Final failure modes for all columns of group No. 1.

In order to assist the effect of end eccentricity on the ultimate capacity of eccentrically loaded columns, an expression is proposed based on the experimental results as given by Eq. (4).

$$P_u = P_o e^{-2.9\left(\frac{e}{b}\right)} \tag{4}$$

where P_u is the ultimate capacity, P_o is the nominal capacity of the column cross-section, which considered in the current study as the ultimate capacity of concentrically loaded column C-0-0, e/b is the ratio between the equal end eccentricity and the column side. However, this expression was derived for columns subjected to equal end eccentricities, i.e., the maximum moment occurs at the mid-height point of the column. For the column subjected to unequal end moments and bent in either single or double curvature mode, the maximum moment may occur at the column's end or somewhere within the column. For such cases, the concept of equivalent moment could be implemented.

For a column subjected to end moments M_1 and M_2, where M_2 is greater than M_1, the magnitude of the equivalent moment, M_{eq}, is such that the maximum moment produced by it will be equal to that produced by the actual end moments M_1 and M_2 as depicted in Fig. 10. Austin (Chen and Lui 1987) proposed a general expression for the equivalent moment that gives the same effect at the mid-height of the column as given by Eq. (5).

$$M_{eq} = 0.6M_2 - 0.4M_1 \geq 0.4M_2 \tag{5}$$

where M_1 has a negative value for column bent in single curvature mode. Since the equivalent end eccentricity can be obtained by dividing the equivalent moment by the acting normal force on the column, the equivalent end eccentricity, e_{eq}, can be obtained from Eq. (6).

$$e_{eq} = 0.6e_2 - 0.4e_1 \geq 0.4e_2 \tag{6}$$

where e_1 and e_2 are the corresponding end eccentricities for moments M_1 and M_2, respectively.

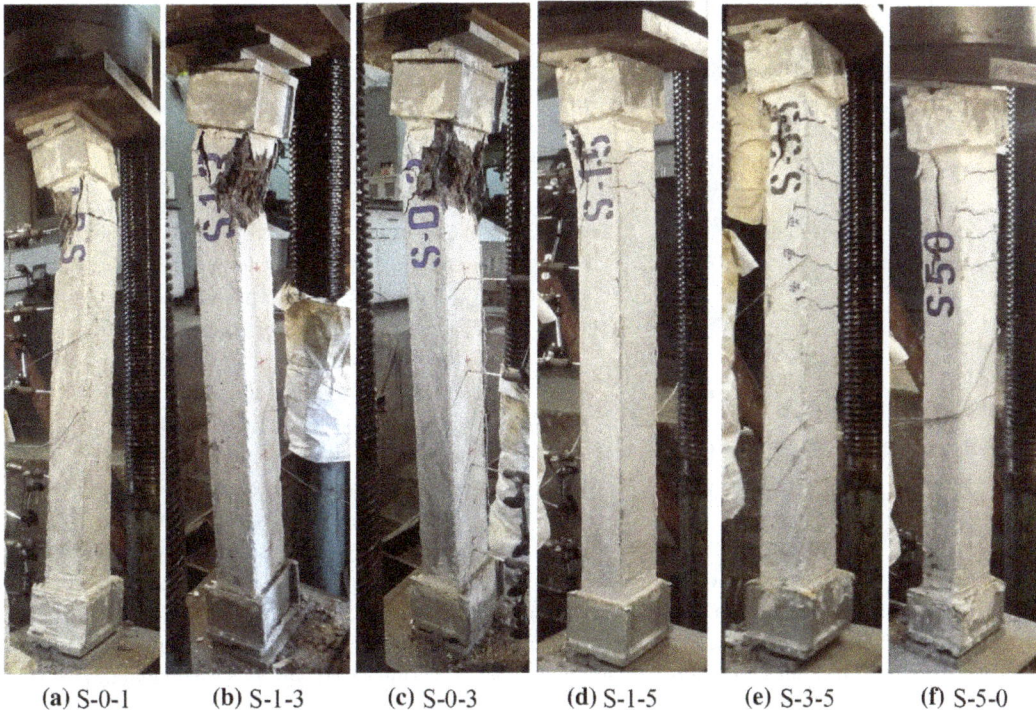

(a) S-0-1   **(b)** S-1-3   **(c)** S-0-3   **(d)** S-1-5   **(e)** S-3-5   **(f)** S-5-0

Fig. 5 Final failure modes for all columns of group No. 2.

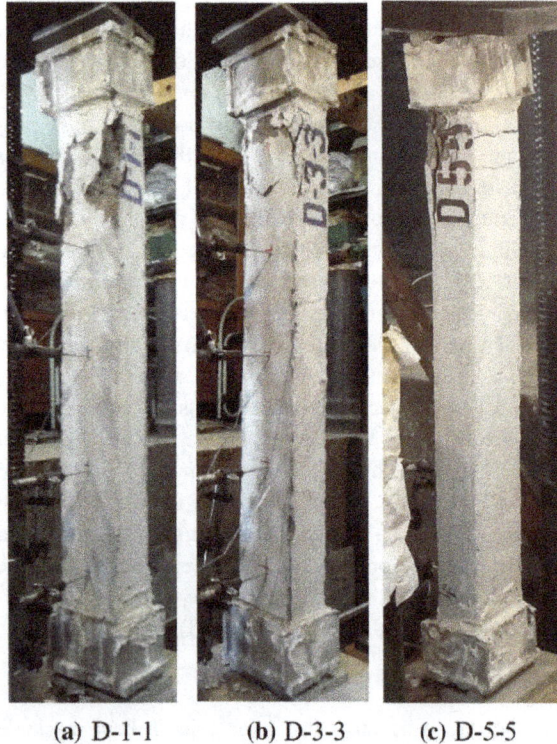

(a) D-1-1   (b) D-3-3   (c) D-5-5

Fig. 6 Final failure modes for all columns of group No. 3.

(a) D-1-3   (b) D-1-5   (c) D-3-5

Fig. 7 Final failure modes for all columns of group No. 4.

Table 3 lists normalized capacities based on both experimental findings and those obtained from the proposed expression. It can be noted that the coefficient of variation was 0.0941. In addition, the maximum variation is ranging from −10 % to +21 %, while in most cases small variations were recorded. This indicated that the proposed expression

can predict well the ultimate capacities of eccentrically loaded columns bent in either single or double curvature modes.

Furthermore, the proposed expression based on experimental tests was compared with the proposed expression by Afefy (2012). Based on about 400 test results from literature, Afefy proposed Eq. (7) in order to correlate the normalized axial capacity and the end eccentricity ratio e/b.

$$P_u = P_o e^{-2.4\left(\frac{e}{t}\right)} \tag{7}$$

Figure 11 shows comparison between both expressions. It can be concluded that the proposed expression based on experimental test results showed more conservative results within about 10 % compared to that presented by Afefy (2012).

3.4 Developed Normal Strain on the Longitudinal Bars at the Mid-height Section

Inspite that the maximum stressed section was not the same for all tested columns depending on the end eccentricities combinations, the developed normal strains on the longitudinal bars were measured at the mid-height section. Based on the used steel type, the yield strain of the longitudinal bars is 2069 micro-strain. Since the column C-0-0 was short in both directions, i.e., the effect of slenderness is minimal, the developed strains along the entire height of the reinforcing bars should reach the yielding point at failure. That happened, as expected, where the measured compressive strain near failure was 2247 micro-strain for column C-0-0.

The application of end eccentricities at column ends changed the strain distribution along the column cross-section at the mid-height point, where tensile strain maybe developed based on the end eccentricity value as well as the curvature mode. For columns bent in single curvature modes, tensile strain could be developed at the mid-height section since this section is the maximum stressed section for the case of equal end eccentricities. While, for unequal end eccentricities, the maximum stressed section could be shifted based on the end eccentricities combinations. On the other hand, for the columns bent in double curvature modes, the mid-height section could develop the lowest strain for the case of equal end eccentricities and higher values but not the maximum ones for the case of unequal end eccentricities.

For group No. 1, only column S-1-1 developed compressive strain along the entire cross-section with a maximum value exceeded the yielding strain (2326 micro-strain). This can be attributed to small end eccentricities, which resulted in subjecting the column cross-section to non-uniform compressive stress. Increasing the end eccentricity to 30 mm, resulted in increasing the acting bending moment. Hence, tensile stress was developed and the measured tensile strain exceeded the yielding point (3453 micro-strain). Increasing the end eccentricity further to 50 mm, showed the same behavior as exhibited by column S-3-3 but the measured tensile strain was lower than that developed by column S-3-3 inspite that the acting moment was greater. That can be attributed to the lower sustained load by column S-5-5

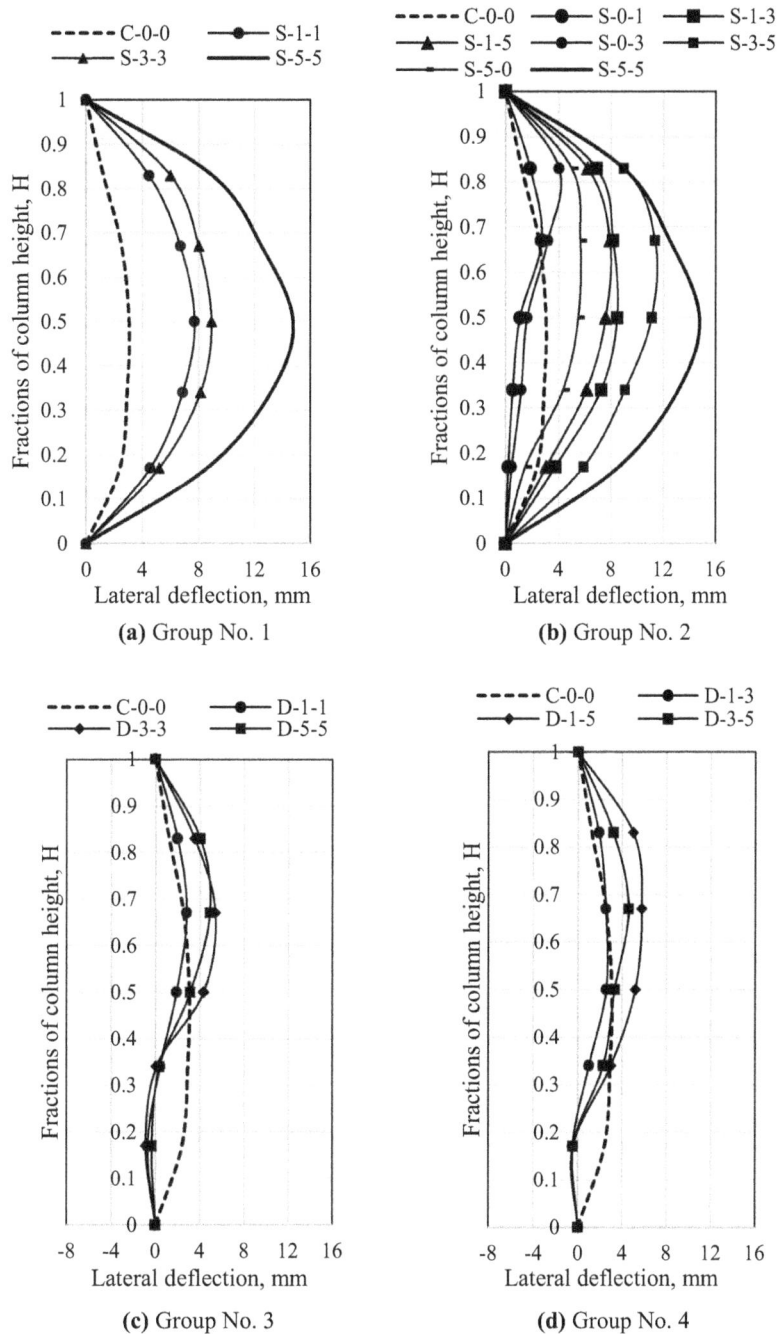

Fig. 8 Deformed shapes of all tested columns.

compared to that of column S-3-3. It can be noted that increasing the end eccentricities resulted in decrease the manifested compressive strains. That is owing to the decrease of the effect of normal force compared to the increased effect of the bending moment due to increased end eccentricities.

For columns having unequal end eccentricities of group No. 2, none of them reached the yielding point of the longitudinal steel bars on either tension or compression sides. That is because the maximum stressed section was shifted away from the measured locations. In all cases, the maximum stressed sections were located at the upper quarter of the tested column as depicted in Fig. 4. As shown in

Table 2, only the columns of 50 mm end eccentricity developed tensile strain on the longitudinal bars at the mid-height point.

As for columns bend in double curvature modes as groups No. 3 and 4, none of them developed tensile strain in the longitudinal bars at the mid-height section. That can be attributed to the minimal effect of the developed bending moment at these sections, where the maximum stressed section were near the supports as depicted in Figs. 5 and 6. As illustrated in Table 2, it can be noted that increasing the end eccentricities resulted in decreasing the developed compressive strain on the longitudinal bars at the mid-height section due to increase the bending moment effect.

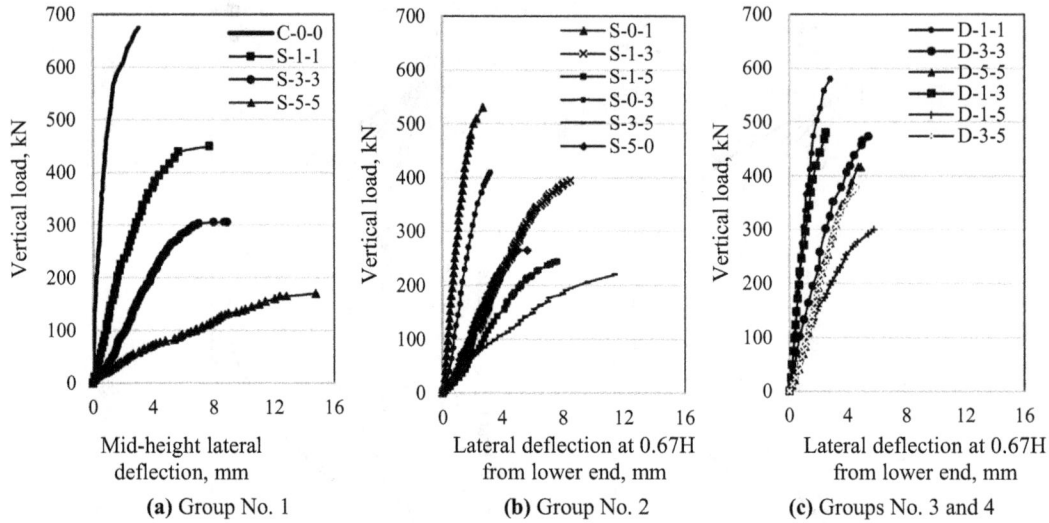

Fig. 9 Vertical load versus lateral deflection for all tested columns.

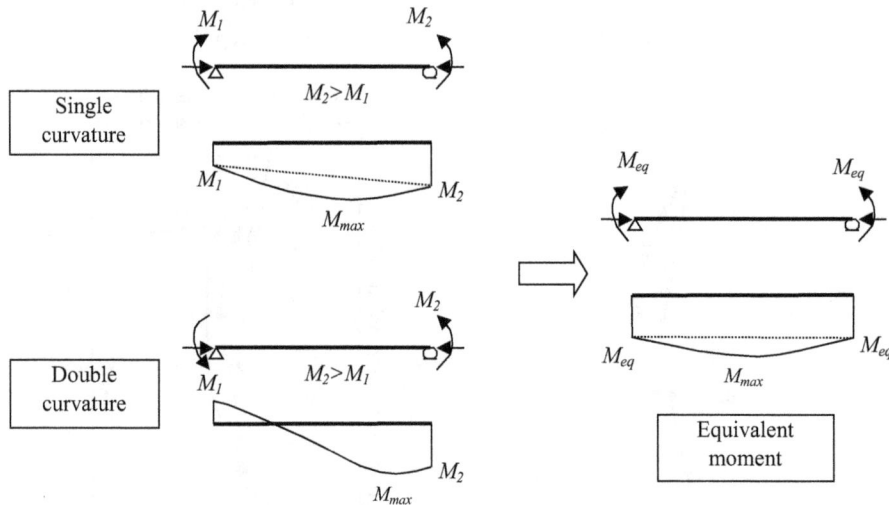

Fig. 10 Schematic representation of the concept of equivalent moment.

3.5 Equivalent Column

The relationship between the equivalent pin-ended column, H^*, and the end eccentricity is given in Eq. (1). Assuming balanced failure of the column, the curvature at the mid-height section of the equivalent column, ϕ_m, can be represented by Eq. (8).

$$\phi_m = \frac{\varepsilon_{cu} + \varepsilon_y}{b - c} \tag{8}$$

where ε_{cu} is the concrete crushing strain $= 0.003$, ε_y is the steel yield strain, equals yield stress divided by steel modulus of elasticity, and c is concrete cover. Hence the maximum mid-height eccentricity can be rewritten as given in Eq. (9).

$$e_o = \frac{\varepsilon_{cu} + \varepsilon_y}{b - c} \times \frac{H^{*2}}{\pi^2} \tag{9}$$

Knowing the end eccentricity value as well as the curvature mode, the equivalent column could be obtained.

3.5.1 Implementation of the Equivalent Column Concept on Column Bend in Single Curvature Mode

Consider column S-3-5 as an example for column bent in single curvature mode, the equivalent pin-ended axially loaded column is determined in the following, refer to Fig. 12a.

$$e = e_o \sin\left(\frac{\pi x}{H^*}\right)$$

$$e_o = \frac{\varepsilon_{cu} + \varepsilon_y}{b - c} \times \frac{H^{*2}}{\pi^2} = 0.00000568 H^{*2},$$

$$e_2 = 50 \text{ mm}, e_1 = 30 \text{ mm}$$

$$e_1 = 0.00000568 H^{*2} \sin\left(\frac{\pi x_1}{H^*}\right) = 30 \to (1)$$

$$e_2 = 0.00000568 H^{*2} \sin\left(\frac{\pi x_2}{H^*}\right) = 50 \to (2)$$

$$x_2 = x_1 + 1200 \to (3)$$

Solving the three equations by trial and error yields

Table 3 Calculated axial load capacities versus experimental results.

Group no.	Specimen	Failure load, kN	Equivalent eccentricity, mm	Equivalent e/t	Normalized load based on experimental results (6)	Normalized load based on Eq. (4) (7)	(7)/(6)
Control	C-0-0	675	0.0	0.00	1.00	1.00	1.00
1	S-1-1	450	10	0.10	0.67	0.75	1.12
	S-3-3	300	30	0.30	0.44	0.42	0.94
	S-5-5	170	50	0.50	0.25	0.23	0.93
2	S-0-1	530	6	0.06	0.79	0.84	1.07
	S-1-3	395	22	0.22	0.59	0.53	0.90
	S-1-5	245	34	0.34	0.36	0.37	1.03
	S-0-3	410	18	0.18	0.61	0.59	0.98
	S-3-5	220	42	0.42	0.33	0.30	0.91
	S-5-0	265	30	0.30	0.39	0.42	1.07
3	D-1-1	580	2	0.02	0.86	0.94	1.10
	D-3-3	473	6	0.06	0.70	0.84	1.20
	D-5-5	416	10	0.10	0.62	0.75	1.21
4	D-1-3	480	14	0.14	0.71	0.67	0.94
	D-1-5	300	26	0.26	0.44	0.47	1.06
	D-3-5	379	18	0.18	0.56	0.59	1.06
Average							1.03
Coefficient of variation							0.0941

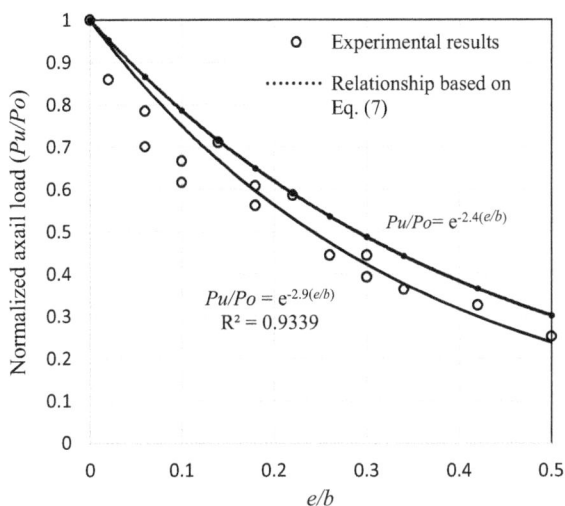

Fig. 11 Relationship between the normalized axial capacity and the end eccentricity ratio to the column side.

$H^* = 3028$ mm,

$x_1 = 0.592$ m

$e_o = 0.00000568\,H^{*2} = 52.1$ mm

3.5.2 Implementation of the Equivalent Column Concept on Column Bend in Double Curvature Mode

Consider column D-3-5 as an example for column bent in double curvature mode, the equivalent pin-ended axially loaded column is determined in the following, refer to Fig. 12b.

$e_2 = 50$ mm, $e_1 = 30$ mm

$$e_o = \frac{\varepsilon_{cu} + \varepsilon_y}{b - c} \times \frac{H^{*2}}{\pi^2} = 0.00000568 H^{*2}$$

$$e = e_o \sin\left(\frac{\pi \times x}{H^*}\right)$$

$$e_1 = 0.00000568 H^{*2} \sin\left(\frac{\pi x_1}{H^*}\right) = 30 \rightarrow (1)$$

$$e_2 = 0.00000568 H^{*2} \sin\left(\frac{\pi x_2}{H^*}\right) = 50 \rightarrow (2)$$

$$x_1 + x_2 = 1200 \rightarrow (3)$$

Assuming the maximum moment occurs at the end column having 50 mm end eccentricity and solving the three equations by trial and error yields $H^* = 1702$ mm, $x_1 = 0.349$ m, $x_2 = 0.851$ m.

(a) Column S-3-5

(b) Column D-3-5

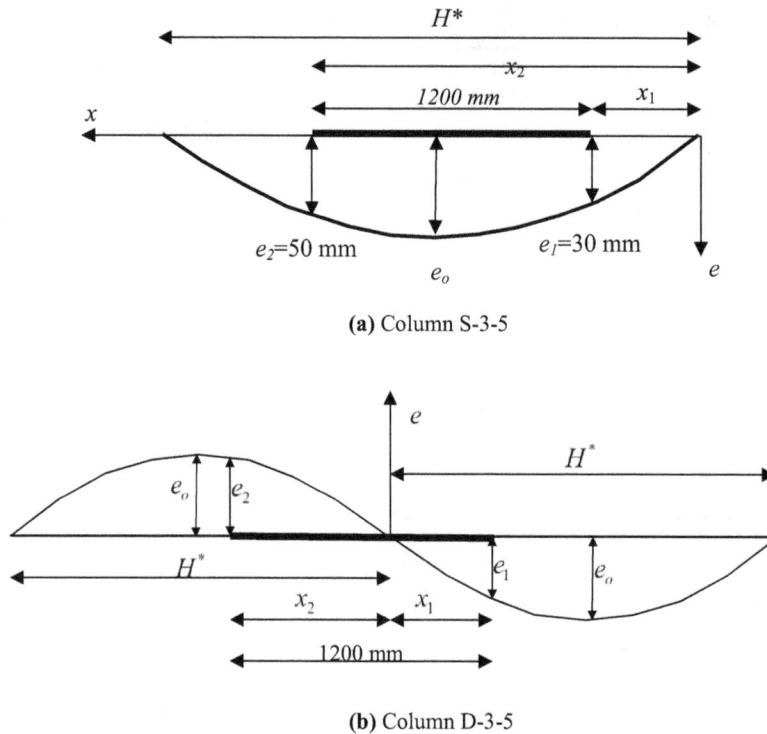

Fig. 12 Representation of the equivalent axially loaded column.

It can be noted that the equivalent column for the case of double curvature mode is lower than that of the single curvature mode. Hence, the slenderness effect of the single curvature mode is higher ($H^*/b = 30.28$), which resulted in a significant decrease in the ultimate capacity as confirmed by the experimental result of such column (S-3-5) where its ultimate capacity was about 33 % of the axial capacity of C-0-0. On the other hand, column bent in double curvature mode has slenderness ratio of 17.02, which resulted in moderate effect on the ultimate capacity. This contact was confirmed by the experimental result where column D-3-5 showed about 56 % of the ultimate capacity of axially loaded column C-0-0.

3.5.3 Relationship Between the End Eccentricity Ratio and the Equivalent Column Length

The same procedure used in clause 3.5.1 was implemented considering different end eccentricities combinations and the corresponding equivalent columns were obtained. Hence, a relationship between the normalized equivalent column length and the end eccentricity ratio was obtained as presented on Fig. 13 and given by Eq. (10).

$$\frac{H^*}{H} = 1 + 5(e/b) - 3.17(e/b)^2 \qquad (10)$$

As a consequence, knowing any end eccentricity combinations and the original column height for the column bent in single curvature mode, the equivalent pin-ended column subjected to axial load can be obtained using Eq. (10). Therefore, the design procedure could be simplified.

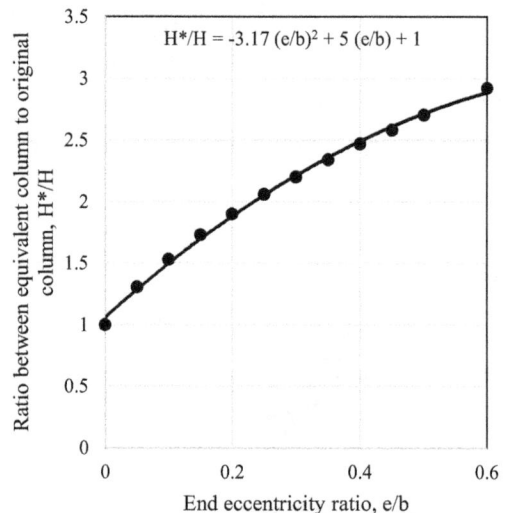

Fig. 13 Relationship between the end eccentricity ratio and the equivalent column length.

For the case of columns bent in double curvature mode, generalize the equivalent column concept maybe led to inaccurate situation and each case should be treated individually. For instance, for column having $e_1 = 5$ mm and $e_2 = 20$ mm, the equivalent column will be 1.58 times the original column length. On the other hand, for column having $e_1 = 30$ mm and $e_2 = 50$ mm, the equivalent column will be 1.42 times the original column length. Therefore, the value of the higher end eccentricity and the ratio between higher end eccentricity and the lower end eccentricity have to be considered.

3.6 Simplified Design Procedure

The measured lateral deformation showed that inspite the column was consider as short column it exhibited lateral deformation. This lateral deformation results in decreased axial capacity of the column due to the resulting bending moment. In addition, the resulting lateral deformation is directly proportional to the column height even if the column still short one where this lateral deformation is neglected. In order to account for such additional moment as well as the acting primary end moments, the column is reduced to an equivalent pin-ended slender column. Hence, the additional moment can be calculated and the column cross-section can be proportionated using any available design charts as explained in the following.

Consider any short column subjected to any end eccentricities combinations, the column can be design as follows:

1. Calculate the equivalent end eccentricity, Eq. (6)
2. Calculate the equivalent pin-ended column, Eq. (10)
3. Check the upper slenderness limit through comparing the acting axial load and the critical buckling load, $P_{critical}$, as calculated from Eq. (11)

$$P_{critical} = \frac{\pi^2 EI}{H^{*2}} \qquad (11)$$

where EI is the flexural rigidity, which can be calculated according to the relevant design standard.

4. If the acting load is more than the critical buckling load then the column is unsafe and the concrete dimensions of the cross-section have to be increased.
5. If the acting load is less than the critical buckling load then calculate the mid-span lateral deformation e_o from Eq. (9).
6. Calculate the additional moment as the multiplication of the acting load and the mid-span lateral deformation.
7. Use any ready-made design charts to obtain the steel reinforcement.

3.6.1 Implementation of the Proposed Procedure

Considered a fixed-ended braced column subjected to an axial ultimate load of 1600 kN and the acting end moments about the minor axis are 133 kN-m and 95 kN-m. The column height is 5 m and it has cross section 300 by 500 mm. Assuming the flexural rigidity of the column cross section as calculated by the ACI 318-14 as 1.04×10^{13} N/mm². The design moment will be calculated by both ACI standard and the proposed procedure herein below.

3.6.1.1 Proposed Procedure

- $e_2 = \frac{M_2}{P_u} = \frac{133}{1600} = 0.083$ m, $e_1 = \frac{M_1}{P_u} = \frac{95}{1600} = 0.059$ m
- Using Eq. (6) the equivalent eccentricity, e_{eq}, equals 0.0737 m, $\frac{e_{eq}}{b} = \frac{0.0737}{0.3} = 0.245$
- Using Eq. (10), $H^* = H\left(1 + 5\left(e_{eq}/b\right) - 3.17\left(e_{eq}/b\right)^2\right)$ $= 7.24$ m
- $P_{critical} = \frac{\pi^2 EI}{H^{*2}} = \frac{\pi^2 * 1.04 * 10^{13}}{7240^2} = 3343$ kN $> P_u \rightarrow OK$
- $e_o = \frac{\varepsilon_{cu} + \varepsilon_y}{b-c} \times \frac{H^{*2}}{\pi^2} = \left(\frac{0.003 + 0.002}{270}\right) \times \frac{7240^2}{\pi^2} = 98.4$ mm $> e_2$

- $M_{design} = P_u * e_o = 1600 * 98.4/1000 = 157.4$ kN-m $> M_2$

3.6.1.2 ACI-318-14 Code

- $M_c = \frac{C_m * M_2}{1 - \frac{P_f}{\emptyset_m P_c}} \geq M_2$
- $\emptyset_m = 0.75$
- $P_{critical} = \frac{\pi^2 EI}{(kl)^2}$
- Consider $kl = 0.7$, since the column is fixed-ended at both ends, $P_{critical} = 8411.4$ kN $> P_u \rightarrow OK$
- $C_m = 0.6 + 0.4\frac{M_1}{M_2} = 0.886$
- $M_c = \frac{C_m * M_2}{1 - \frac{P_f}{\emptyset_m P_c}} = \frac{0.886 * 133}{1 - \frac{1600}{0.75 * 8411.4}} = 157.8$ kN-m $> M_2$

It can noted that both methods give approximately the same design moment value; 157.4 and 157.8 kN-m. That means the proposed simplified design procedure based on the equivalent column concept gives a comparable result against the results of the ACI 318-14.

4. Conclusions

Based on the studied end eccentricities combinations for reinforced concrete columns bent in either single or double curvature mode and according to the used concrete dimensions and adopted material properties, the following conclusions maybe drawn:

1. Providing end eccentricities resulted in decrease the axial capacity proportionally with respect to the value of the end eccentricity. For equal end eccentricities ratio of $0.5b$, the column had lost about 75 % of its axial capacity. In addition, columns bent in double curvature modes can sustain higher load than those bent in single curvature modes having the same end eccentricities combinations.
2. Considering the second order effect, the deformed shapes of columns bent in double curvature mode were approximately symmetric about the deformed shape of axially loaded column not about the original undeformed axis of the column.
3. The proposed expression correlating the axial capacity and the end eccentricity ratio showed good results against the experimental data and showed more conservative results when compared with the available formula.
4. The equivalent column concept can be generalized to simplify columns bent in single curvature modes with different end eccentricities combinations to pin-ended axially loaded columns. On the other hand, the equivalent column concept can be implemented for a particular case of a column bent in double curvature mode.
5. The results of the proposed design procedure was comparable for those obtained by the ACI 318-14 for braced columns. Therefore, as a first step, the proposed design procedure could be applied for braced columns

and an additional work could be done to cover unbraced columns.

Acknowledgments

The experimental work had been conducted at Alexandria University's Reinforced Concrete laboratory.

References

Afefy H. M. (2007) Experimental and numerical instability analysis of high strength reinforced concrete systems. Ph.D. thesis, Faculty of Engineering, Tanta University, Tanta, Egypt, 244 pp.

Afefy, H. M. (2012). Ultimate flexural rigidity for stability analysis of RC beam column members. *Structural and Building, 165*, 299–308.

Afefy, H. M., Taher, S. F., & El-Metwally, S. E. (2009). A new design procedure for braced reinforced high strength concrete columns under uniaxial and biaxial compression. *Arabian Journal for Science and Engineering, KFUPM, KSA, 34*, 349–377.

American Concrete Institute. (2014). *ACI, Building Code Requirements for Structural Concrete (ACI 318-14) and Commentary.* Farmington Hills, MI: ACI.

Barrera, A. C., Bonet, J. L., Romero, M. L., & Miguel, P. F. (2011). Experimental tests of slender reinforced concrete columns under combined axial load and lateral force. *Engineering Structures, 33*, 3676–3689.

CAN/CSA-A23.3-04 (R2010) (2010) Design of concrete structures. Canadian Standard Association, Toronto, Ontario, Canada, 258 p.

Chen W. F., Lui E. M. (1987). Structural stability: Theory and implementation, New York, NY: Elsevier Science Publishing Co., Inc., 483 pp.

Chuang H., Kong F. K. (1997) Large scale tests on slender reinforced concrete columns. Journal of the Institution of the Structural Engineers, Nanyang Technological University, Singapore, *75*, 410–416.

Cranston W. B. (1972) Analysis and design of reinforced concrete columns. Cement and Concrete Association, No. 20, 54 p.

Egyptian Code for Design and Construction of Reinforced Concrete Structures, (ECP 203-2007) 2007.

El-Metwally, S. E. (1994). Method of segment length for instability analysis of reinforced concrete beam-columns. *ACI Structural Journal, 91*, 666–677.

Ivanov, A. (2004). Special features for the design of slender columns in monolithic multistory buildings with the consideration of longitudinal bending. *Journal of Concrete and Reinforced Concrete, 5*, 27–29.

MacGregor, J. G., Breen, J. E., & Pfrang, E. O. (1970). Design of slender columns. *ACI Structural Journal, 67*, 6–28.

MacGregor, J. K., & Wight, J. G. (2009). *Reinforced concrete: Mechanics and design* (5th ed.). Upper Saddle River, NJ: Prentice Hall.

Mari, R., & Hellesland, J. (2005). Lower slenderness limits for rectangular concrete columns. *Journal of Structural Engineering, ASCE, 131*, 85–95.

McCormac, J. C. (1998). *Design of reinforced concrete* (4th ed.). Menlo Park, CA: Addison-Wesley Longman, Inc.

Milner, D. M., Spacone, E., & Frangopol, D. M. (2001). New light on performance of short and slender reinforced concrete columns under random loads. *Engineering Structures, 23*, 147–157.

Mirza, S. A., & MacGregor, J. G. (1982). Probabilistic study of strength of reinforced concrete members. *Canadian Journal of Civil Engineering, 9*, 431–448.

Nilson, A. H. (2004). *Design of concrete structures* (10th ed.). New York, NY: McGw-Hill book Company.

Park, H., & Paulay, T. (1975). *Reinforced concrete structures.* New York, NY: Wiley.

Yalcin, C., & Saatcioglu, M. (2000). Inelastic analysis of reinforced concrete columns. *Computers & Structures, 77*, 539–555.

12

Seismic Analysis on Recycled Aggregate Concrete Frame Considering Strain Rate Effect

Changqing Wang[1,2,3], Jianzhuang Xiao[1],*, and Zhenping Sun[1,4]

Abstract: The nonlinear behaviors of recycled aggregate concrete (RAC) frame structure are investigated by numerical simulation method with 3-D finite fiber elements. The dynamic characteristics and the seismic performance of the RAC frame structure are analyzed and validated with the shaking table test results. Specifically, the natural frequency and the typical responses (e.g., storey deformation, capacity curve, etc.) from Model 1 (exclusion of strain rate effect) and Model 2 (inclusion of strain rate effect) are analyzed and compared. It is revealed that Model 2 is more likely to provide a better match between the numerical simulation and the shaking table test as key attributes of seismic behaviors of the frame structure are captured by this model. For the purpose to examine how seismic behaviors of the RAC frame structure vary under different strain rates in a real seismic situation, a numerical simulation is performed by varying the strain rate. The storey displacement response and the base shear for the RAC frame structure under different strain rates are investigated and analyzed. It is implied that the structural behavior of the RAC frame structure is significantly influenced by the strain rate effect. On one hand, the storey displacements vary slightly in the trend of decreasing with the increasing strain rate. On the other hand, the base shear of the RAC frame structure under dynamic loading conditions increases with gradually increasing amplitude of the strain rate.

Keywords: recycled aggregate concrete (RAC), frame structure, seismic analysis, strain rate effect, finite element model, shaking table test.

1. Introduction

In civil engineering, almost all concrete structures will inevitably encounter dynamic loads during their design lifetime. For example, structures may suffer from earthquake loading. Because of their unpredictability and destructive nature, these kinds of loading always become important factors in dominating structural design. Concrete is a typical rate-dependent material. Therefore, the strength, stiffness and ductility (or brittleness) of concrete are affected by loading rates. The strain rate at critical sections may be up to 10^{-1}/s for reinforced concrete structures subjected to strong earthquake ground motion excitations (Bischoff and Perry 1991). The properties of structural materials at dynamic loading conditions will be different from those at static loading conditions (Wakabayashi et al. 1984; Shing and Mahin 1988). The research of rate-dependency of concrete started in 1917 with Abrams' dynamic compressive test (1917). Based on the experimental results, Norris et al. (1959) proposed an empirical formula and predicted that the compressive strengths should increase by up to 33, 24 and 17 % as compared with the static strength when the strain rate was 3, 0.3 and 0.1/s, respectively. Atchley and Furr (1967) reported that the dynamic compressive strength of concrete increased by between 25 and 38 %. A number of research efforts have been devoted to the effects of high strain rate (>1/s) on structural materials under impact loading in the past few decades (Le and Bailly 2000; Lin et al. 2008). Recent investigation by Cotsovos and Pavlovic (2006) indicated that the application of high rates of uniaxial compressive loading on concrete prisms results in these specimens exhibiting high rates of axial and lateral deformation which, in turn, triggers the development of significant inertia forces.

The dynamic tensile tests of concrete are more difficult to perform and there are limited results available. Zielinski et al. (1981) studied the behaviour of concrete subjected to uniaxial impact tensile loading and found that the ratios of impact to static tensile strengths were between 1.33 and 2.34 for various concrete mixes. Oh (1987) presented a realistic non-linear stress–strain model that can describe the dynamic tensile behaviour of concrete. In that model, an equation was proposed to predict the increase of tensile strengths. Cadoni et al. (2001) studied the effect of strain rate on the tensile

---

[1]College of Civil Engineering, Tongji University, Shanghai 200092, China.

*Corresponding Author; E-mail: jzx@tongji.edu.cn

[2]Postdoctoral Mobile Research Station, College of Material Science and Engineering, Tongji University, Shanghai, China.

[3]Nanyang Normal University, Nanyang 473000, China.

[4]College of Material Science and Engineering, Tongji University, Shanghai, China.

behaviour of concrete at different relative humidity levels. The numerical analysis performed by Cotsovos and Pavlovic (2008) reveals that at high rates of tensile loading only the upper region of the concrete specimen deforms whereas the rest remains practically unaffected by the application of the external load. Furthermore, the behaviour of the concrete prisms under high rates of uniaxial tensile loading is affected by a number parameters related to the structural characteristics of the prisms such as mass, strength, and stiffness. Experimental investigation on dynamic tensile and compressive tests by Xiao et al. (2008) indicate that the tensile and compressive strengths of concrete increase with the loading rate. The initial tangential modulus and the critical strain of concrete in tension are independent of strain rate but those in compression slightly increase with the strain rate. Poisson's ratio of concrete in both tension and compression is not obviously dependent of loading rate. Some researches (Soroushian and Choi 1987; Chang and Lee 1987; Restrepo-Posada et al. 1994; Malvar and Crawford 1998) on the effects of strain rate on the steel material have been conducted. The conclusions drawn from their studies are that the yield strength and ultimate strength enhance as the strain rate increases, the elastic modulus is not influenced by the strain rate variations, and the strain-rate effects are inversely proportional to the strength of steel. The effects of strain rate on the reinforcing steel were investigated experimentally by Li and Li (2012). Based on the test results, the dynamic increasing factors which are functions of strain rate and quasi-static yield strength of steel are built. In order to study the dynamic behaviour of reinforced concrete structures affected by strain rate when subjected to seismic loading, a strain rate dependent material model for concrete was proposed by Pandey et al. (2006) for analysis of 3-D reinforced concrete structures under transient dynamic loads.

The material property of recycled aggregate (Hansen 1986), the mechanical behavior of recycled aggregate concrete (RAC) (ACI Committee 555 2002), and the constitutive relationship of stress–strain of RAC under static loadings (Xiao et al. 2005) have been experimentally studied and theoretically analyzed. It has been found that there is slight difference in mechanical properties between RAC and natural aggregate concrete. A large number of research works indicate that RAC can be used in civil engineering applications; however, knowledge of the seismic behavior of RAC is insufficient and incomplete. This is further confirmed by recent experimental studies of the mechanical behavior of structural members made of RAC (Fathifazl et al. 2009; Xiao et al. 2012a, b; Ajdukiewicz and Kliszczewicz 2007). It may be noted that most of the above studies on RAC were carried out under static or quasi-static loadings. A long-standing interest in the response of concrete subjected to dynamic loads stems from the widespread use of concrete structures in seismic region. However, only a few works have been reported on the mechanical behavior of RAC under dynamic loading conditions. Xiao et al. (2014) performed a series of experiments on modeled RAC under uniaxial compressive loading condition, and observed compressive strength and elastic modulus to increase with

the increase of strain rate. The split Hopkinson pressure bar tests of RAC under compression loading condition were carried out by Lu et al. (2014) and Xiao et al. (2015). Results show that, impact properties of RAC exhibit strong strain rate dependency, and increase approximately linearly with the strain rate. The compressive behaviour of RAC with different recycled coarse aggregate (RCA) replacement percentages was experimentally investigated under quasi-static to high strain rate loading by Xiao et al. (2015), and the strain rate effects on the failure pattern, compressive strength, initial elastic modulus and peak strain were studied.

The dynamic mechanical behaviors of RAC structures at high strain rate representative of seismic conditions were also investigated experimentally in recent years. Zhang et al. (2014) performed shaking table tests on four 1/5 scaled RAC frame-shear wall structures with concealed bracing detail. The dynamic characteristics, dynamic response and failure mode of each model were compared and analyzed. Shaking table tests on a full scale model of RAC block masonry building with the tie column + ring beam + cast-in-place slab system was carried out by Wang and Xiao (2012). The dynamic characteristics, the seismic performance, and the damage assessment of RAC frame structure were analyzed under different earthquake levels.

Economic considerations and the seismic design philosophy indicate that building structures are able to resist major earthquakes without collapse but with some structural damage. Therefore it is imperative that seismic design is based on nonlinear analysis of structures. For the nonlinear analysis of reinforced concrete structures a variety of models have been considered. For the early research on the nonlinear analysis, the material models which were obtained at low strain rates without strain rate effect included were generally adopted to predict behavior of the structures under seismic conditions which were characterized by high strain rates. There is little evidence of the stress–strain behavior considering strain rate dependency of RAC for large scale loaded specimens finding a place in the seismic nonlinear analysis of RAC structures. This paper examines the influence of the strain rate effect on the seismic response of an RAC frame structure.

In this study, the tested structure applied to evaluate the seismic behavior with the influence of the strain rate of material is a one-fourth scaled, two-bay, two-span and six-storey RAC frame. The RAC members are modelled using the flexibility-based fiber beam–column element with the material models of Kent–Scott–Park concrete model and the hysteretic material steel model. To examine the influence of the strain rate on dynamic response of the structure the strain rate effects for RAC and reinforcing steel have been taken into account in the Kent–Scott–Park model and the hysteretic model, respectively, by applying the dynamic increase factors (DIFs) which were derived by a test as well as from CEB (1993) on the test results basis. The dynamic characteristics (e.g., natural frequency) and the seismic behaviors (e.g., storey deformation, capacity curve, etc.) of the RAC frame structure are analyzed and discussed systematically by examining both the model with the strain rate effect included and the model with the strain rate effect excluded, and compared with the shaking table test results. For the purpose to examine how the seismic behaviors of the RAC frame structure vary under different strain rates in a real seismic situation, a numerical simulation is performed by varying the parameters of strain rates. The storey displacement response and the base shear for RAC frame structure under different strain rates are investigated and analyzed. It is implied that the structural behavior of RAC frame structure is significantly influenced by the strain rate effect. Firstly, the storey displacements vary slightly in the trend of decreasing with the increasing strain rate. Secondly, the base shear of the RAC frame structure under dynamic loading conditions increases with gradually increasing amplitude of strain rate.

2. Strain Rate-Dependent Material Model

2.1 Strain Rate-Dependent Model of Concrete

In this study, the dynamic mechanic tests of RAC under dynamic loading conditions with strain rates of 1×10^{-5}, 1×10^{-3}, and 1×10^{-2}/s were performed at the test setup of MTS 815 concrete test system. For the specimen, shown in Fig. 1, the dimension of cross section is 150 mm × 150 mm, and the longitudinal size is 450 mm. The test setup in the dynamic tests is shown in Fig. 2 and the test model is exhibited in Fig. 3.

Through the dynamic tests, the complete curves of stress–strain relationship for RAC with replacement ratios (R) of 0 and 100 % are plotted in Fig. 4a, b, respectively. Based on these stress–strain curves, the characteristic parameters such as the compressive peak stress and critical peak strain, etc., can be easily identified.

Based on regression analysis of the experimental test data from dynamic tests of RAC, the relationship between the DIF of compressive strength and the strain rate is established

Fig. 2 Test setup.

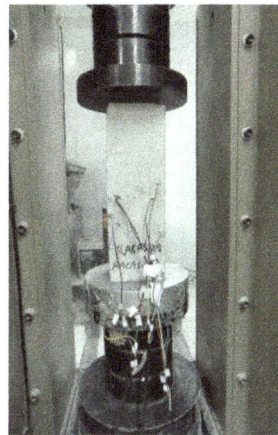

Fig. 3 Overview of test model.

and illustrated in Fig. 5, and the corresponding model of DIF is preliminary proposed and expressed in the following:

$$k_{f_c} = \frac{f'_{cd}}{f'_c} = \left(\frac{\dot{\varepsilon}_c}{\dot{\varepsilon}_{c0}} \right)^{\alpha} \tag{1}$$

with

$$\alpha = 6.664 \times \frac{1}{6.943 + 8.656 f_c} \tag{2}$$

The DIF formula proposed in this study for RAC is valid for strain rates at a constant range of approximately $1 \times 10^{-5} < \dot{\varepsilon}_c < 1 \times 10^{-1}$/s under compression. Where, k_{f_c} is the DIF for the compressive strength of concrete, f'_{cd} and f'_c represent the prism compressive strength of concrete under dynamic loading and quasi-static loading conditions in MPa, respectively, f_c stands for the nominal compressive strength of RAC (MPa), which is equal to 30 MPa in this study, $\dot{\varepsilon}_c$ is the compressive strain rate of concrete (1/s), $\dot{\varepsilon}_{co}$ is the quasi-static compressive strain rate of RAC, which is equal to 3.04×10^{-5} in this study.

The discreteness can be observed in testing mechanical indexes of concrete, especially for RAC because of the different sources of RCA which has been proved in previous conclusions by researchers (Xiao 2008), although 81 test units are carried out in the dynamic tests by the authors. The

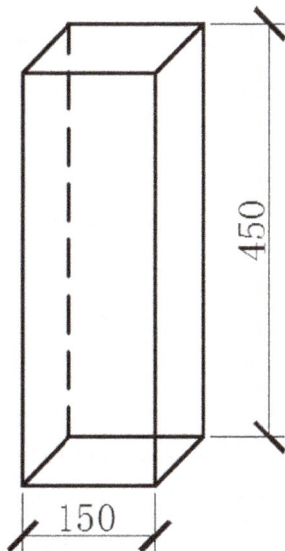

Fig. 1 Dimension of specimen.

(a) R=0

(b) R=100%

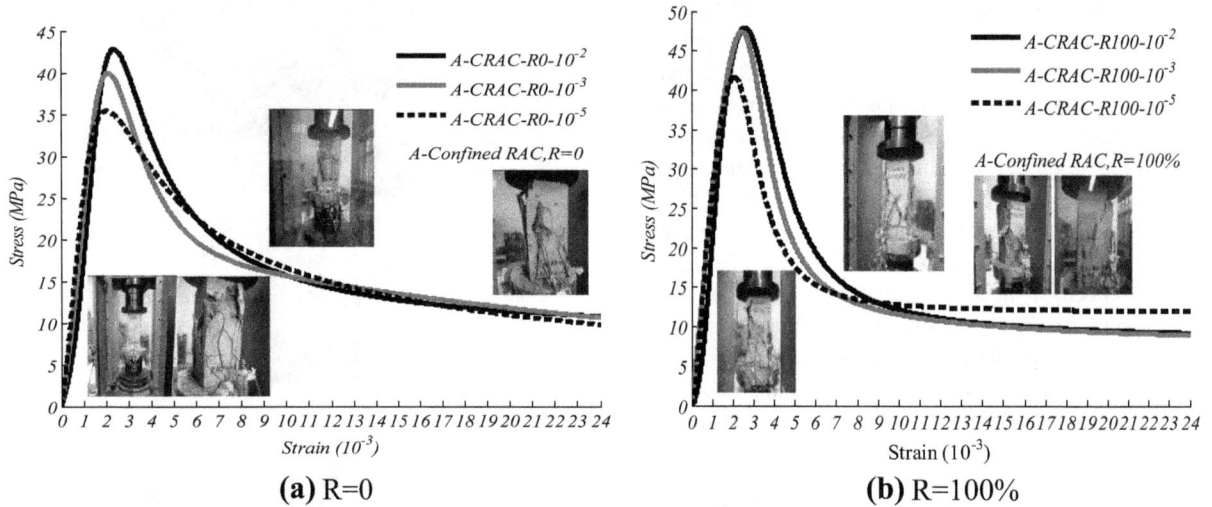

Fig. 4 Completed *curve* of stress–strain relationship for RAC under compression.

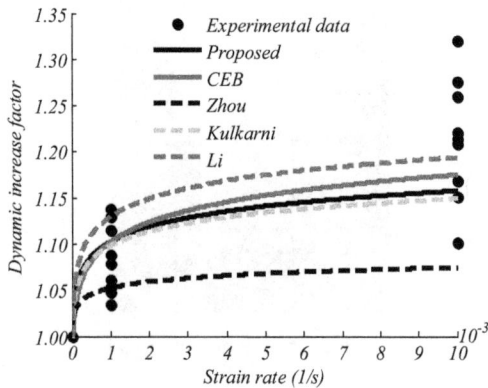

Fig. 5 DIF of compressive strength of RAC.

proposed DIF function model is compared to those from other researchers (Li and Li 2012; The Euro-International Committee for Concrete (CEB) 1993; Zhou and Hao 2008; Kulkarni and Shah 1998) and shown in Fig. 5. Comparing the enhancement values of the compressive peak stress of RAC calculated from the empirical formulas of DIF under different strain rates, it is implied that the distribution of the calculation curve of DIF from the proposed model is consistent with those from other function models. The calculation values of DIF given by CEB are more close to those from the proposed model. Therefore, the proposed formula of DIF of the compressive peak stress for RAC is reasonable and applicable to describe the changing law of compression strength of RAC subjected to the excitation of the dynamic loads.

For a given strain rate the tensile strength under the dynamic loading condition advised by CEB (1993) is adopted in this study and defined as follows

$$k_{ft} = \frac{f_{td}}{f_t} = \left(\frac{\dot{\varepsilon}_t}{\dot{\varepsilon}_{to}}\right)^{1.016\delta_s} \tag{3}$$

with

$$\delta_s = \frac{1}{10 + 6\frac{f_c'}{10}} \tag{4}$$

The DIF formula proposed in this study for RAC is valid for strain rates at a constant range of approximately $1 \times 10^{-5} < \dot{\varepsilon}_c < 1 \times 10^{-1}/s$ under tension. Where, k_{ft} is the DIF for the tensile strength of concrete, f_{td} and f_t represent the tensile strength of concrete under the dynamic and quasi-static loading conditions in MPa, respectively, f_c' represents the compressive strength of concrete under quasi-static loading conditions (MPa), $\dot{\varepsilon}_t$ is the tensile strain rate of concrete (1/s), $\dot{\varepsilon}_{to}$ is the quasi-static tensile strain rate of concrete, which is equal to 3.04×10^{-5} in this study.

Critical strain is defined as the strain when the stress reaches the peak value. Bischoff and Perry (1991) summarized a wide range of concrete with various static strengths and strain rates, showing the significant increases in compressive critical strain were observed during impact loading, although these increases were generally less than those observed for strength in the literature by Chen et al. (2013). According to Bischoff and Perry (1991), the extension of cracking required for failure increases with the strain rate. The increased compressive critical strain may also be explained by the lateral confinement which results in the formation of significant amounts of microcracks, but prevents the formation of macrocracks (Lai and Sun 2009). In this study, based on regression analysis of experimental test data from dynamic tests of RAC, the relationship between the DIF of the compressive critical strain of RAC and the strain rate is displayed in Fig. 6, and the corresponding model of DIF is preliminary proposed in the following form:

$$k_{\varepsilon_c} = \frac{\varepsilon_{d0}}{\varepsilon_{c0}} = \left(\frac{\dot{\varepsilon}_c}{\dot{\varepsilon}_{c0}}\right)^{\phi} \tag{5}$$

$$\phi = 0.01597 \tag{6}$$

where k_{ε_c} is the DIF for the compressive critical strain of concrete, ε_{d0} and ε_{c0} are the compressive critical strain of concrete under the dynamic and quasi-static loading conditions, respectively, other symbols are the same as in Eq. (1).

The model implemented in this study takes into account the effect of concrete confinement on the monotonic envelope curve in compression is illustrated in Fig. 7. The successive degradation of the stiffness of both reloading and the unloading curves is included, because of the increasing values in compressive strain, the tension stiffness, and the hysteretic response under seismic conditions (Yassin and Hisham 1994). The monotonic envelope curve of concrete in compression follows the model by Kent and Park (1971) that was later extended by Scott et al. (1982). In this study, the effect of the strain rate has been taken into account in this Kent–Scott–Park model by applying the DIFs (k_f, k_ε) which were derived by CEB (1993). Thus, the RAC compressive stress–strain relation, in which the strain rate effect is considered, is proposed as follows:

Region OA: $\sigma_c = K_d f'_{cd} \left[2\left(\dfrac{\varepsilon_c}{\varepsilon_{do}}\right) - \left(\dfrac{\varepsilon_c}{\varepsilon_{do}}\right)^2 \right]$ $(\varepsilon_c \leq \varepsilon_{d0})$

$$(7)$$

Region AB: $\sigma_c = K_d f'_{cd} [1 - Z_d(\varepsilon_c - \varepsilon_{do})]$
$(\varepsilon_{d0} < \varepsilon_c \leq \varepsilon_{2d0})$

$$(8)$$

Region BC: $\sigma_c = 0.2 K_d f'_{cd}$ $(\varepsilon_c > \varepsilon_{2d0})$

$$(9)$$

The corresponding tangent modulus for each region is listed as follows:

$$E_{td} = \frac{2 K_d f'_{cd}}{\varepsilon_{do}} \left(1 - \frac{\varepsilon_c}{\varepsilon_{do}} \right) \quad (\varepsilon_c \leq \varepsilon_{d0}) \tag{10}$$

$$E_{td} = -Z_d K_d f'_{cd} \quad (\varepsilon_{d0} < \varepsilon_c \leq \varepsilon_{2d0}) \tag{11}$$

$$E_{td} = 0 \quad (\varepsilon_c > \varepsilon_{2d0}) \tag{12}$$

where

$$f'_{cd} = k_{fc} f'_c \tag{13}$$

$$\varepsilon_{d0} = 0.002 K_d k_{\varepsilon c} \tag{14}$$

$$K_d = \left(1 + \frac{\rho_s f_{yh}}{f'_{cd}} \right) \tag{15}$$

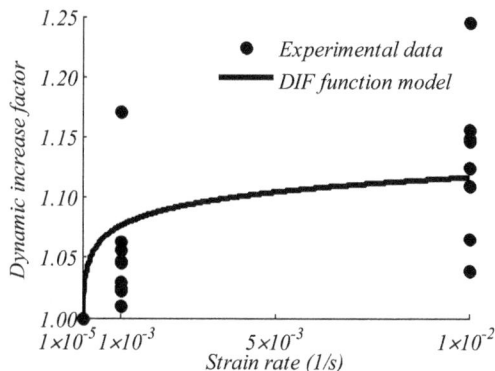

Fig. 6 DIF of the compressive critical strain of RAC.

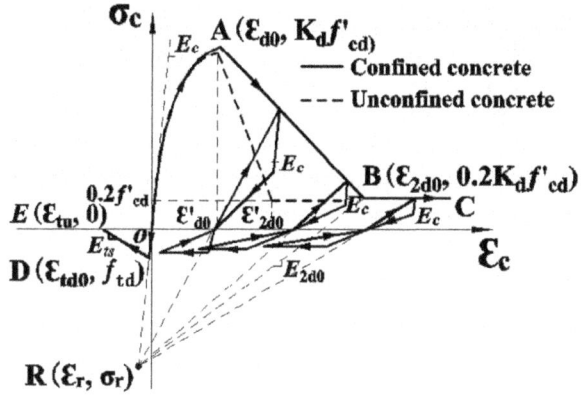

Fig. 7 Modified Kent–Scott–Park model.

$$Z_d = \frac{0.5}{\frac{3 + 0.29 f'_{cd}}{145 f'_{cd} - 1000} + 0.75 \rho_s \sqrt{\frac{h'}{s_h}} - 0.002 K_d} \tag{16}$$

σ_c and ε_c are the compressive stress and the corresponding strain, respectively, ε_{d0} is the compressive critical strain under the dynamic loads, ε_{2d0} is the strain corresponding to 20 % of the compressive peak stress at the descending branch of the stress–strain curve under the dynamic loads, f'_c and f'_{cd} stand for the quasi-static and dynamic axial compressive strength of concrete in MPa, respectively, K_d is a factor which accounts for the strength increase due to the confinement and the strain rate effect of concrete, Z_d is the strain softening slope considering the strain rate effect of concrete, f_{yh} is the dynamic yield strength of stirrups in MPa, ρ_s is the volume ratio of the hoop reinforcement to the concrete core measured to outside of stirrups, h' is the width of concrete core measured to outside of stirrups, and s_h is the center to center spacing of stirrups or hoop sets. The cyclic unloading and reloading behavior is represented by a set of straight lines as shown in Fig. 7 which shows that hysteretic behavior occurs under both tension and compression stress.

The ultimate strain of concrete confined by stirrup-ties ε_u is given by Eq. (17) which was suggested by Scott et al. (1982).

$$\varepsilon_u = 0.004 + 0.9 \rho_s \left(\frac{f_{yh}}{300} \right) \tag{17}$$

In the model the rate-dependent monotonic tensile stress–strain relation of concrete is describe by the following equations:

$$\sigma_t = E_{td} \varepsilon_t \quad \varepsilon_t \geq \varepsilon_{td0} \tag{18}$$

$$\sigma_t = f_{td} + E_{ts}(\varepsilon_t - \varepsilon_{td0}) \quad \varepsilon_{tu} \leq \varepsilon_t < \varepsilon_{td0} \tag{19}$$

with

$$f_{td} = -0.6228 \sqrt{k_{ft} f'_c} \tag{20}$$

$$E_{td} = \frac{f_{td}}{\varepsilon_{td0}} \quad (\varepsilon_t \leq \varepsilon_{td0}) \tag{21}$$

where σ_t and ε_t are the tension stress and the corresponding strain, respectively, ε_{td0} is the tensile critical strain of concrete, ε_{tu} is the strain at the point where the tensile stress is reduced to zero, which is assumed to be kept unchanged under the dynamic and quasi-static loadings, f_{td} is the tensile strength of concrete under the dynamic condition in MPa, f_c' stands for the quasi-static axial compressive strength of concrete in MPa, E_{ts} is the tension stiffening modulus that depends on numerical and physical parameters. The modulus E_{ts} controls the degree of tension stiffening.

According to the results by earlier researchers (Xiao et al. 2008; Malvar and Ross 1998), it is observed that, in the range of strain rate from 10^{-5} to 10^{-2}/s, the effect of strain rate on the critical strain value of concrete in tension is limited. Therefore, DIF of the tensile peak strain, i.e., the ratio of the tensile peak strain at dynamic loading rate to that at quasi-static loading rate, is assumed to be equal to 1.0 in the present work.

In this study, the quasi-static mechanical properties of RAC were measured. Thus, the dynamic stress–strain relation of the modified Kent and Park concrete model can be determined by the above equations, and the tested mechanical properties of RAC under the quasi-static loadings related to the model are listed in Table 1.

2.2 Strain Rate-Dependent Model of Steel

Based on experimental studies and numerical analyses, some empirical formulae of DIF of reinforcing steel are developed. In this study, the DIF formula for reinforcing steel derived by CEB (1993) is adopted and recalled here.

The DIF formula proposed by CEB (1993) for reinforcing steel is valid for strain rates at a constant range of approximately $5 \times 10^{-5} < \dot{\varepsilon}_s < 10$/s, which is given as follows

$$k_{sy} = \frac{f_{yd}}{f_y} = 1.0 + \left(\frac{6.0}{f_y}\right) \ln\left(\frac{\dot{\varepsilon}_s}{\dot{\varepsilon}_{s0}}\right) \tag{22}$$

$$k_{su} = \frac{f_{ud}}{f_u} = 1.0 + \left(\frac{6.0}{f_u}\right) \ln\left(\frac{\dot{\varepsilon}_s}{\dot{\varepsilon}_{s0}}\right) \tag{23}$$

where k_{sy} and k_{su} are the DIF of the yield and ultimate strength of reinforcing steel, respectively, f_y and f_{yd} represent the quasi-static and dynamic yield strength of steel in MPa, respectively, f_u and f_{ud} denote the quasi-static and dynamic

ultimate strength of steel in MPa, respectively, $\dot{\varepsilon}_s$ is the strain rate of steel, $\dot{\varepsilon}_{s0}$ means the quasi-static strain rate of steel, which is equal to 3.04×10^{-5} in this study.

Several models were proposed to represent the stress–strain relationship of steel reinforcements (Soroushian and Choi 1987; Chang and Lee 1987; Restrepo-Posada et al. 1994; Malvar and Crawford 1998). In this study, the hardening and softening of strain, the pinching effect of force and deformation, and unloading stiffness degradation are considered in this material model. The material model (Hognestad 1951; Filippou et al. 1992) for the steel bar is shown in Fig. 8, and the effect of the strain rate has been taken into account in this hysteretic model by applying the DIFs k_{sy} and k_{su} which are derived by CEB (1993). Thus, the stress–strain relation of the hysteretic model is modified by considering the DIF k_{sy} in the yield strengths (s_{1p} and s_{1n}) and the critical strains at the yield strengths (e_{1p} and e_{1n}), as well as k_{su} in the ultimate strengths (s_{3p} and s_{3n}). The other characteristic parameters are assumed to be unchanged. The slope of each branch of the hysteretic model of the reinforcing steel is given by the above equations.

In this study, the quasi-static mechanical properties of fine iron wires including the yield strength and the initial elastic modulus were tested. The stress–strain relation of the modified hysteretic model can be determined by the above equations, and the tested material properties of the fine iron wires under the quasi-static loadings related to the steel model are listed in Table 2.

In this study, the models of DIF are applied to the stress–strain relationships of reinforcement and RAC. In the numerical simulation, these constitutive models are used to simulate stirrups, longitudinal reinforcement, core concrete, and concrete cover, respectively.

3. Numerical Modelling of RAC Frame Structure

The present study concentrates on the discrete finite element models (FEMs), which are the best compromise between simplicity and accuracy in nonlinear seismic response studies and represent the simplest class of model that still provides significant insight into the seismic response of members and the whole structure.

Table 1 Measured material properties of RAC under the quasi-static loading.

Properties		Prism compressive strength f_c' (MPa)	Elastic modulus E_c (GPa)
Specimens	Floor 1	35.31	24.38
	Floor 2	42.36	26.18
	Floor 3	35.96	24.25
	Floor 4	31.86	23.24
	Floor 5	27.89	21.13
	Floor 6	35.82	23.16

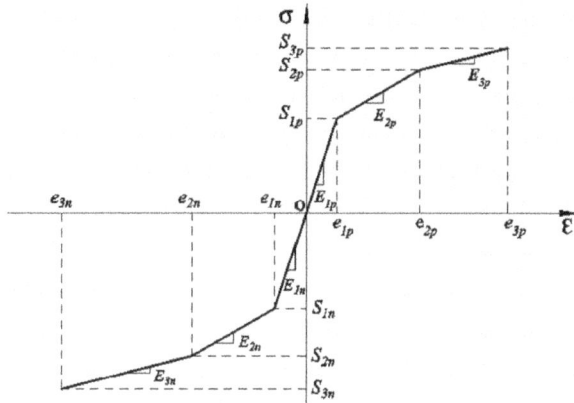

Fig. 8 Steel material model.

3.1 Description of the Tested RAC Frame Model

The tested model is a six-storey frame structure and was designed according to Chinese Standard GB 50011 (2010). The details of the general geometry are given in Fig. 9. An overview of the model after installation on the shaking table facility and the experimental set-up are shown in Fig. 10 (Xiao et al. 2012a). The Wenchuan earthquake wave (WCW, 2008, N–S) belonging to Type-II site soil was chosen. Considering the spectral density properties of Type-II site soil, El Centro earthquake wave (ELW, 1940, N–S) and Shanghai artificial wave (SHW) were also selected. The test procedure consisted of nine phases, namely, tests for peak ground acceleration (PGA) of 0.066, 0.13 g (frequently occurring earthquake of intensity 8), 0.185, 0.264, 0.370 g (basic occurring earthquake of intensity 8), 0.415, 0.55, 0.75 g (rarely occurring earthquake of intensity 8) and 1.17 g (rarely occurring earthquake of intensity 9).

3.2 Strain Rate Calculation

It is indicated from the damage pattern and the dynamic characteristics of the tested model (Xiao et al. 2012a) that the tested model suffered slight damage firstly and stepped into elastic–plastic state. With increasing intensity of shaking, the strain response of RAC becomes stronger. The maximum compressive strain reaches 1854×10^{-6} in the 0.415 g test phase, and is close to the critical strain of RAC. It is inferred the structure suffers severe damage. The strain rate of RAC is calculated by numerical differentiation to the strain response history. With increasing input acceleration amplitudes, the stain rate of RAC increases progressively. The maximum concrete strain rate of RAC at the bottom of the corner column (Fig. 11) under three different earthquake excitations with PGA of 0.415–1.170 g is listed in Table 3.

The orders of the maximum strain rate of RAC at different peaks of acceleration are all at 10^{-2}/s, which is consistent with the results reported by Bertero et al. (1984). To simply the calculation, the mean value of the maximum strain rates which are to induce the most significant effect on mechanical behaviors of the structures is considered in this study, although the strain rate varies with the time. Therefore, the strain rates of 3.04×10^{-5}, 3.04×10^{-3}, 3.04×10^{-2} and 3.04×10^{-1}/s, which are regarded as being indicative of the strain rate expected during the response of reinforced concrete to earthquakes, are taken into account in the numerical simulation, respectively. The investigators (Bischoff and Perry 1991) found that there is no clear increase in strength of concrete or steel up to a strain rate of about 5×10^{-5}/s. In this study the typical seismic response for the model with strain rate of 3.04×10^{-5}/s are defined as the quasi-static response and used as the reference values of the structural responses.

3.3 Finite Element Model (FEM) of RAC Frame Structure

The most promising models for the nonlinear analysis of reinforced concrete members are, presently, flexibility-based fiber elements. In this study, the spread plasticity fiber model is developed. The flexibility-based fiber element is subdivided into longitudinal fibers, which has two inherent advantages: (a) the reinforced concrete section behavior is derived from the uniaxial stress–strain behavior of the fibers and three-dimensional effects, such as concrete confinement by transverse steel can be incorporated into the uniaxial stress–strain relation, and (b) the interaction between bending moment and axial force can be described in a rational way. The fiber beam–column element originally formulated by Taucer et al. (1991) is a two-node, cubic, three-dimensional element with multiple fiber control sections. The formulation of the flexibility-based fiber beam–column element is based on the assumption of linear geometry, where plane sections remain plane and normal to the longitudinal axis during the element deformation history. While this hypothesis is acceptable for small deformation of elements composed of homogeneous material, it does not properly account for phenomena such as cracking and bond-slip which are characteristic of reinforced concrete elements. Investigation on effect of the bond-slip in the nonlinear analysis of RAC structures will be carried out in the future research work.

In order to study the effect of strain rate on the dynamic response of RAC structures, the dynamic responses of the RAC spatial frame are analyzed with the FE fiber models.

Table 2 Measured material properties of fine iron wires under the quasi-static loading.

Properties		Diameter (mm)	Yield strength f_y (MPa)	Ultimate strength f_u (MPa)	Elastic modulus E_s (GPa)
Specimens	8#	4.01	274.11	377.81	182.01
	10#	3.53	247.00	365.05	148.00
	14#	2.21	261.84	368.74	134.01

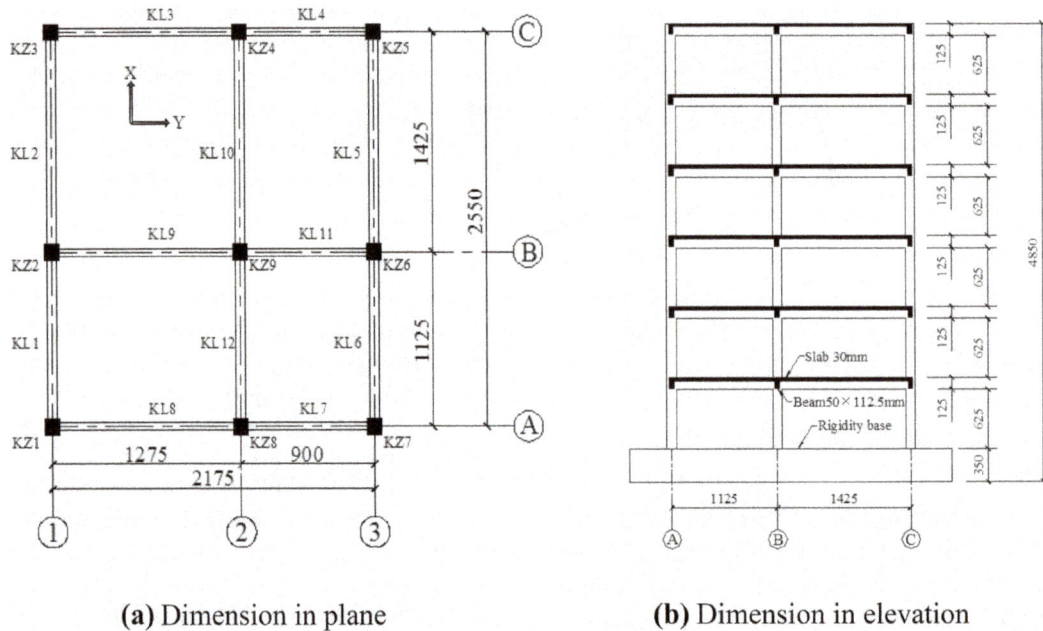

(a) Dimension in plane   **(b)** Dimension in elevation

Fig. 9 Dimension of the six-storey frame.

Fig. 10 RAC frame tested.

The RAC frame specimen is idealized as a three-dimensional discrete numerical model as shown in Fig. 12.

The beam and column members are modeled using the flexibility-based distributed-plasticity nonlinear fiber beam–column elements. For each element it is subdivided into several control sections and each section is composed of a number of fibers. The number of sections and their locations depend on the integration scheme and the desired level of accuracy. The analysis results show that, an indistinguishable indicating convergence to the analytical solution of the problem is observed for eight to ten integration points, a very good agreement with the response is obtained for four to six integration points, and the maximum force discrepancy is observed for only two integration points. In this study the Gauss–Lobatto integration scheme is used since it allows for two integration points to coincide with the end sections of the element, where significant inelastic deformations typically take place. In the numerical implementation of the RAC numerical model each nonlinear fiber beam–column element is subdivided into five integration points in order to

achieve good agreement with the experimental results. The material behavior of the element depends entirely on the stress–strain relation of the fibers, which follows the unconfined and confined concrete as well as reinforcing steel models involving the strain rate effect mentioned in Sect. 2. Different concrete and steel material types can be specified for the fibers by varying the values of material parameters. Figure 13 shows the details of the section modeling for the beam and column members. Since attention is focused on the behavior of bare frames, it is assumed that floor diaphragms are infinitely rigid. The RAC frame is modeled based on the OpenSees computational platform (Mazzoni et al. 2006), a general purpose nonlinear analysis program for the static and/or dynamic analysis of complete three-dimensional structural systems.

4. Seismic Analysis of RAC Frame Structure

The typical responses of the structure subjected to WCW, ELW and SHW excitations are analyzed in this section. The dynamic characteristics and seismic responses obtained from the numerical simulation with and without strain rate effect are compared with the shaking table test results.

In this study, *Model 1* represents RAC frame numerical model without the strain rate effect, while *Model 2* represents RAC frame numerical model, in which the material models are adopted for the strain rate of 3.04×10^{-2}/s representative of seismic conditions by applying the corresponding DIF to the peak stress and the critical strain. The tested structure model is shown in Fig. 5. The total mass including live load for the frame is 17,000 kg. The columns are assumed to be fixed at the base. The FEM used two-node fiber beam–column elements. Nine elements are used to discretize the individual columns in each storey and 12

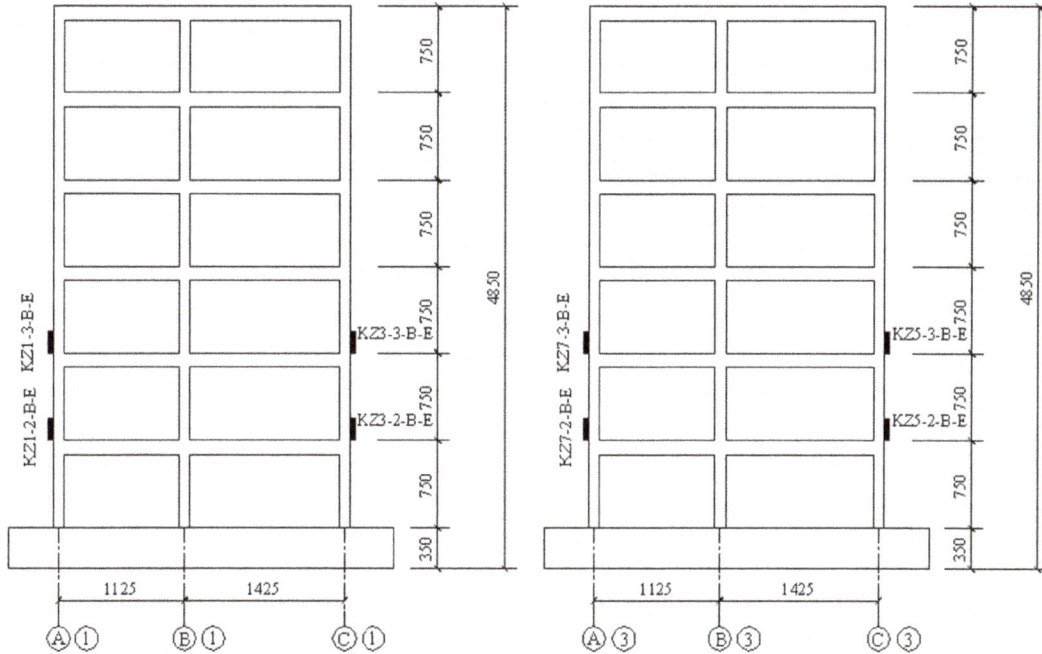

Fig. 11 Arrangement of strain sensors.

Table 3 Maximum strain rate of concrete under different test phases (10^{-2}/s).

PGA (g)	Storey 2				Storey 3		
	Corner column				Corner column		
	KZ3	KZ1	KZ7	KZ5	KZ3	KZ1	KZ7
0.415	0.59	0.56	0.67	1.10	0.76	3.96	4.76
0.550	0.69	4.39	2.07	1.40	0.88	3.51	5.16
0.750	0.71	8.25	3.72	3.40	0.86	5.53	6.91
1.170	1.20	0.65	2.97	3.73	1.57	5.52	9.51
Mean value	3.04						

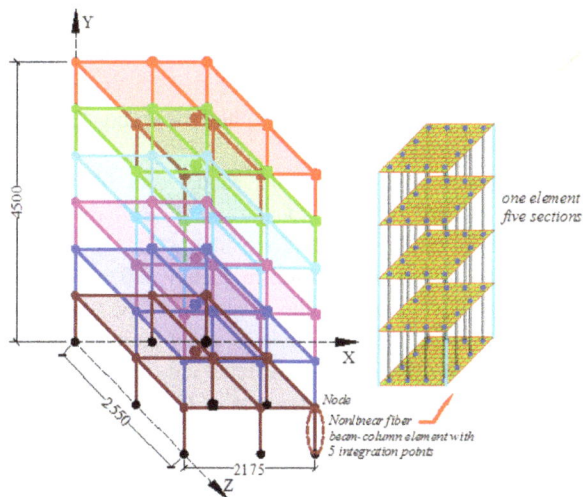

Fig. 12 Proposed modeling of RAC spatial frame.

elements the individual beams at each floor level. The same input earthquake waves and loading procedure used in the shaking table tests (Wang 2012) are followed in numerical modeling, so that the calculated and tested results can be directly compared. WCW, ELW and SHW are input horizontally in sequence in the different test phases. During earthquake motions with various intensities, typical responses of the structure are influenced more significantly for the excitation SHW than for others. Therefore, the structural responses under SHW excitation are mainly analyzed and discussed in this paper. The gradually increasing amplitudes of base excitation are input successively in a manner of time-scaled earthquake waves with 0.00736 s intervals. Before and after each dynamic response time history analysis, a modal analysis is performed to capture the natural frequency and equivalent lateral stiffness.

4.1 Dynamic Characteristics Analysis

The natural vibration frequency of the structure is obtained with the white noise scanning for the shaking table tests and the modal analysis for the numerical modelling at different earthquake motions of various intensities. Table 4 lists the initial frequency of the structure before the test.

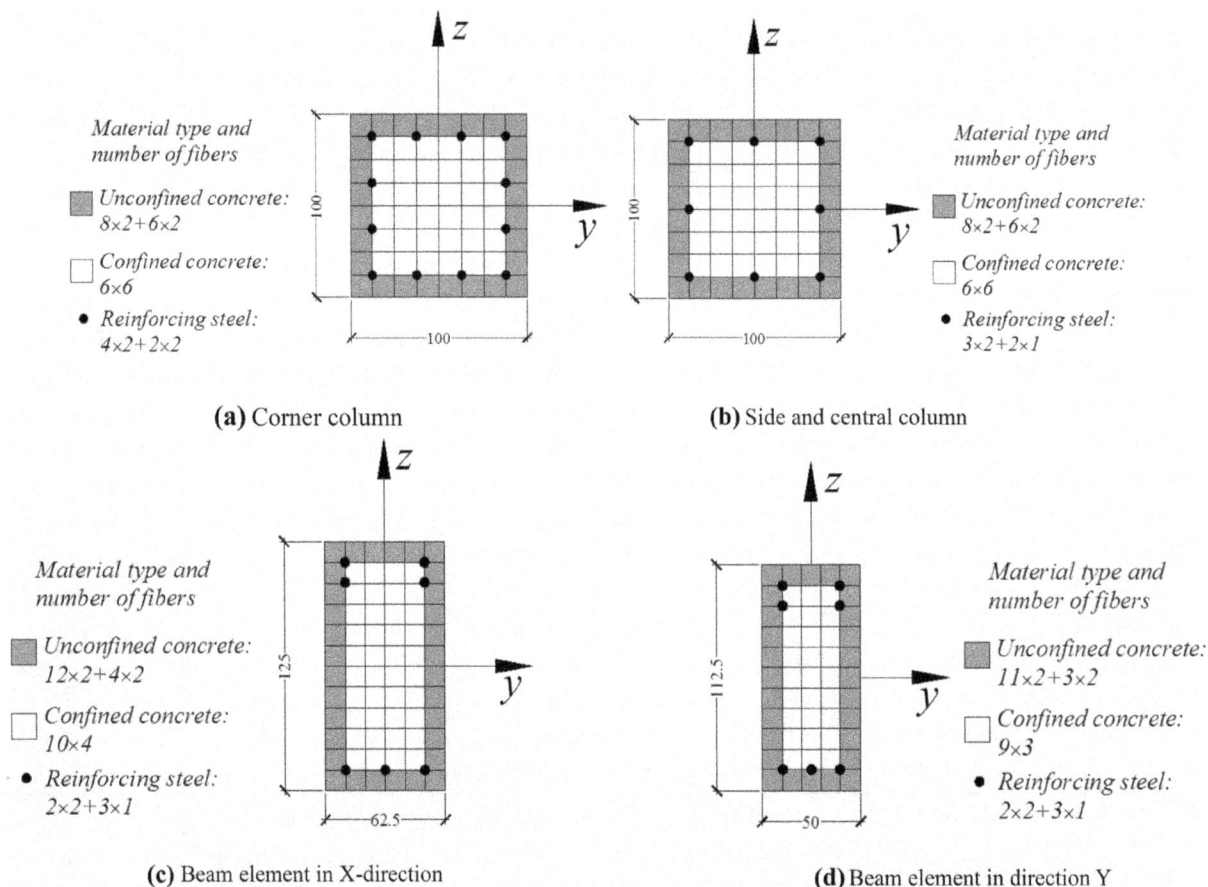

(a) Corner column

(b) Side and central column

(c) Beam element in X-direction

(d) Beam element in direction Y

Fig. 13 Subdivision of cross section into fibers for fiber beam–column element.

For the experimental results the first two modes are of translation in Y and X directions with initial natural frequency of 3.450 and 3.715 Hz, respectively. Based on Model 1, the calculated initial natural frequencies in the Y- and X-direction are 3.392 and 3.756 Hz, respectively. Using Model 2 with strain rate effect the obtained initial natural frequencies in direction Y and X are 3.419 and 3.669 Hz, respectively. It can be seen that the frequency from numerical simulations match test results very well.

Choosing the natural frequency f_0 under the test phase with 0.130 g peak acceleration amplitude as the reference, the variation of the natural frequency under WCW, ELW and SHW is shown in Fig. 14a–c, respectively. It can be seen that the natural frequency variation values obtained by Models 1 and 2 agree well with the tested model values, more closed for Model 2 than for Model 1 without the strain rate effect. The error of the natural frequency variation value increases in the severe elastic–plastic stage due to the complex attributes of the materials such as strain

softening in compression and tension of RAC and tensile degradation of reinforcement steel that cannot be captured very well by the analysis procedure in the post-peak stress regime. The variation values of the simulated natural frequency for RAC frame structure under the excitations WCW, ELW and SHW are listed in Table 5. Table 5 shows that, under the 0.415 g test phase, the errors in the variation values (f/f_0) for the strain rate included and exclude are respectively 5 and 18 % for the excitation WCW, about 15 and 34 % for the excitation ELW, about 16 and 42 % for the excitation SHW.

On the other hand, when either Model 1 or Model 2 is subjected to different earthquake wave excitations, an obvious discrepancy is found. For example, for Model 1 the variation of the natural vibration frequency is 0.695 for the excitation WCW, 0.789 for the excitation ELW, and 0.853 for the excitation SHW. It is indicated that the nature of an input earthquake wave can sometimes induce a significant effect on the dynamic characteristics of the structures.

Table 4 Initial natural frequency comparison between calculated and tested results.

Models	First order frequency in the X direction (Hz)	First order frequency in the Y direction (Hz)
Model 1	3.756	3.392
Model 2	3.669	3.419
Tested model	3.715	3.450

(a) Under WCW

(b) Under ELW

(c) Under SHW

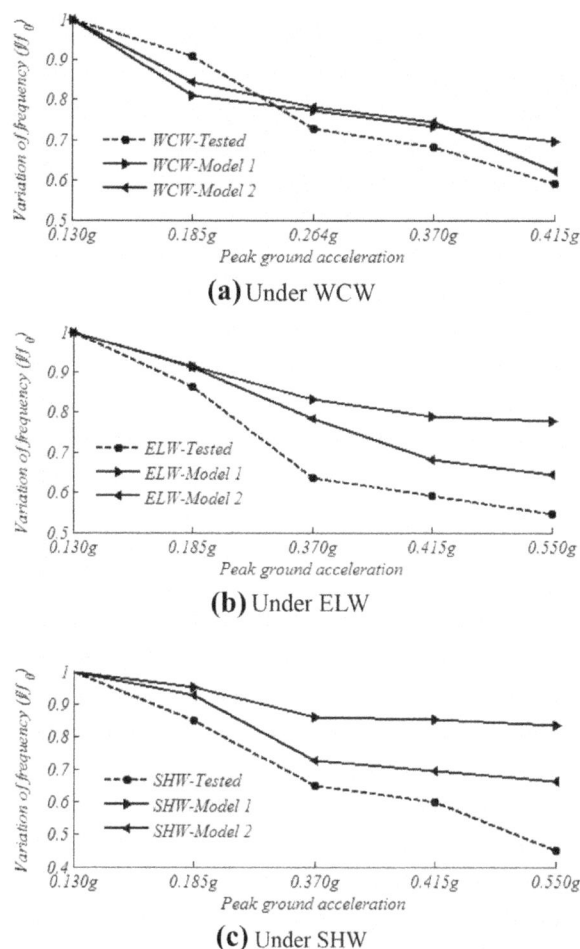

Fig. 14 Variation of natural vibration frequency from different excitation histories.

4.2 Seismic Response of RAC Frame Structure
4.2.1 Storey Displacement

The simulated top displacement time histories of the RAC frame structure subjected to excitation SHW using Models 1 and 2 are compared with that of the experimental results as shown in Fig. 15. For the purpose of easier comparison, only the response history from 4 to 15 s is presented in the figure.

In general the inclusion or exclusion of strain rate makes little difference to the overall frequency content of the displacement response. The peak displacement for Model 1 is slightly higher than that of Model 2 and the tested model. It is also interesting to see that the peak and troughs for the three models are similarly located. It can be seen that the direction of the peak response is basically the same for the three models. For example, the maximum top displacements in the numerical models and the tested model are all negative. However, the peak values occur at different times.

The simulated storey displacement envelops of the RAC frame structure subjected to excitation SHW model are shown in Figs. 16a to 16d along with that obtained from experimental results. Examining Fig. 16 it can be seen that the storey displacement distribution for Models 1 and 2 is similar to that from shaking table tests along the height, more closely for Model 2 than for Model 1. In general, the storey displacement curve along the height is found to show a shear-type feature, which is in accordance with the distribution feature of the first order vibration mode curve of the model as presented by Wang (2012). While inputting different earthquake waves, it has little influence on the shape of the structural displacement response curve but has obvious influence on the amplitude of structural displacement. The maximum storey displacement curves are relatively smooth without obvious inflexion, which means that the distribution of the equivalent rigidity along the height of the structure is well proportioned.

From Fig. 16, it can be seen that the relative errors of the storey displacement obtained using Model 2 are mostly smaller than those using Model 1. For example, the maximum relative error of the storey displacement for floors 1–6 is found to be around −8.9 % from Model 2 due to the strain rate effect, but the maximum relative error for the exclusion of the strain rate effect of Model 1 is around 42.8 %, under the 0.130 g test phase. The maximum relative error of the storey displace for floors 1–6 is about −16.0 % from Model 2 and around −23.2 % from Model 1, under the 0.370 g test phase. The maximum relative error of the storey displace for floors 1–6 is around −20.6 % from Model 2 and about −32.3 % from Model 1, under the 0.415 g test phase. The

Table 5 Natural vibration frequency variation under different test phases (f/f_0).

Wave types	Model	0.130 g	0.185 g	0.264 g	0.370 g	0.415 g	0.550 g
WCW	Model 1	1.000	0.809	0.771	0.733	0.695	0.685
	Model 2	1.000	0.842	0.780	0.745	0.622	0.590
	Test model	1.000	0.909	0.727	0.682	0.591	0.545
ELW	Model 1	1.000	0.914	0.872	0.832	0.789	0.779
	Model 2	1.000	0.911	0.831	0.784	0.682	0.644
	Test model	1.000	0.864	0.727	0.637	0.591	0.545
SHW	Model 1	1.000	0.954	–	0.861	0.853	0.834
	Model 2	1.000	0.928	–	0.725	0.694	0.664
	Test model	1.000	0.8500	–	0.650	0.600	0.450

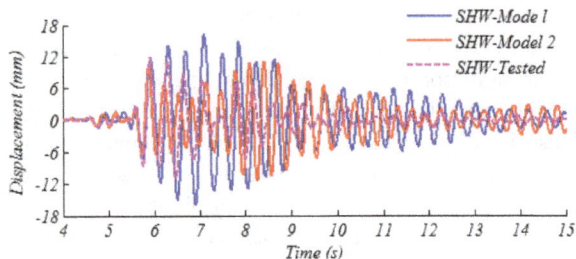

Fig. 15 Top displacement history of RAC frame structure under 0.130 g test phase.

maximum relative error of the storey displacement for floors 1–6 is about −11.9 % from Model 2 and around −28.0 % from Model 1, under the 0.550 g test phase. It can be seen that while the relative errors of the storey displacement for floors 4–6 from Model 2 are smaller than these from Model 1, the relative errors of the storey displacement for the floors 1–3 from Model 2 are slightly larger than these from Model 1. In general, it is indicated that Model 2 considering the

strain rate effect of the material is more likely to provide a better match between the numerical simulation and the dynamic tests as the key attributes of the seismic behavior of the structure are captured by the model. On the other hand, it is worthy to note that the storey displacement response for the tested model is slightly larger than that the other two numerical models in the post-elastic regime. This can be explained that in the degradation stage of the lateral stiffness for the real structure is more significant than due to inclusion or exclusion of the strain rate for the numerical simulation.

4.2.2 Capacity Curves

The base shear versus top deformation capacity curves for both cases with and without strain rate effect are compared and discussed along with the experimental results as shown in Fig. 17. By examining Fig. 17, it can be seen that the capacity curves during earthquake motion excitation with PGAs from 0.066 to 1.170 g for both Models 1 and 2 are similar to that from shaking table tests, more closely for Model 2 than for Model 1. Simultaneously, while the

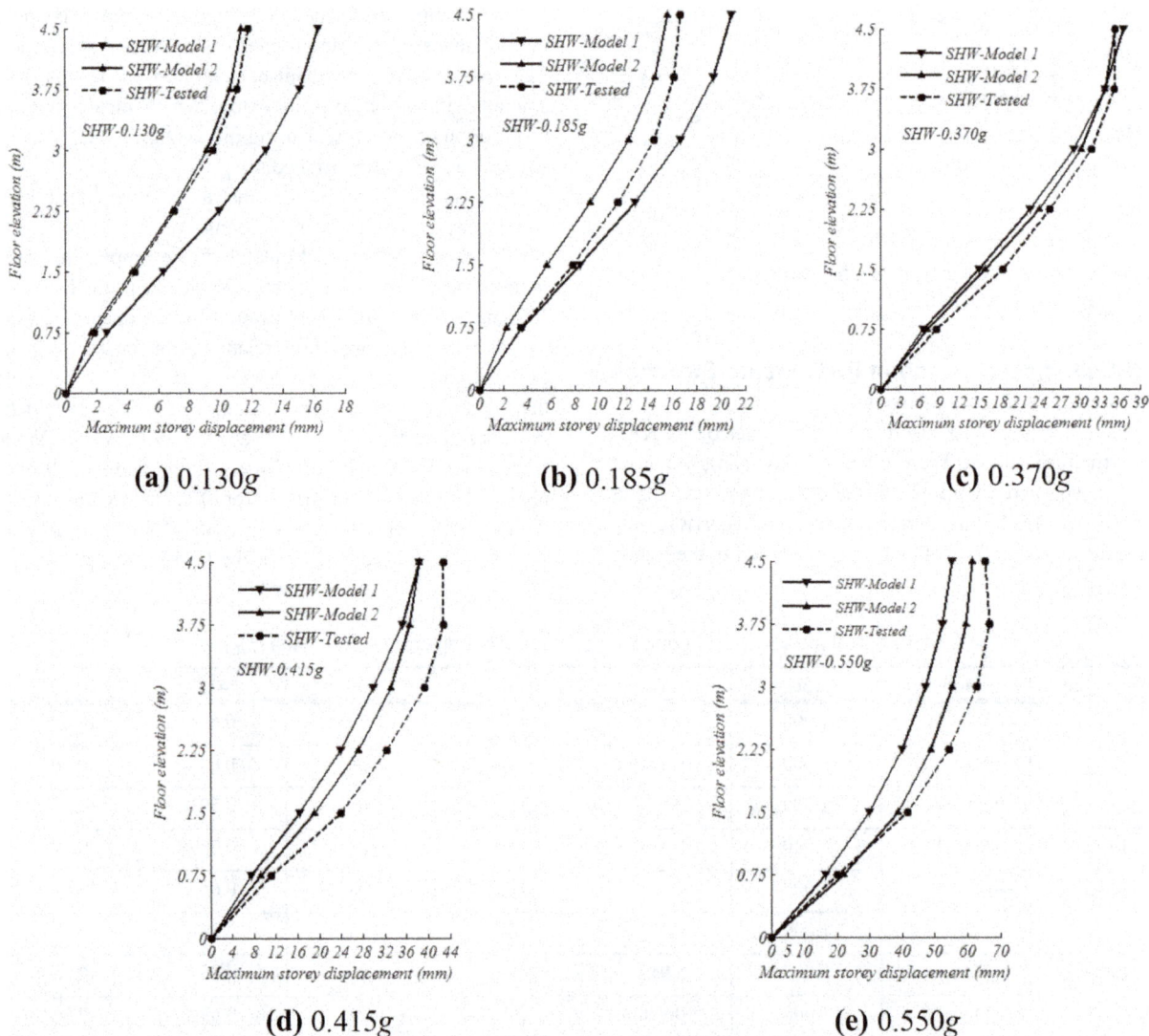

(a) 0.130g

(b) 0.185g

(c) 0.370g

(d) 0.415g

(e) 0.550g

Fig. 16 Envelope of storey displacement with various intensities.

capacity curves from Models 1 and 2 match the experimental curves closely in particular for the latter with the inclusion of the strain rate effect in the early test phase, the error increases for the capacity curves in the post-yield regime of the structures for both Models 1 and 2.

5. Parameter Study for Varying Strain Rate

In this section, for the motivation to examine how the seismic behaviors of the RAC structure vary in a real seismic situation for different strain rates, the strain rates of 3.04×10^{-5}, 3.04×10^{-3}, 3.04×10^{-2} and 3.04×10^{-1}/s, which are regarded as being indicative of the strain rate expected during the response of reinforced concrete to earthquakes, are taken into account in the numerical simulation, respectively. Here, the typical seismic response for the model with strain rate of 3.04×10^{-5}/s are defined as the quasi-static response and used as the reference values of the structural responses. Once again the six-storey two-span and two-bay RAC frame structure discussed earlier is used. For easy comparison of the seismic responses under dynamic loads with different strain rates included in this section the DIF is defined as the ratio of dynamic response value to quasi-static response value.

5.1 Storey Displacement

The storey displacement response of the structure subjected to earthquake excitations and modeled using the verified model proposed with strain rates effect are shown in Figs. 18a to 18f. The figures show that the inclusion of strain rate makes little difference to the overall storey displacement distribution of the structure along the height. However, the amplitude quantities for different cases obviously suffer a significant influence.

Comparisons for the storey displacements under different strain rates show two important features as follows: firstly, the storey displacements vary slightly in the trend of decreasing with the increasing strain rate. For example, in the 0.066 g test phase the DIFs of the top deformation are 0.964, 0.949 and 0.935 under the earthquake excitation history WCW with the strain rate of 3.04×10^{-3},

3.04×10^{-2} and 3.04×10^{-1}/s, respectively. In the 0.130 g test phase the DIFs of the top deformation are 0.948, 0.946 and 0.884 under the earthquake excitation history SHW with the strain rate of 3.04×10^{-3}, 3.04×10^{-2} and 3.04×10^{-1}/s, respectively. In the 0.185 g test phase the DIFs of the top deformation are 0.916, 0.877 and 0.788 under the earthquake excitation history WCW with the strain rate of 3.04×10^{-3}, 3.04×10^{-2} and 3.04×10^{-1}/s, respectively. Secondly, the storey displacements vary obviously in the trend of decreasing with the increasing input acceleration amplitudes. For example, for the strain rate of 3.04×10^{-1}/s the DIFs of the top deformation are 0.935, 0.884, 0.788, and 0.702 under the test phases with PGA of 0.066, 0.130, 0.185 and 0.370 g, respectively. On the other hand, it is interesting to note that, in the post-yield regime (e.g., the test phases from 0.415 to 0.550 g) variation of the maximum storey deformation response due to the strain rate is unconspicuous with the increasing input acceleration amplitudes.

5.2 Storey Shear Force

The base shear response of the structure subjected to earthquake excitation history SHW and modeled using the proposed model in this study are shown in Fig. 19. The figure shows that the inclusion of strain rate makes little difference to the curve distribution of the base shear against the PGA for the RAC structure. However, the magnitude of the base shear for different loading cases by varying strain rates obviously suffers a significant influence, i.e., the base shear of the RAC structure under dynamic loading conditions increases with gradually increasing amplitude of strain rate. Analyzing the base shear versus PGA curves, it can be demonstrated that the base shear augments steeply with the increasing input acceleration amplitudes at the cracking regime with the PGAs from 0.130 to 0.370 g. The base shear increases flatly with the increasing input acceleration amplitudes at the yield regime with the PGAs from 0.370 to 0.550 g, at the yield regime the strain of the RAC in compression steps into hardening stage. For the ultimate regime, with the PGAs from 0.550 to 0.750 g, the base shear decreases flatly with the increasing input acceleration amplitudes, then the base shear decreases steeply up to the PGA of 1.170 g, at the ultimate regime the strain of the RAC in comparison steps into softening stage.

Based on the information presented in Fig. 19, the fitting curves for the base shear versus PGA of the RAC frame structure under dynamic loads with different strain rates included are plotted in Fig. 20. The corresponding fitting polynomial function model derived from the information shown in Fig. 20 for each fitting curve is expressed as follows:

$$BS_i(PGA) = a_i \cdot (PGA)^3 - b_i \cdot (PGA)^2 + c_i \cdot (PGA) + d_i \tag{24}$$

Here, the PGA as the independent variable of the function is assigned at the interval [0.130 1.170 g] with g (9.8 m/s^2) as the fundamental unit. BS_i ($i = 1, 2, 3, 4$) denotes the base

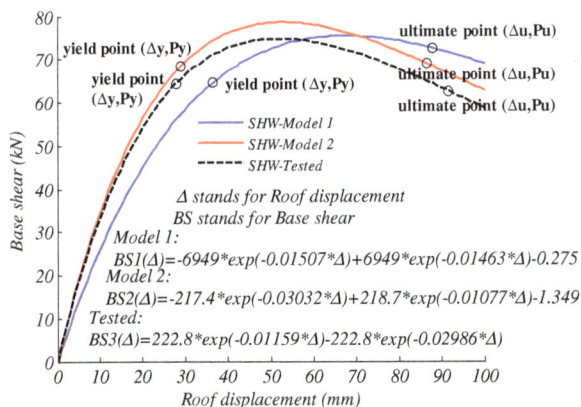

Fig. 17 Fitting function model for the capacity curves.

(a) 0.066g   **(b)** 0.130g   **(c)** 0.185g

(d) 0.370g   **(e)** 0.415g   **(f)** 0.550g

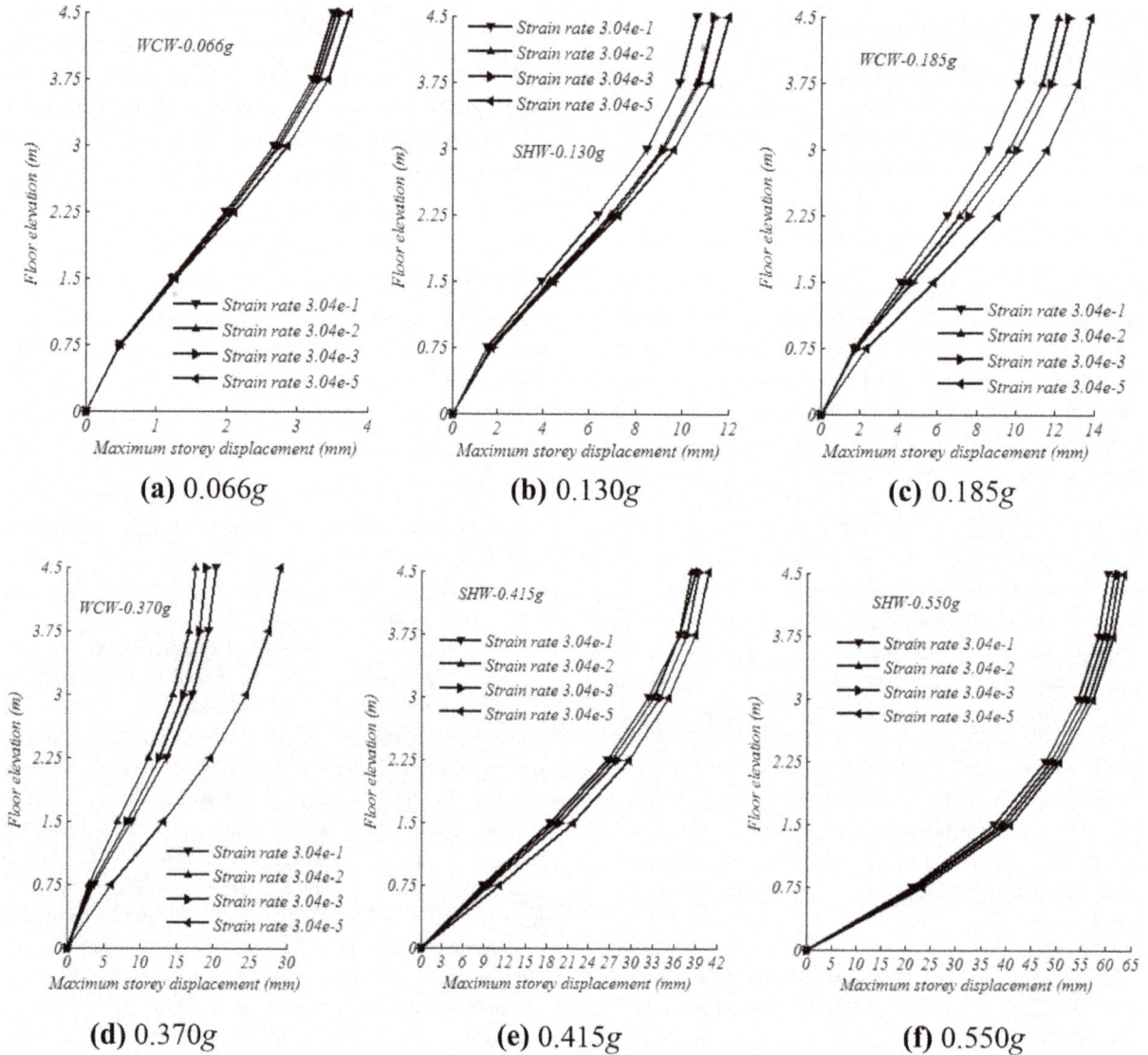

Fig. 18 Storey displacement envelop of the RAC with different strain rates included.

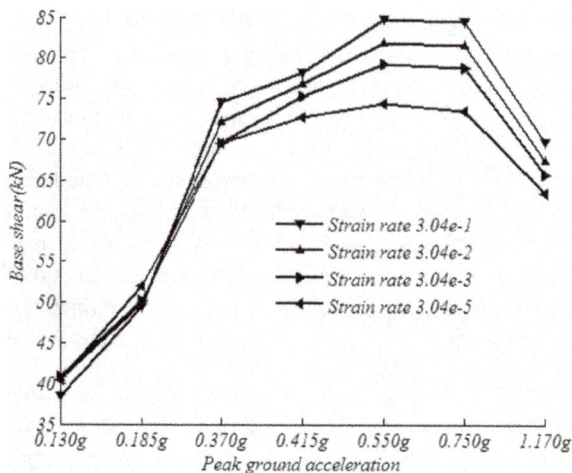

Fig. 19 Base shear distribution with different strain rates included.

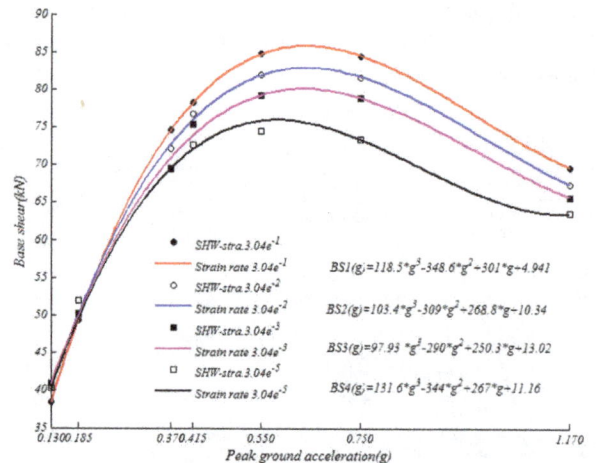

Fig. 20 Fitting function curves for the base shear with different strain rates included.

Table 6 Parameters of the equations used for fitting the base shear versus PGA curves.

Parameters				Curve equations
a_i	b_i	c_i	d_i	
118.5	−348.6	301	4.941	BS_1
103.4	−309	268.8	10.34	BS_2
97.93	−290	250.3	13.02	BS_3
131.6	−344	267	11.16	BS_4

shear with strain rate of 3.04×10^{-1}, 3.04×10^{-2}, 3.04×10^{-3} and 3.04×10^{-5}/s, respectively. a_i, b_i, c_i and d_i represent the function model parameters ($i = 1, 2, 3, 4$). The results of the fitting are presented in Table 6.

6. Summary and Conclusions

The dynamic characteristics and the seismic behaviors of the RAC frame structure are analyzed and discussed by examining the FE models with and without considering the strain rate effect, and compared with the shaking table test results. For the purpose to examine how the seismic behaviors of the RAC frame structure vary under different strain rates in a real seismic situation, a numerical simulation is performed by varying the parameters of strain rate. Several typical conclusions are summarized and presented as follows:

(1) The natural frequency variation values obtained from Models 1 and 2 agree well with the experimentally measured values, and Model 2 with strain rate effect performs better than Model 1 without strain rate effect. The error increases for the natural frequency variation value in the severe elastic–plastic stage. This can be partially attributed to the complex attributes of the materials such as strain softening in compression and tension of RAC and tensile degradation of reinforcement steel that cannot be captured very well by the analysis procedure in the post-elastic regime.

(2) The seismic responses for Models 1 and 2 agree well with the tested model values, with Model 2 giving better predictions than Model 1. Model 2 considering the strain rate effect of the material is more likely to provide a better match between the numerical simulation and the dynamic tests as the key attributes of the seismic behavior of the structure are captured by the model.

(3) The storey deformations of the RAC frame structure subjected to dynamic loads of the lower strain rate are consistently larger than those subjected to dynamic loads of the higher strain rate, because of the inclusion of the strain rate effect on the peak strain of the RAC and the yield strain of reinforcement steel.

(4) The base shear decreases slightly with the increasing strain rate of RAC in the cracking regime. However, the base shear increases slightly with the increasing strain rate of RAC in the yield and ultimate regime.

This can be attributed to the inclusion of the strain rate effect on the peak strength of the RAC and yield strength of the reinforcement steel.

Acknowledgments

The authors wish to acknowledge the financial support from Project funded by China Postdoctoral Science Foundation through Grant Nos. 2014M550247 and 2015T80449, the National Natural Science Foundation of China (51438007), and the Key Projects of Science and Technology Pillar Program of Henan Province (152102310027).

References

Abrams, D. A. (1917). Effect of rate of application of load on the compressive strength of concrete. *ASTM Journal, 17*, 364–377.

ACI Committee 555. (2002). Removal and reuse of hardened concrete. *ACI Material Journal, 99*(3), 300–325.

Ajdukiewicz, A. B., & Kliszczewicz, A. T. (2007). Comparative tests of beams and columns made of recycled aggregate concrete and natural aggregate concrete. *Journal of Advanced Concrete Technology, 5*(2), 259–273.

Atchley, B. L., & Furr, H. L. (1967). Strength and energy-absorption capabilities of plain concrete under dynamic and static loading. *ACI Journal Proceedings, 64*(11), 745–756.

Bertero, V. V., Aktan, A. E., Charney, F. A., & Sause, R. (1984). *Earthquake simulation tests and associated studies of a 1/5th scale model of a 7-story RC test structure*. U.S.–Japan cooperative earthquake research program, report no. UCB/EERC-84/05. Berkeley, CA: Earthquake Engineering Research Center, University of California.

Bischoff, P. H., & Perry, S. H. (1991). Compressive behavior of concrete at high strain rates. *Materials and Structures, 24*(6), 425–450.

Cadoni, E., Labibes, K., & Albertini, C. (2001). Strain-rate effect on the tensile behaviour of concrete at different relative humidity levels. *Materials and Structures, 235*(34), 21–26.

Chang, K. C., & Lee, G. C. (1987). Strain rate effect on structural steel under cyclic loading. *Journal of Engineering Mechanics, ASCE, 113*(9), 1292–1301.

Chen, X. D., Wu, S. X., & Zhou, J. K. (2013). Experimental and modeling study of dynamic mechanical properties of cement paste, mortar and concrete. *Construction and Building Materials, 47*, 419–430.

Cotsovos, D. M., & Pavlovic, M. N. (2006). Simplified FE model for RC structures under earthquakes. *Structure Buildings, 159*, 87–102.

Cotsovos, D. M., & Pavlovic, M. N. (2008). Numerical investigation of concrete subjected to high rates of uniaxial tensile loading. *International Journal of Impact Engineering, 35*, 319–335.

Fathifazl, G., Razaqpur, A. G., Isgor, O. B., Abbas, A., Fournier, B., & Foo, S. (2009). Flexural performance of steel-reinforced recycled concrete beams. *ACI Structural Journal, 106*(6), 858–867.

Filippou, F. C., D'Ambrisi, A., & Issa, A. (1992). *Nonlinear static and dynamic analysis of reinforced concrete sub-assemblages*. Report No. UCB/EERC-92/08. Berkeley, CA: Earthquake Engineering Research Center, University of California.

GB 50011. (2010). *Code for seismic design of buildings*. Beijing, China: Chinese Building Press.

Hansen, T. C. (1986). Recycled aggregate and recycled aggregate concrete, second state-of-the-art report, developments from 1945–1985. *Materials and Structures, 111*, 201–246.

Hognestad, E. (1951). A study of combined bending and axial load in reinforced concrete. *Bulletin series 339*, Illinois (USA), University of Illinois Experiment Station.

Kent, D. C., & Park, R. (1971). Flexural members with confined concrete. *Journal of the Structural Division, ASCE, 97*(ST7), 1969–1990.

Kulkarni, S. M., & Shah, S. P. (1998). Response of reinforced concrete beams at high strain rates. *ACI Structural Journal, 95*(6), 705–715.

Lai, J., & Sun, W. (2009). Dynamic behavior and visco-elastic modeling of ultra-high performance cementitious composite. *Cement Concrete Research, 39*, 1044–1051.

Le, N. H., & Bailly, P. (2000). Dynamic behaviour of concrete: The structural effects on compressive strength increase. *Mechanics of Cohesive-Frictional Materials, 5*(6), 491–510.

Li, M., & Li, H. N. (2012). Effects of strain rate on reinforced concrete structure under seismic loading. *Advances in Structural Engineering, 15*(3), 461–475.

Lin, F., Gu, X. L., Kuang, X. X., & Yin, X. J. (2008). Constitutive models for reinforcing steel bars under high strain rates. *Journal of Building Materials, 11*(1), 14–20.

Lu, Y. B., Chen, X., Teng, X., & Zhang, S. (2014). Dynamic compressive behavior of recycled aggregate concrete based on split Hopkinson pressure bar tests. *Latin American Journal of Solids and Structures, 11*(1), 131–141.

Malvar, L. J., & Crawford, J. E. (1998). Dynamic increase factors for steel reinforcing bars. In *Proceedings of the twenty-eighth DoD explosives safety seminar*, Orlando, FL, USA.

Malvar, L. J., & Ross, C. A. (1998). Review of strain rate effects for concrete in tension. *ACI Materials Journal, 95*(6), 435–439.

Mazzoni, S., Mckenna, F., Scott, M. H., & Fenves, G. L. (2006). *Open system for earthquake engineering simulation, user command-language manual*. Berkeley, CA: Pacific Earthquake Engineering Research Center, University of California.

Norris, C. H., Hansen, R. J., Holley, M. J., Biggs, J. M., Namyet, S., & Minasmi, J. K. (1959). *Structural design for dynamic loads*. New York, NY: McGraw-Hill.

Oh, B. H. (1987). Behavior of concrete under dynamic tensile loads. *ACI Materials Journal, 84*(1), 8–13.

Pandey, A. K., Kumar, R., Paul, D. K., & Trikha, D. N. (2006). Strain rate model for dynamic analysis of reinforced concrete structures. *Journal of Structural Engineering, ASCE, 132*(9), 1393–1401.

Restrepo-Posada, J. I., Dodd, L. L., Park, R., & Cooko, N. (1994). Variables affecting cyclic behavior of reinforcing steel. *Journal of Structural Engineering, ASCE, 120*(11), 3178–3196.

Scott, B. D., Park, R., & Priestley, M. J. N. (1982). Stress–strain behavior of concrete confined by overlapping hoops at low and high strain rates. *ACI Journal, 79*(1), 13–27.

Shing, P. S. B., & Mahin, S. A. (1988). Rate of loading effects on pseudo dynamic tests. *Journal of Structural Engineering, ASCE, 114*(11), 2403–2420.

Soroushian, P., & Choi, K. B. (1987). Steel mechanical properties at different strain rates. *Journal of Structural Engineering, ASCE, 113*(4), 663–672.

Taucer, F. F., Spacone, E., & Filippou, F. C. (1991). *A fiber beam–column element for seismic response analysis of reinforced concrete structures*. Report no. UCB/EERC-91/17. Berkeley, CA: Earthquake Engineering Research Center, University of California.

The Euro-International Committee for Concrete (CEB). (1993). *CEB-FIP model code 1990*. Lausanne: Thomas Telford Ltd.

Wakabayashi, M., Nakamura, T., Iwai, S., & Hayashi, Y. (1984). Effect of strain rate on the behavior of structural members subjected to earthquake force. In *Proceeding 8th world conference on earthquake engineer*, San Francisco, CA (pp. 491–498).

Wang, C. Q. (2012). Research on shaking table test and nonlinear analysis for recycled aggregate concrete frame structure. PhD Thesis, College of Civil Engineering, Tongji University, Shanghai, China.

Wang, C. Q., & Xiao, J. Z. (2012). Shaking table tests on a recycled concrete block masonry building. *Advances in Structural Engineering, 15*(10), 1843–1860.

Xiao, J. (2008). *Recycled concrete*. Beijing, China: Chinese Building Press. (in Chinese).

Xiao, S., Li, H., & Lin, G. (2008). Dynamic behaviour and constitutive model of concrete at different strain rates. *Magazine of Concrete Research, 60*(4), 271–278.

Xiao, J. Z., Li, L., Shen, L. M., & Poon, C. S. (2015). Compressive behaviour of recycled aggregate concrete under impact loading. *Cement and Concrete Research, 71*, 46–55.

Xiao, J. Z., Li, J. B., & Zhang, C. (2005). Mechanical properties of recycled aggregate concrete under uniaxial loading. *Cement and Concrete Research, 35*, 1187–1194.

Xiao, J. Z., Wang, C. Q., Li, J., & Tawana, M. M. (2012a). Shake-table model tests on recycled aggregate concrete frame structure. *ACI Structural Journal, 109*(6), 777–786.

Xiao, J. Z., Xie, H., & Yang, Z. (2012b). Shear transfer across a crack in recycled aggregate concrete. *Cement and Concrete Research, 42*(5), 700–709.

Xiao, J. Z., Yuan, J. Q., & Li, L. (2014). Experimental study on dynamic mechanical behavior of modeled recycled aggregate concrete under uniaxial compression. *Journal of Building Structures, 35*(3), 201–207. (in Chinese).

Yassin, M., & Hisham, M. (1994). Nonlinear analysis of prestressed concrete structures under monotonic and cyclic load. PhD Thesis, University of California, Berkeley, CA.

Zhang, J. W., Cao, W. L., Meng, S. B., Yu, C., & Dong, H. Y. (2014). Shaking table experimental study of recycled concrete frame-shear wall structures. *Earthquake Engineering and Engineering Vibration, 13*(2), 257–267.

Zhou, X. Q., & Hao, H. (2008). Modelling of compressive behaviour of concrete-like materials at high strain rate. *International Journal of Solids and Structures, 45*, 4648–4661.

Zielinski, A. J., Reinhardt, H. W., & Körmeling, H. A. (1981). Experiments on concrete under uniaxial impact tensile loading. *Materials and Structures, 80*(14), 103–112.

13

Shear Tests for Ultra-High Performance Fiber Reinforced Concrete (UHPFRC) Beams with Shear Reinforcement

Woo-Young Lim[1], and Sung-Gul Hong[2],*

Abstract: One of the primary concerns about the design aspects is that how to deal with the shear reinforcement in the ultra-high performance fiber reinforced concrete (UHPFRC) beam. This study aims to investigate the shear behavior of UHPFRC rectangular cross sectional beams with fiber volume fraction of 1.5 % considering a spacing of shear reinforcement. Shear tests for simply supported UHPFRC beams were performed. Test results showed that the steel fibers substantially improved of the shear resistance of the UHPFRC beams. Also, shear reinforcement had a synergetic effect on enhancement of ductility. Even though the spacing of shear reinforcement exceeds the spacing limit recommended by current design codes (ACI 318-14), shear strength of UHPFRC beam was noticeably greater than current design codes. Therefore, the spacing limit of $0.75d$ can be allowed for UHPFRC beams.

Keywords: spacing limit, shear reinforcement, ultra-high performance fiber-reinforced concrete (UHPFRC), shear strength, shear test, failure modes.

1. Introduction

Recently, the steel fiber-reinforced concrete (SFRC) has been widely used as structural material due to its remarkable mechanical properties compared to conventional concrete. Through the numerous experimental studies, it turns out that the addition of steel fibers can improve the structural capability of concrete (Fanella and Naaman 1985; Sharma 1986; Narayanan and Darwish 1987; Wafa and Ashour 1992; Ashour et al. 1992; Ezeldin and Balaguru 1997; Kwak et al. 2002). Even though SFRC has many advantages as structural material, some limitations still exist in the construction of the large-scale structures that requires very high compressive and tensile strength.

To overcome these limitations, ultra-high performance fiber-reinforced concrete (UHPFRC) has been developed. The UHPFRC has a compressive strength of about 150–200 MPa and a tensile strength of 10 MPa or more (Rossi et al. 2005; Farhat et al. 2007; Wille et al. 2011a, b; Park et al. 2012). In addition, shear resistance of UHPFRC beam is outstanding. Previous research on shear tests for UHPFRC beam has focused on the I-shaped beam or girder without shear reinforcement because UHPFRC can reduce a web thickness of the beam due to its great compressive and tensile strength.

According to Baby et al. (2014), the presence of shear reinforcement has increased the shear capacity of the beams. Voo et al. (2010) found that a significant distribution of shear cracking occurs prior to the formation of the critical failure crack. Due to its superior mechanical properties, the UHPFRC has been successfully applied in the construction of bridges and also used for retrofitting and strengthening existing concrete structures in building structures (Alaee and Karihaloo 2003; Meda et al. 2014).

One of the primary concerns about the design aspects is that how to deal with the shear reinforcement in the UHPFRC beams. The formation of inclined shear cracking might lead directly to critical failure without warning. To avoid sudden failure in beams, shear reinforcement is required in a proper spacing so that the shear reinforcement should intersect with the diagonal shear cracks, even when shear reinforcement is not necessary according to the computation. Current design codes for reinforced concrete (RC) beams (ACI 318-14 2015; EC2 2004; CSA A23.3-04 2004; AASHTO-LRFD 2004; MC2010 2012) requires a minimum shear reinforcement in beams to ensure adequate reserve shear strength and to prevent possible sudden shear failure, when the factored shear force (V_u) exceeds $0.5\phi V_c$. Here, ϕ is the strength reduction factor for shear and V_c is the shear strength provided by concrete. Also, a spacing limit of shear reinforcement is served in design codes (ACI 318-14 2014; CSA A23.3-04 2004).

For SFRC beams, ACI 544 (1988) reported that the steel fibers show potential advantages as shear reinforcement. Previous studies have identified the synergetic effect of fiber volume fraction and presence of shear reinforcement on shear behavior of beams (Mansur et al. 1986; Narayanan 1987; Li et al. 1992; Khuntia et al. 1999; Noghabai 2000).

[1]Institute of Engineering Research, Seoul National University, Seoul 08826, Korea.

[2]Department of Architecture and Architectural Engineering, Seoul National University, Seoul 08826, Korea.

*Corresponding Author; E-mail: sglhong@snu.ac.kr

They found that the combination of steel fibers and shear reinforcement depicted slow and controlled cracking and better distribution of tensile cracks, and minimized the penetration of shear cracks into the compression zone. According to Parra-Montesinos (2006), SFRC beams that contained fiber volume fraction (V_f) more than 0.75 % exhibited a shear stress at failure greater than the conservative lower bound value of $0.3\sqrt{f_c'}$. Also, the use of a minimum V_f of 0.75 % has been recommended by ACI Subcommittee 318-F.

However, the effect of shear reinforcement in a rectangular UHPFRC beam section has not been recognized even though the design shear strength for the UHPFRC structural member is obtained by summing the shear strengths provided by cement matrix, steel fibers, and shear reinforcement (JSCE 2004; K-UHPC 2012; AGFC 2013). Especially, a spacing limit of shear reinforcement have not been provided due to the lack of previous test data. Thus, it is necessary to investigate the shear behaviour of the UHPFRC beams regarding the spacing of shear reinforcement because the rectangular beam section in building structures might require sufficient beam width to provide the shear reinforcement.

In this study, shear tests for simply supported rectangular UHPFRC beam sections with and without shear reinforcement were performed to characterize the shear behavior depending on the spacing of shear reinforcement. Also, the shear contribution for the spacing of shear reinforcement is discussed.

2. Current Design Guidelines for Shear

2.1 Shear Strength

The JSCE (2004) and K-UHPC (2012) design guidelines provide the shear strength of UHPFRC beam with or without shear reinforcement.

The design shear strength (V_d) is obtained by summation of the shear strength provided by cement matrix, steel fiber, and shear reinforcement as follows:

$$V_d = V_c + V_{fb} + V_s \tag{1}$$

where V_c, V_{fb}, and V_s are shear strength provided by cement matrix, steel fibers, and shear reinforcement, respectively.

The shear strength provided by cement matrix is obtained as given:

$$V_c = \phi_b 0.18\sqrt{f_c'}b_w d \tag{2}$$

where ϕ_b is the member reduction factor and is recommended as 0.77, f_c' is the compressive strength, b_w is the beam width, and d is the effective depth of the beam.

The shear strength by steel fibers can be determined as follows:

$$V_{fb} = \phi_b \left(\frac{f_{vd}}{\tan \beta_u} \right) b_w z \tag{3}$$

where f_{vd} is the design average tensile strength in the direction perpendicular to diagonal tensile crack; β_u is the angle occurring between axial direction and diagonal tensile crack plane. This angle shall be larger than 30°. The value of z is distance from the position of the resultant of the compressive stresses to the centroid of tensile steel (mm), generally $d/1.15$.

In these guidelines, the design average tensile strength in the direction perpendicular to diagonal tensile crack can be expressed as Eq. (4) since the material reduction factor considers the orientation of the steel fibers. Thus, the value of f_{vd} is obtained as follows:

$$f_{vd} = \frac{1}{w_v} \int_0^{w_v} \phi_c \sigma_k(w)dw = \frac{1}{w_v} \int_0^{w_v} \sigma_d(w)dw \tag{4}$$

where $w_v = \max(w_u, 0.3\text{ mm})$; w_u is the ultimate crack width corresponding to peak stress on the outer fiber; ϕ_c is material reduction factor (= 0.8); $\sigma_k(w)$ is the tension softening curve; and $\sigma_d(w)$ is equal to $\phi_c\sigma_k(w)$.

Shear strength by the shear reinforcement is provided in K-UHPC recommendations (2012) and it can be determined as follows:

$$V_s = \phi_b \frac{A_v f_{yt}(\sin \alpha_s + \cos \alpha_s)}{s} d \tag{5}$$

where A_v is the cross sectional area of shear reinforcement; f_{yt} is the design yield strength of shear reinforcement; α_s is the angle between longitudinal axis of beam and shear reinforcement; and s is the spacing of shear reinforcement. It should be noted that JSCE guidelines does not provide this term.

In AFGC design guidelines (2013), shear strength of UHPFRC members is computed by summing ($V_d = V_c + V_{fb} + V_s$) of the shear strength provided by cement matrices; steel fibers; and shear reinforcements in the same manner as other design recommendations assuming the web shear failure.

For a reinforced section, the term of shear strength provided by cement matrices is given by:

$$V_c = \frac{0.21}{\gamma_{cf}\gamma_E} k \sqrt{f_c'}b_w d \tag{6}$$

where γ_{cf} is the partial safety factor on fibers and is assumed to be a value of 1.3; γ_E is a safety coefficient; $\gamma_{cf}\gamma_E$ is equal to 1.5, k is determined by $1 + 3\sigma_{cp}/f_c'$ for $\sigma_{cp} \geq 0$; and $1 + 0.7$ $\sigma_{cp}/f_{c,0.05}'$ for $\sigma_{cp} < 0$; σ_{cp} is calculated by the equation of N_{ed}/A_c; N_{ed} is the axial force in the cross section due to prestressing; and A_c is the area of concrete cross section.

The part of shear strength provided by the fiber is determined as follows:

$$V_{fb} = \frac{A_{fv}\sigma_{Rd,f}}{\tan \theta} \tag{7}$$

where A_{fv} is the area of fiber effect and is assumed to be $b_w z$ for rectangular section; z is the inner lever arm and is approximately equal to $0.9d$; and θ is the angle between the principal compression stress and the beam axis, which a minimum value of $30°$ is recommended; and $\sigma_{Rd,f}$ is residual tensile strength and can be computed as follows:

$$\sigma_{Rd,f} = \frac{1}{K\gamma_{cf}} \frac{1}{w_{\lim}} \int_0^{w_{\lim}} \sigma_f(w)dw, \text{ where } w_{\lim} = \max(w_u, w_{\max}) \tag{8}$$

where K is the fiber orientation factor assuming to be a value of 1.25; $\sigma_f(w)$ is a function of the tensile stress and crack width; and w_{max} is the maximum crack width.

The shear strength by the vertical shear reinforcement is as follows:

$$V_s = \frac{A_v}{s} z f_{yt} \cot\theta \tag{9}$$

Meanwhile, ACI 544 (1988) provides shear strength for fiber-reinforced concrete proposed by Sharma (1986) as follows:

$$V_{cf} = \frac{2}{3} f_{ct} \left(\frac{d}{a}\right)^{0.25} b_w d \tag{10}$$

where f_{ct} is splitting tensile strength of FRC.

2.2 Minimum Shear Reinforcement

Current design provisions (ACI 318-14 2014; EC2 2004; CSA A23.3-04 2004; AASHTO-LRFD 2004; MC2010 2012) for reinforced concrete (RC) beam provide the minimum and maximum shear reinforcement as shown in Table 1. According to Section 9.6.3 of ACI 318-14 (2015), a minimum area of shear reinforcement should be provided in beams where the factored shear force (V_u) exceeds $0.5\phi_c V_c$. In Section 9.2.2 of EC2 (2004), when design shear force (V_d) is higher than design shear resistance (V_{dc}) provided by concrete ($V_d > V_{dc}$), sufficient shear reinforcement should be provided in order that shear resistance (V_{Rd}) is larger than design shear force ($V_d \leq V_{Rd}$).

For fiber-reinforced concrete beams, when compressive strength (f_c') is not exceeding 40 MPa, an overall height (h) not > 600 mm, and the factored shear force not larger than $\phi 0.17\sqrt{f_c'} b_w d$, the minimum shear reinforcement would not be required. Parra-Montesinos (2006) suggested that

shear strength of FRC with hooked or crimped steel fibers exhibits greater than $0.29\sqrt{f_c'} b_w d$.

2.3 Spacing Limits for Shear Reinforcement

ACI 318-14 (2014) prescribes the spacing limitation of shear reinforcement in Section 9.7.6.2.2. Spacing of shear reinforcement installed perpendicular to the axis of the member should not exceed $d/2$ in beams nor 600 mm. Where shear strength contributed by shear reinforcement (V_s) exceeds $0.33\sqrt{f_c'} b_w d$, maximum spacing should be reduced by one-half. EC2 suggests the spacing limits as $0.75d$ or 600 mm. In Section 11.3.8.1 of CSA A23.3-04 (2004), the spacing of shear reinforcement shall not exceed $0.7d_v$ ($d_v = \max(0.9d, 0.72h)$) or 600 mm in case of beams with an overall thickness greater than 750 mm. According to MC2010 (Section 7.13.5.2), shear reinforcement generally is provided in their spacing not exceed $0.75d$ or 500 mm. However, current design guidelines for UHPFRC members does not provide the spacing limits for shear reinforcement.

3. Experimental Program

3.1 Specimen Description

Test specimens which had a same dimension were designed in accordance with K-UHPC (2012) guidelines as shown in Fig. 1. As shown in Table 2, primary test parameter is the spacing (s) of shear reinforcement. Figure 1a shows the cross-section of the test specimens. Rectangular cross-sectional specimens had a dimension ($b_w \times h$) of 150×290 mm, where, b_w is the beam width and h is the overall height of the beam. The effective depth (d) of the beam is 220 mm and the shear span to depth ratio (a/d) is 3.0. Concrete cover is 30 mm. To induce shear failure, four D29 ($d_b = 29$ mm) high-strength reinforcements ($f_y = 600$ MPa) were used, where d_b is a diameter of reinforcing bars and f_y is a design yield strength of reinforcement. The longitudinal reinforcement ratio (ρ) is equal to the value of 0.078. Shear reinforcement [D10 ($d_b = 10$ mm), $f_{yt} = 400$ MPa] was designed in accordance with ACI 318-14 (2014). Therefore, the moment capacities (M_n) of all specimens were 338.3 kN m and shear strength corresponding to moment capacity was 512.6 kN.

The SB1 specimen is a control test specimen without shear reinforcement. (see Fig. 1b) This specimen was designed as

Table 1 Minimum shear reinforcement for RC beam in current design codes.

Design codes	Minimum shear reinforcement
ACI 318-14 (2014)	$\rho_{v,\min} = 0.062\sqrt{f_c'}/f_{yt} \geq 0.35/f_{yt}$
EC2 (2004)	$\rho_{v,\min} = 0.08\sqrt{f_c'}/f_{yt}$
CSA A23.3-04 (2004)	$\rho_{v,\min} = 0.06\sqrt{f_c'}/f_{yt}$
AASHTO-LRFD (2004)	$\rho_{v,\min} = 0.083\sqrt{f_c'}/f_{yt}$
MC2010 (2012)	$\rho_{v,\min} = 0.08\sqrt{f_c'}/f_{yt}$

f_c' is the compressive strength of concrete (in MPa) and f_{yt} is the design yield strength of shear reinforcement (in MPa).

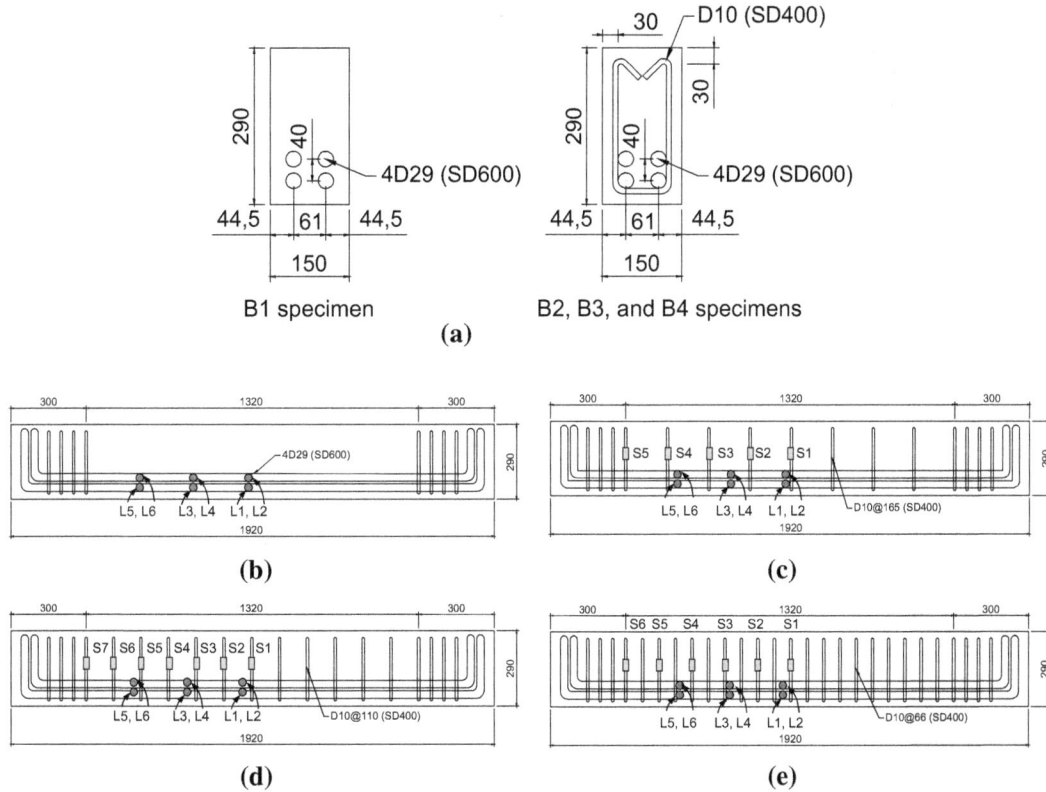

Fig. 1 Details of the test specimens (unit: mm). **a** Cross section, **b** SB1 specimen, **c** SB2 specimen ($s = 0.75d$), **d** SB3 specimen ($s = 0.5d$), **e** SB4 specimen ($s = 0.3d$).

Table 2 Test variables.

Specimens	f_{ct} (MPa)	V_f (%)	a/d	ρ_l (%)	ρ_v (%)	f_y (MPa)	f_{yv} (MPa)	s (mm)	M_n (kN-m)	$V_{@Mn}$ (kN)	V_n (kN)	$V_{@Mn}/V_n$
SB1	11.5	1.5	3	0.78	–	617.7	537.5	–	338.3	512.6	347.6	1.47
SB2	11.5	1.5	3	0.78	0.6	617.7	537.5	165	338.3	512.6	449.8	1.14
SB3	11.5	1.5	3	0.78	0.9	617.7	537.5	110	338.3	512.6	501.0	1.02
SB4	11.5	1.5	3	0.78	1.4	617.7	537.5	66	338.3	512.6	603.2	0.85

f_{ct} is measured tensile strength obtained using direct tension test; V_f is fiber volume fraction; a/d is the shear span-to-depth ratio; ρ_l is the longitudinal reinforcement ratio ($A_s/b_w d$); ρ_v is the shear reinforcement ratio ($A_{sv}/b_w s$); A_s is the area of the longitudinal reinforcement; A_{sv} is the area of the shear reinforcement; f_y is measured yield strength of longitudinal reinforcement; f_{yv} is measured yield strength of shear reinforcement; s is the spacing of shear reinforcement; M_n is the flexural moment strength; $V_{@Mn}$ is the shear force at flexural moment strength; V_n is the shear strength determined in accordance with JSCE and K-UHPC recommendations.

the specimen failed by diagonal tension failure ($V_{@Mn} > V_n$). The SB2, SB3, and SB4 specimen has shear reinforcement with a spacing of $0.75d$ (165 mm), $0.5d$ (110 mm) and $0.3d$ (66 mm), respectively (Figs. 1c to 1e). Here, the spacing of $0.5d$ is a spacing limit provided in ACI 318-14 (2014). In SB3 and SB4 specimens, shear reinforcements were provided at a spacing. Thus, the shear reinforcement ratios of SB2, SB3, and SB4 specimens were 0.6, 0.9 and 1.4 %, respectively.

3.2 Test Set-Up and Instrumentation

Figure 2 shows the test set-up and instrumentation. Simply supported beams were loaded with a capacity of 1000 kN actuator by displacement control. Deflection of the beam

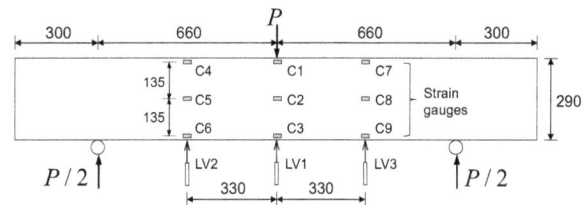

Fig. 2 Test set up (unit: mm).

was measured using three Linear Vertical Displacement Transducers (LVDTs). One is installed at the mid-span of the beam and others are at one-half distance (330 mm) of both sides with respect to mid-span.

Strains of the longitudinal and shear reinforcing bars was measured by using strain gauges during the tests. The location of the strain gauges is presented in Fig. 1. Strain distribution of concrete was obtained at top, mid-height, and bottom of the beam using strain gauges.

4. Material Properties

4.1 Materials and Mix Design of UHPFRC
The UHPFRC is a kind of reactive powder concrete that coarse aggregates were not included. Fine aggregates consist of sand with a diameter of < 0.5 mm, which is the largest component of the UHPFRC. Portland cement is used as the binder, and the filler material is crushed quartz with an average diameter of 10 lm and a density of 2600 kg/m^3. The workability provided by the low water-to-cement ratio of the concrete is maintained by the addition of a high-performance water reducing agent, a polycarboxylate superplasticizer with a density of 1060 kg/m^3. In Table 3, the proportions of the components are shown in terms of weight ratios.

Two different straight-shaped steel fibers with a diameter of 0.2 mm are used to produce the UHPFRC containing steel fibers. According to Park et al. (2012), the overall shape of tensile stress–strain curves of the UHPFRC was substantially dependent on the type of macro fibers. The addition of micro fibers had an effect on the strain hardening and multiple cracking behaviors. For each batch, UHPFRC includes both steel fibers with different lengths of 16 and 19 mm. The fibers had a yield strength of 2500 MPa. Test specimens were produced after adding in a volume of 1.5 % of the total mix volume.

4.2 Compressive Behavior of UHPFRC
Compression tests for cylindrical test specimens with a diameter of 100 mm and a height of 200 mm were performed to obtain the compressive strength of UHPFRC in accordance with ASTM C39/C39M (2005). Figure 3a shows stress–strain curves of the test specimens with a fiber volume fraction of 1.5 %. Compressive strength was measured using universal testing machine controlling by displacement and axial strain (ε_c) was obtained using two strain gauges on the opposite surface of the test specimen. Loading rate was 0.3 mm/min during the tests. The cylindrical test specimens were produced with each batch simultaneously and were cured by steam curing at a temperature above 90 °C for 48 h, and then they cured at room temperature for 60 days until testing.

The UHPFRC showed a linear-elastic behavior until the end of the test. After reaching the peak strength, a brittle failure occurred as shown in Fig. 3b. However, a post-peak behavior was not observed in all of the test specimens. The average compressive strength (σ_{cu}) and ultimate strain (ε_{cu}) were determined to be 166.9 MPa and 0.0041 mm/mm, respectively. The modulus of elasticity (E_c) was a value of 41.1 GPa, where it was calculated using ultimate stress and strain corresponding to ultimate stress under stress–strain relationship in accordance to AFGC design recommendations (2013).

4.3 Tensile Behavior of UHPFRC
The tensile strength of UHPFRC was obtained using direct tension tests for dog-bone shaped specimens in accordance with K-UHPC (2012) guidelines as shown in Fig. 4a. Test specimens had an overall width of 125 mm, a height of 300 mm, and a thickness of 25 mm, but an effective width and a height are 75 and 150 mm, respectively. To induce critical crack at the center of the specimen, notches were installed at both sides of the specimen. The length (a_0) and width of the notches was 12.5 and 2 mm, respectively.

Test specimens are loaded with 100 kN actuator by displacement control. During the test, a loading speed is 0.3 mm/min. The tensile stress was computed with the load divided by an effective cross-sectional area of the specimen, which is equal to $(75 - 2 \times 12.5) \times 25$ mm = 1250 mm^2. The effective cross-sectional area is defined as the area considering the width except for the overall notch length.

Figure 4b shows tensile strength-crack opening relationship of the notched specimens. Crack opening was measured using clip gauges with a capacity of 10 mm installing at both notches. As shown in Fig. 4b, after reaching the peak tensile stress, the stress gradually decreased as increasing the crack opening. The significant variation of the peak tensile stress is because the non-uniform distribution of the steel fibers at the notch tip. Test results showed that the average tensile stress (f_{ct}) was 11.5 MPa.

4.4 Tensile Behavior of Reinforcing Bars
Uniaxial tension tests for D29 ($d_b = 29$ mm, $f_y = 600$ MPa) and D10 ($d_b = 10$ mm, $f_{yt} = 400$ MPa) reinforcing bars were also carried out in accordance with ASTM A370-14 (2014). The average tensile stresses of longitudinal (D29) and shear (D10) reinforcement were 617.7 and 537.5 MPa, respectively.

5. Test Results

5.1 Damage and Crack Patterns
The amount of shear reinforcement greatly affected the damage and crack patterns for UHPFRC rectangular cross-sectional beams ($V_f = 1.5$ %). Figure 5 shows damage and

Table 3 Mix proportion (weight ratio).

Water-binder ratio	Cement	Zirconium	Filler	Fine aggregate	Water-reducing admixture
0.2	1.0	0.25	0.3	1.1	0.02

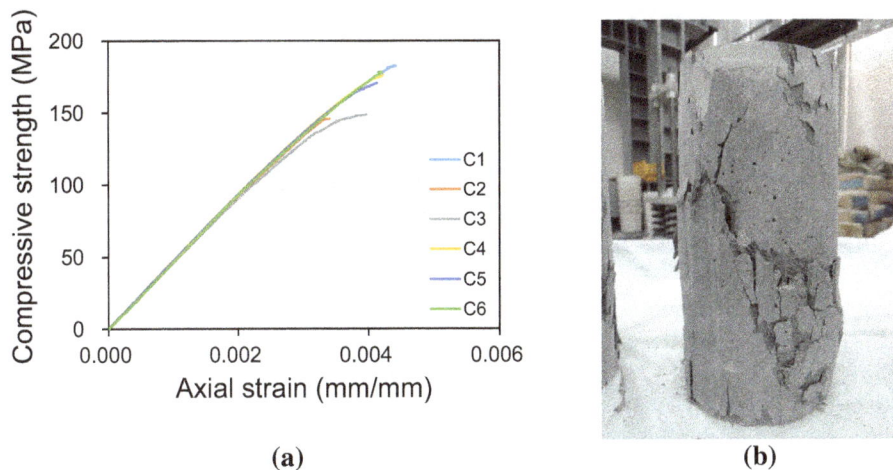

Fig. 3 Uniaxial compression test for UHPFRC. **a** Stress–strain curves, **b** failure mode.

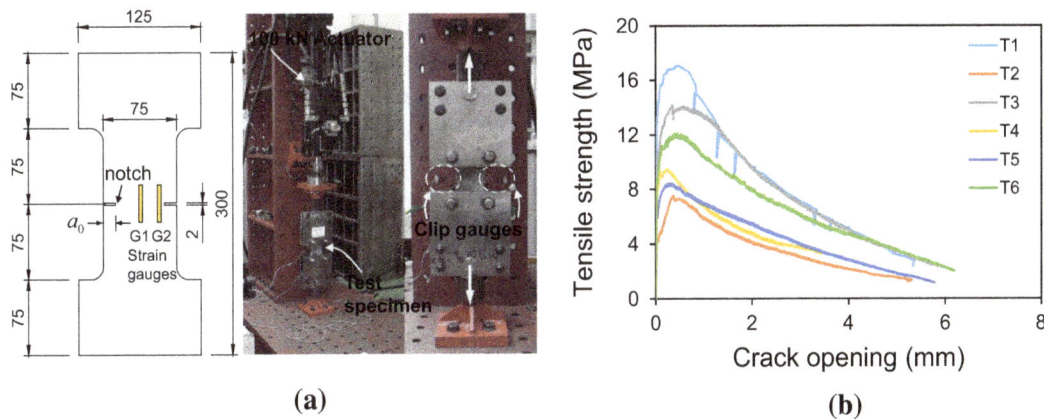

Fig. 4 Direct tensile test for UHPFRC. **a** Dog-bone shaped specimen, **b** stress–strain curves.

crack patters at the end of the tests. For control specimen SB1 which does not contain the shear reinforcement, flexural cracks initiated at the bottom of beam at the mid-span, and then the diagonal cracks occurred at the end of the flexural cracks. Finally the diagonal tension failure occurred after the yielding of longitudinal reinforcing bars. In this specimen,

compression failure at the compression zone was also observed with shear cracks.

In case of SB2 specimen, a diagonal tension failure as well as the compression failure of concrete occurred and shear reinforcement yielded prior to the yielding of longitudinal reinforcement. In this specimen, the compression failure and the yielding of longitudinal reinforcement occurred almost simultaneously. The specimen SB3 adopted the minimum shear reinforcement ($s = 0.5d$) in accordance with ACI 318-14 (2014) showed a compression failure of concrete at the compression zone occurred prior to shear failure. The inclined shear cracks were developed subsequently after the flexural yielding of longitudinal reinforcing bars. For SB4 specimen installing the shear reinforcement at the spacing of $0.3d$, flexural failure occurred without observation of critical shear cracks due to the excessive amount of shear reinforcement. After the compression failure of concrete, the yielding of longitudinal and shear reinforcement was followed. Test results indicated that if the minimum shear reinforcement is installed at a spacing of $0.5d$ presented in ACI 318-14 (2014), the flexural failure may occur prior to shear failure. On the other hand, for beams with the spacing

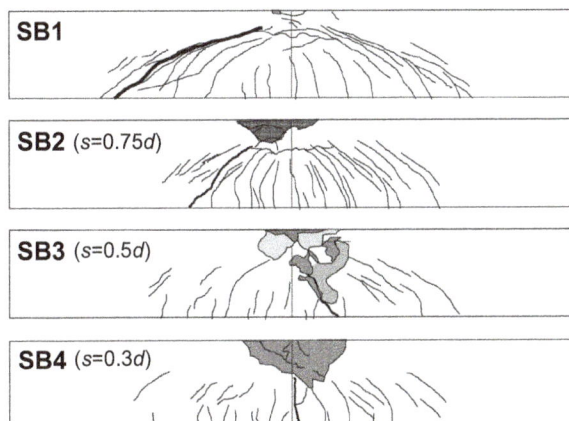

Fig. 5 Damage and crack patterns at the end of the test.

which is greater than minimum values in current design codes, the yielding of shear reinforcement might be observed prior to the yielding of flexural reinforcement and compression failure.

5.2 Load–Displacement Relationship

Figure 6a shows the load–displacement relationship of test specimens. Here, the displacement is a deflection measured at the mid-span of the beam. Figure 6b depicts definition of yielding point and ductility. The value of V_y is the yield strength, V_{peak} is the peak shear strength, $V_{failure}$ is the strength at failure, Δ_y, $\Delta_{@Vpeak}$ and $\Delta_{failure}$ are the displacement corresponding to the strength of V_y, V_{peak} and $V_{failure}$, respectively. The displacement at failure is defined as the deflection when the load dropped to 80 % of the peak load. The secant stiffness at a point of two-third of the measured peak strength is used to idealize the elastoplastic curve that passes through the peak point of the load–displacement curve, and the displacement at an intersecting point between the two lines is used to determine the yield point on the curve (Pan and Moehle 1989). The ultimate shear strengths of the UHPFRC beams are reported in Table 4 in terms of the average shear stress which is defined as the peak shear force divided by the beam width and effective depth ($v_u = V_u/b_w d$). The ductility (μ) is defined as the ability of the structure or parts of it to sustain large deformations beyond the yield point, which is obtained in terms of displacements, as the maximum displacement ($V_{failure}$) divided with the yield displacement (Δ_y).

As shown in Fig. 6, the peak load of the beams with shear reinforcement was greater than the beams without shear reinforcement. However, initial stiffness was very similar regardless of the presence of shear reinforcement and their spacing. For the control specimen (SB1), non-linear behavior showed after reaching the yielding point due to the yielding of longitudinal reinforcing bars and flexural cracks. Eventually the load suddenly dropped due to the diagonal tension failure after reaching the peak load. In case of the specimen SB2, SB3, and SB4, the strength was maintained almost being constantly at the peak strength, and then the strength dropped abruptly due to the compression failure at the compression zone without critical shear cracks even though several inclined cracks occurred. Unlike the control specimen, the strength gradually decreased due to the shear reinforcement after the compression failure of concrete. However, the peak strength of the beams with shear reinforcement was very similar. These results indicated that the shear reinforcement ratio might not influence on the peak strength of UHPFRC beams with shear reinforcement.

Shear reinforcement also had an effect on improvement of deformation capacity. Ductility (μ) of beams with shear reinforcement also appeared to be somewhat higher than the control specimen. The ductility of the control specimen was 2.04 and in case of the specimens with shear reinforcement (SB2, SB3, and SB4) were between 2.15 and 2.23.

5.3 Strain Response

Figure 7 shows the strain response of the test specimens. To measure the strains, strain gauges were used. Flexural yielding and shear yielding are defined as the point when the strain of the reinforcing bars reaches a yield strain (=0.002). Also, a concrete failure is defined as a failure at the compression zone of the beam, that is, the compressive strain at the extreme fiber of the beam reaches an ultimate limit state of the UHPFRC. To define the ultimate state of the UHPFRC, the ultimate strain determined using material tests was used. Material tests showed that the ultimate strain of UHPFRC was between 0.003 and 0.0032. The specimen SB1 shows the diagonal tension failure after flexural yielding. In this specimen, the flexural yielding occurred prior to concrete failure. For SB2 specimen installed shear reinforcement at the spacing of $0.75d$, shear reinforcement yielded before flexural yielding and concrete failure. In case of the SB3 specimen, a flexural-shear failure occurred. The strain of shear reinforcement reaches the yield strain after flexural yielding and concrete failure. On the other hand, the SB4 specimen which is over-reinforced beam shows that shear yielding occurred before the flexural yielding, but after concrete compressive failure. As shown in Fig. 5, the ultimate failure mode was the compressive concrete failure. In this study, a critical shear crack was not observed during the tests.

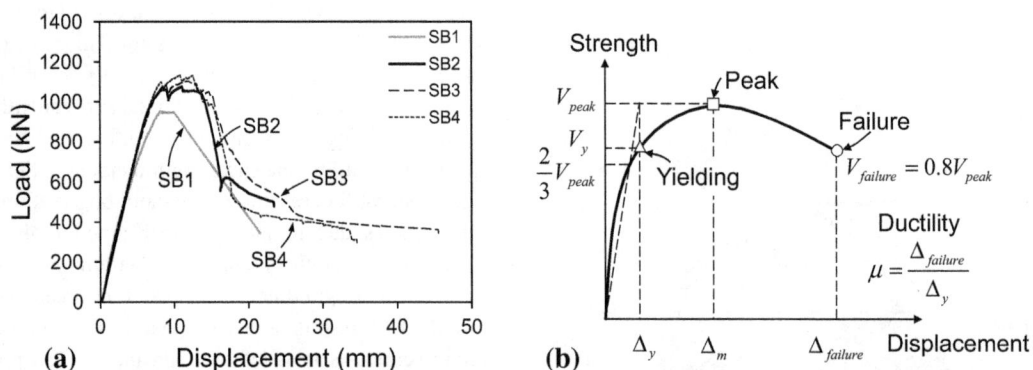

Fig. 6 Load–displacement relationship and definition of yielding point. **a** Load–displacement relationship, **b** definition of yielding point and ductility.

Table 4 Summary of test results.

Specimens	Failure mode	At initial cracking		At yielding		At peak		At failure		V_{test} (MPa)	$\dfrac{v_{test}}{\sqrt{f_{cf}}}$ (MPa)	μ ($\Delta_{failure}/\Delta_y$)
		Δ_{cr} (mm)	V_{cr} (kN)	Δ_y (mm)	V_y (kN)	$\Delta_{@Vpeak}$ (mm)	V_{peak} (kN)	$\Delta_{failure}$ (mm)	$V_{failure}$ (kN)			
SB1	S	2.1	339.7	6.7	347.8	8.2	475.8	26.4	172.0	14.4	1.12	2.04
SB2	SY	1.1	150.2	7.1	479.1	11.1	537.3	15.3	408.9	16.3	1.26	2.15
SB3	C	3.6	555.6	7.3	359.8	11.8	551.7	16.3	441.0	16.7	1.29	2.23
SB4	F	1.2	190.5	7.3	296.1	10.9	567.0	16.0	436.1	17.2	1.33	2.19

V_{cr} is the initial cracking strength; V_y, V_{peak}, and $V_{failure}$ are the yield strength, peak strength, shear strength at failure, respectively. Δ_{cr}, Δ_y, $\Delta_{@Vpeak}$, and $\Delta_{failure}$ are the measured displacement corresponding to the strength of V_{cr}, V_y, V_{peak}, and $V_{failure}$ at the mid-span of the test specimen, respectively. μ is the ductility obtained by the equations of $\Delta_{failure}/\Delta_y$. It should be note that S means the diagonal tension failure; SY is the shear yielding; C is the compression failure of concrete; F is the flexural yielding.

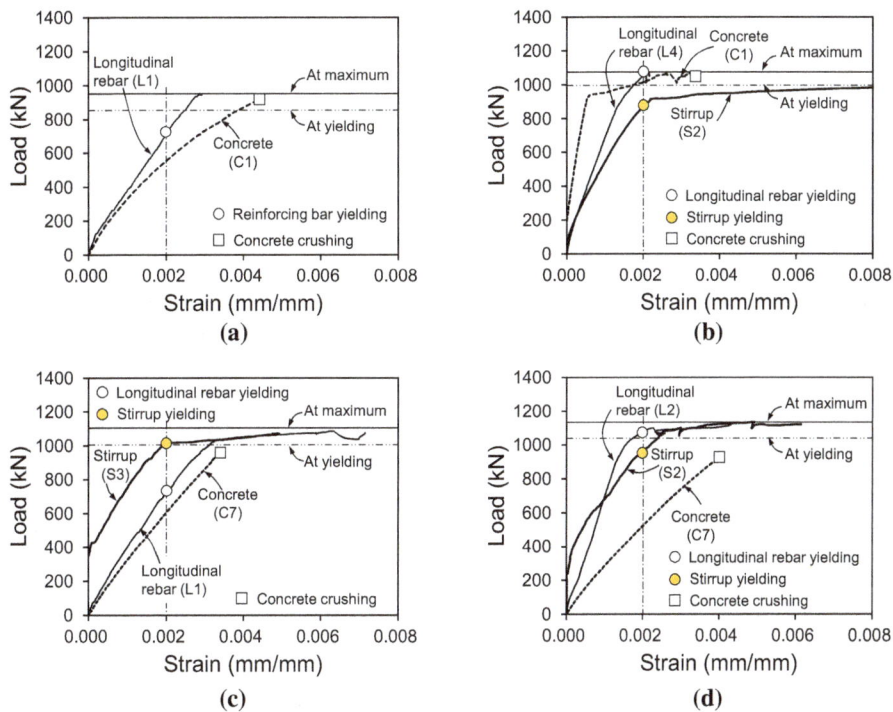

Fig. 7 Measured strain values of concrete, longitudinal and shear reinforcement. **a** SB1, **b** SB2 ($s = 0.75d$), **c** SB3 ($s = 0.5d$), **d** SB4 ($s = 0.3d$).

6. Discussion of Test Results

6.1 Effect of Shear Reinforcement on Shear Strength

The ultimate shear strength of the UHPFRC beams was dependent on the presence of shear reinforcement. The shear strength of the beams with shear reinforcement was larger than that of control specimen and was improved about 13–19 %. However, the effect of the amount of shear reinforcement was insignificant. Although the area of shear reinforcement increases about 55.6 % with respect to the current design codes, the increase of shear strength was only about 2.6 % (in case of SB3 and SB4). In addition, provided that the amount of shear reinforcement decreases about 37.5 % regarding the minimum shear reinforcement, deterioration of the shear strength was in about 2.6 %. These

results indicated that the steel fibers substantially contribute to enhancement of the shear resistance of UHPFRC beams. The shear contributions of UHPFRC are reported in Table 5, where the shear resistances (V_c, V_{fb}, and V_s) are obtained using AFGC recommendations. As shown in Table 5, current design guidelines had some conservatism to the test results. Among the component of shear resistance for the beams, shear strength provided by steel fibers was determined to be the greatest value regardless of shear reinforcement. Especially, in case of the control specimen, the shear strength was more largely affected by the shear contribution of steel fibers ($V_{test}/V_{fb} = 1.76$) than cement matrix ($V_{test}/V_c = 6.2$). On the other hand, in case of the specimens with shear reinforcement, the shear resistances were affected by shear contributions by steel fibers and shear reinforcement. As increasing the amount of shear reinforcement,

Table 5 Shear contributions of UHPFRC.

Specimens	s (mm)	V_c (kN)	V_{fb} (kN)	V_s (kN)	V_{test} (kN)	$\dfrac{V_{test}}{V_c}$	$\dfrac{V_{test}}{V_{fb}}$	$\dfrac{V_{test}}{V_s}$	$\dfrac{V_{test}}{V_c+V_{fb}}$	$\dfrac{V_{test}}{V_{fb}+V_s}$	$\dfrac{V_{test}}{V_c+V_{fb}+V_s}$
SB1	–	76.7	270.9	–	475.8	6.20	1.76	–	1.37	1.76	1.37
SB2	$0.75d$	76.7	270.9	102.2	537.3	7.01	1.98	5.26	1.55	1.44	1.19
SB3	$0.5d$	76.7	270.9	153.4	551.7	7.19	2.04	3.60	1.59	1.30	1.10
SB4	$0.3d$	76.7	270.9	255.6	567.0	7.39	2.09	2.22	1.63	1.08	0.94

s is the spacing of shear reinforcement; d is the effective depth; V_c, V_{fb}, and V_s are shear strength provided by cement matrices, steel fiber and shear reinforcement obtained in accordance with AFGC design guidelines (2013); and V_{test} is the peak shear force determined from the tests.

shear contribution provided by shear reinforcement decreased. However, if the spacing of shear reinforcement is $0.75d$ ($\rho_v = 0.9$ %), the effect of shear resistance provided by shear reinforcement decreased while the shear contribution by steel fibers increased. If the shear reinforcement provides in about 1.4 % ($s = 0.3d$), the shear contributions by steel fibers and shear reinforcement was very similar ($V_{test}/V_{fb} = 2.09$ and $V_{test}/V_s = 2.22$) even though the shear resistance is larger than test results. On the contrary to this, in case of the UHPFRC beam with a shear reinforcement ratio of 0.6 %, the effect of the shear reinforcement was less significant than other reinforced beams.

From these results, it is found that the steel fibers irregularly distributed on the diagonal cracked section play a key role to restrain the shear crack along with the shear reinforcement.

6.2 Evaluation of Shear Strength

The shear strength predictions of FRC beams were evaluated as to whether or not they are applicable to UHPFRC beams. For comparisons, the existing shear strength models for SFRC beams proposed by Sharma (1986), Narayanan et al. (1987), Ashour (1992), ACI 544 (1997), Kwak et al. (2002) were used. They are summarized in Table 6.

Sharma (1986) investigated the effect of steel fibers on shear strength performing seven SFRC beams with a compressive strength of about 45 MPa. From their shear tests, it is found that steel fibers are effective in increasing the shear strength and SFRC beams have a high post-cracking strength. Narayanan and Darwish (1987) carried out shear tests for forty-nine SFRC rectangular cross-sectional beams with a compressive strength of 40–79.5 MPa regarding shear span to depth ratio (a/d), longitudinal and shear reinforcement, presence of shear reinforcement, and the fiber factor ($F = (L/D)\rho_f d_f$). Based on the observations of first cracks in shear, empirical shear strength equation was suggested for the evaluation of cracking shear strength. Ashour et al. (1992) tested eighteen HSFRC beams ($f_c' = 93$ MPa) with or without shear reinforcement. Test variables were shear span-to-depth (a/d), longitudinal reinforcement ratio, fiber volume fraction. They found that shear strength of beams increase with an increase of fiber volume fraction and a decrease in a/d. On the basis of test results, predictions of shear strength for high-strength SFRC beams without shear reinforcement. ACI 544 (1997) adopted the shear strength equations proposed by Sharma (1986) based on the test results. The proposed equations follows the method of ACI 318 for calculating the contribution of stirrups to the shear

Table 6 Existing shear strength models.

Authors	Shear strength models
Sharma (1986)	$v_u = kf_t'(d/a)^{0.25}$
	where $k = 2/3$; a/d is the shear span-to-depth ratio; $f_t' = 0.17\sqrt{f_{cf}}$, if the tensile strength is unknown, and f_{cf} is the concrete cylinder compressive strength
Narayanan et al. (1987)	$v_u = e\left[0.24f_{spfc} + 80\rho\frac{d}{a}\right] + v_b$
	where f_{spfc} is the computed split-cylinder strength of fiber concrete ($= f_{cuf}/(20 - \sqrt{F}) + 0.7 + 1.0\sqrt{F}$); ρ is the longitudinal reinforcement ratio; F is the fiber factor ($=(L_f/D_f)V_f d_f$; e is the arch action factor, 1.0 for $a/d > 2.8$ and $2.8d/a$ for $a/d \leq 2.8$; f_{cuf} is the cube strength of fiber concrete; V_f is the fiber volume fraction; d_f is a bond factor, 0.5 for round fibers, 0.75 for crimped fibers, and 1.0 for indented fibers; v_b is equal to the equations of $0.41\tau F$, and τ is the average fiber matrix interfacial bond stress, taken as 4.15 MPa
Ashour et al. (1992)	For $a/d \geq 2.5$ $v_u = \left(2.11\sqrt[3]{f_{cf}} + 7F\right)\left(\rho\frac{d}{a}\right)^{1/3}$
Kwak et al. (2002)	$v_u = 3.7ef_{spfc}^{2/3}\left(\rho\frac{d}{a}\right)^{1/3} + 0.8v_b$
	where e is the arch action factor, 1 for $a/d > 3.4$, and $3.4d/a$ for $a/d \leq 3.4$

capacity, to which is added the resisting force of the concrete calculated from the shear stress. Kwak et al. (2002) performed twelve four-point shear tests for normal— (30.8 MPa) and high-strength (68. 6 MPa) SFRC beams without shear reinforcement considering fiber volume fraction (V_f = 0, 0.5, 0.75 %) and shear span to depth ratio (a/d = 2, 3, and 4). Shear strength equations for shear cracking was proposed to improve the accuracy of existing procedures suggested by Narayanan and Darwish (1987).

As shown in Table 7, the existing shear strength equations for SFRC beams were very conservative compared to the experimental data. This means that they would be unreasonable to predict the shear strength of the UHPFRC beams with a compressive strength more than 160 MPa. On the other hand, AFGC recommendations (2013) showed a relatively accurate evaluations of UHPFRC beams with and without shear reinforcement.

6.3 Steel Fibers as Shear Reinforcement

According to ACI 318-14 (2014), when the normalized shear strength ($v_{test}/\sqrt{f_c'}$) defined as divided the average shear stress by the square root of the compressive strength is greater than $0.29\sqrt{f_c'}$ (MPa), the steel fibers can use as the shear reinforcement for SFRC beam ($f_c' \leq 40$ MPa, $d \leq 600$ mm). Parra-Montesinos found that the shear strength of SFRC beam strength was larger than $0.3\sqrt{f_c'}$ (MPa) when fiber content (V_f) is equal to or greater than 0.75 %.

Normalized shear strengths were evaluated whether or not current design codes are applicable to UHPFRC beams with shear reinforcement. As reported in Table 4, normalized shear strengths of all the specimens with V_f = 1.5 % were larger than $1.12\sqrt{f_c'}$ (MPa) regardless of the presence of shear reinforcement and its spacing as shown in Fig. 8.

These results indicate that if the rectangular beam contains UHPFRC with fiber volume fraction of 1.5 %, shear reinforcement need not be provided.

6.4 Spacing Limit of Shear Reinforcement for UHPFRC Beam

As aforementioned, current design codes for reinforced concrete beam provide the spacing limit of shear reinforcement as $0.5d$ in ACI 318-14 (2014) when the factored shear force V_u exceeds $0.5\phi V_c$. Also, CSA A23.3-04 (2004) suggests its distance as $0.7d_v$, where d_v is a maximum value between $0.9d$ and

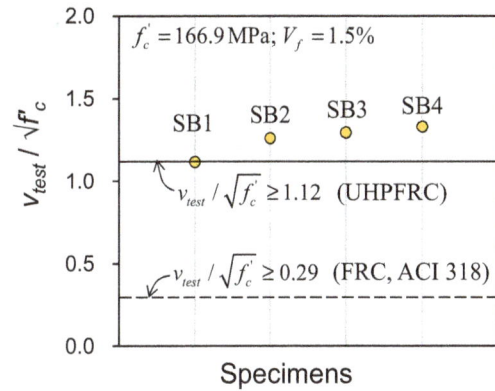

Fig. 8 Lower bound of normalized shear strength for UHPFRC.

$0.72h$. To investigate the effect of spacing limit, this study considered the distance of $0.75d$, $0.5d$, and $0.3d$.

Test results showed that even though the spacing of shear reinforcement exceeds the spacing limit recommended by ACI 318-14 (2014), shear strength of UHPFRC beam was substantially greater than current design codes. Based on the test results, it is concluded that the spacing limit of $0.75d$ can be allowed for UHPFRC beams.

7. Summary and Conclusions

In this study, shear tests on simply supported UHPFRC rectangular beam sections with and without shear reinforcement were carried out to investigate the shear behaviour considering the spacing of shear reinforcement. The main test parameter was the spacing of shear reinforcement. Findings obtained through the experiments are as follows:

1. Compression and direct tension tests were carried out to investigate the material properties of UHFRC. The UHPFRC used in this study showed a linear-elastic behavior until the end of the test and a brittle failure occurred after reaching the peak strength, not observing a post-peak behavior in all of the test specimens. The average compressive strength was 166.9 MPa and the modulus of elasticity was about 41.1 GPa. Also, tensile strength of UHPFRC obtained using direct tension tests was determined to be about 11.5 MPa.

Table 7 Comparison between the predicted strength and test data.

Specimens	Sharma (1986)	Narayanan and Darwish (1987)	Ashour et al. (1992)	Kwak et al. (2002)	AFGC (2013)
SB1	2.79	3.71	6.30	3.12	1.37
SB2	2.09	2.56	3.58	2.27	1.19
SB3	2.28	2.86	4.18	2.49	1.10
SB4	2.46	3.15	4.83	2.71	0.94
Mean	2.41	3.07	4.72	2.65	1.15
SD	0.30	0.49	1.17	0.36	0.08

2. The steel fibers substantially contributes to enhancement of the shear resistance of UHPFRC beams. The shear strength of the beams with shear reinforcement was larger than that of control specimen and was improved about 13–19 %. In addition, the steel fibers in UHPFRC beam play a key role to restrain the shear crack along with the shear reinforcement.

3. Shear reinforcement also had an effect on improvement of deformation capacity. The ductility of beams with shear reinforcement also appeared to be higher than the control specimen. The ductility of the control specimen was 2.04 and in case of the specimens with shear reinforcement (SB2, SB3, and SB4) were between 2.15 and 2.23.

4. The AFGC recommendations (2013) showed a relatively accurate evaluations of UHPFRC beams with and without shear reinforcement compared to the existing shear strength equations for SFRC beams.

5. Even though the spacing of shear reinforcement exceeds the spacing limit suggested by current design code (ACI 318-14), shear strength of UHPFRC beam was substantially greater than current design codes. Therefore, the spacing limit of $0.75d$ can be allowed for UHPFRC beams.

Acknowledgments

This research was supported by a grant (13SCIPA02) from Smart Civil Infrastructure Research Program funded by Ministry of Land, Infrastructure and Transport (MOLIT) of Korean government and Korea Agency for Infrastructure Technology Advancement (KAIA).

References

ACI Committee 318. (2014). *Building Code Requirements for Structural Concrete (ACI 318M-14) and Commentary (318R-14)*. Farmington Hills, MI: American Concrete Institute.

ACI Committee 544. (1988). Design considerations for steel fiber reinforced concrete. *ACI Structural Journal, 85*(5), 1–18.

Alaee, F. J., & Karihaloo, B. L. (2003). Retrofitting of reinforced-concrete beams with CARDIFRC. *Journal of Composites for Construction ASCE, 7*(3), 174–186.

American Association of State Highway and Transportation Officials. (2004). *AASHTO LRFD Bridge Design Specification* (3rd ed.). Washington, DC: AASHTO.

Ashour, S. A., Hasanain, G. S., & Wafa, F. F. (1992). Shear behavior of high-strength fiber reinforced concrete beams. *ACI Structural Journal, 89*(2), 176–184.

Association Française du Génil Civil (AFGC). (2013). *Bétons fibrés àultra-hautes performances*, Association Française du Génil Civil.

ASTM A370-14. (2014). *Standard test methods and definitions for mechanical testing of steel products*. West Conshohocken, PA: ASTM International.

ASTM C39/C39M-05. (2005). *Standard test method for compressive strength of cylindrical concrete specimens*. West Conshohocken, PA: ASTM International.

Baby, F., Marchand, P., & Toutlemonde, F. (2014). Shear behavior of ultrahigh performance fiber-reinforced concrete beams. I: Experimental investigation. *Journal of Structural Engineering ASCE, 140*(5), 04013111.

CSA A23.3-04. (2004). *Design of concrete structures*. Rexdale, ON: Canadian Standard Association.

EN 1992-1-1. (2004). *Eurocode 2: Design of Concrete Structures—Part 1-1: General Rules and Rules for Buildings*. British Standards Institution.

Ezeldin, S., & Balaguru, P. N. (1997). Normal- and high-strength fiber-reinforced concrete under compression. *ACI Material Journal, 94*(4), 286–290.

Fanella, D. A., & Naaman, A. E. (1985). Stress–strain properties of fiber reinforced mortar in compression. *ACI Journal, 82*(4), 475–483.

Farhat, F. A., Nicolaides, D., Kanellopoulos, A., & Karihaloo, B. L. (2007). High performance fiber-reinforced cementitious composite (CARDIFRC)—performance and application to retrofitting. *Engineering Fracture Mechanics, 74*, 151–167.

Japan Society of Civil Engineers (JSCE). (2004). *Recommendations for Design and Construction of Ultra-High Strength Fiber Reinforced Concrete Structures*, draft.

Khuntia, M., Stojadinovic, B., & Goel, S. C. (1999). Shear strength of normal and high-strength fiber reinforced concrete beams without stirrups. *ACI Structural Journal, 96*(2), 282–289.

Korea Concrete Institute. (2012). *Design recommendations for ultra-high performance concrete (K-UHPC)*, KCI-M-12-003, Korea (in Korean).

Kwak, Y. K., Eberhard, M. O., Kim, W. S., & Kim, J. B. (2002). Shear strength of steel fiber-reinforced concrete beams without stirrups. *ACI Structural Journal, 99*(4), 530–538.

Li, V. C., Ward, R., & Hamza, A. M. (1992). Steel and synthetic fibers as shear reinforcement. *ACI Materials Journal, 89*(5), 499–508.

Mansur, M. A., Ong, K. C. G., & Paramsivam, P. (1986). Shear strength of fibrous concrete beams without stirrups. *Journal of Structural Engineering ASCE, 112*(9), 2066–2079.

MC2010. (2012). *fib Model Code for Concrete Structures 2010*, fédération internationale du béton, Lausanne, Switzerland: Ernst & Sohn.

Meda, A., Mostosi, S., & Riva, P. (2014). Sehar strengthening of reinforced concrete beam with high-performance fiber-reinforced cementitious composite jacketing. *ACI Structural Journal, 111*(5), 1059–1067.

Narayanan, R., & Darwish, I. Y. S. (1987). Use of steel fibers as shear reinforcement. *ACI Structural Journal, 84*(3), 216–227.

Noghabai, K. (2000). Beams of fibrous concrete in shear and bending: Experiment and model. *Journal of Structural Engineering ASCE, 126*(2), 243–251.

Pan, A., & Moehle, J. P. (1989). Lateral displacement ductility of reinforced concrete flat plates. *ACI Structural Journal, 86*(3), 250–258.

Park, S. H., Kim, D. J., Ryu, G. S., & Koh, K. T. (2012). Tensile Behavior of ultra high performance hybrid fiber reinforced concrete. *Cement and Concrete Composites, 34*, 172–184.

Parra-Montesinos, G. J. (2006). Shear strength of beams with deformed steel fibers. *Concrete International, 28*(11), 57–66.

Rossi, P., Arca, A., Parant, E., & Fakhri, P. (2005). Bending and compressive behaviors of a new cement composite. *Cement and Concrete Research, 35*, 27–33.

Sharma, A. K. (1986). Shear strength of steel fiber reinforced concrete beams. *ACI Journal, 83*(4), 624–628.

Voo, Y. L., Poon, W. K., & Foster, S. J. (2010). Sheear strength of steel fiber-reinforced ultrahigh-performance concrete beams without stirrups. *Journal of Structural Engineering ASCE, 136*(11), 1393–1400.

Wafa, F. F., & Ashour, S. A. (1992). Mechanical properties of high-strength fiber reinforced concrete. *ACI Materials Journal, 89*(5), 440–455.

Wille, K., Kim, D. J., & Naaman, A. E. (2011a). Strain hardening UHP-FRC with low fiber contents. *Materials and Structures, 44*, 583–598.

Wille, K., Naaman, A. E., & Parra-Montesinos, G. J. (2011b). Ultra-high performance concrete with compressive strength exceeding 150 MPa (22 ksi): A simpler way. *ACI Materials Journal, 108*(6), 46–54.

14

A Review on Structural Behavior, Design, and Application of Ultra-High-Performance Fiber-Reinforced Concrete

Doo-Yeol Yoo[1], and Young-Soo Yoon[2],*

Abstract: An overall review of the structural behaviors of ultra-high-performance fiber-reinforced concrete (UHPFRC) elements subjected to various loading conditions needs to be conducted to prevent duplicate research and to promote its practical applications. Thus, in this study, the behavior of various UHPFRC structures under different loading conditions, such as flexure, shear, torsion, and high-rate loads (impacts and blasts), were synthetically reviewed. In addition, the bond performance between UHPFRC and reinforcements, which is fundamental information for the structural performance of reinforced concrete structures, was investigated. The most widely used international recommendations for structural design with UHPFRC throughout the world (AFGC-SETRA and JSCE) were specifically introduced in terms of material models and flexural and shear design. Lastly, examples of practical applications of UHPFRC for both architectural and civil structures were examined.

Keywords: ultra-high-performance fiber-reinforced concrete, bond performance, structural behavior, design code, application.

1. Introduction

Ultra-high-performance fiber-reinforced concrete (UHPFRC), which was developed in the mid-1990s, has attracted much attention from researchers and engineers for practical applications in architectural and civil structures, because of its excellent mechanical performance, i.e., compressive strength is greater than 150 MPa and a design value of tensile strength is 8 MPa (AFGC-SETRA 2002), durability, energy absorption capacity, and fatigue performance (Farhat et al. 2007; Graybeal and Tanesi 2007; Yoo et al. 2014c; Li and Liu 2016). In particular, its very high strength properties result in a significant decrease in the structural weight, i.e., the weight of UHPFRC structures is about 1/3 (or 1/2) of the weight of general reinforced concrete (RC) structures at identical external loads (Tam et al. 2012). Consequently, slender structures, which are applicable for long-span bridges, can be fabricated with UHPFRC, leading to low overall construction costs.

In order to apply such a newly developed innovative material to real structures, numerous studies have been carried out in many countries in Europe, North America, and Asia. Since UHPFRC was first developed by France's research group, the first technical recommendation on UHPFRC for both material properties and structural design was introduced in France in 2002 and was called the AFGC-SETRA recommendation (AFGC-SETRA 2002). Thereafter, a state-of-the-art report on UHPFRC covering all material and design aspects was published in Germany in 2003 (DAfStB 2003). Then, in 2004, the Japan Society of Civil Engineers (JSCE) published their own design recommendations for UHPFRC based on Ductal® (Orange et al. 1999), commercial UHPFRC available in the world, (JSCE 2004). Lastly, in recent years, the Korea Concrete Institute (KCI) also developed a design code for UHPFRC (KCI 2012), similar to those in France and Japan, by using K-UHPC, another UHPFRC material developed by Korea Institute of Civil Engineering and Building Technology (Kim et al. 2008).

Due to the superb fiber bridging capacities of UHPFRC at cracked surfaces, leading to a special strain-hardening (or deflection-hardening) response with multiple micro-cracks, many researchers have focused on using it to the structures dominated by flexure, shear, and torsion. Furthermore, UHPFRC has also been considered as one of the promising materials for impact- and blast-resistant structures, because of its enhanced strength and energy absorption capacity, along with strain-hardening cementitious composites containing polymeric fibers (Astarlioglu and Krauthammer 2014; Choi et al. 2014). These properties can help to overcome the brittle failure of plain concrete, which has poor energy absorption capacity for impacts and blasts. Since the structural behaviors of UHPFRC under flexure, shear, and torsion, and when subjected to high-rate loadings, such as impacts and blasts, are highly sensitive to numerous factors,

[1]Department of Architectural Engineering, Hanyang University, Seoul 133-791, Korea.
[2]School of Civil, Environmental and Architectural Engineering, Korea University, Seoul 136-713, Korea.
*Corresponding Author; E-mail: ysyoon@korea.ac.kr

i.e., structural shape, loading condition, strain-rate, casting method, reinforcement ratio, etc., it is necessary to synthetically review the scattered studies.

The purpose of this research is to analyze the current state of knowledge of the structural behavior, design techniques, and applications of UHPFRC under various loading conditions. As explained above, the attention of this paper is focused on (1) bond performance between UHPFRC and various reinforcements, which is basic information needed for the design of reinforced structures, (2) structural behavior of UHPFRC under flexure, shear, torsion, and high-rate loading, (3) the most widely used UHPFRC design recommendations in the world, and (4) examples of practical applications in both architectural and civil structures.

2. Historical Development of UHPFRC

Roy et al. (1972) and Yudenfreund et al. (1972) first introduced ultra-high-strength cementitious paste with low porosity in the early 1970s. With special curing methods using heat (250 °C) and pressure (50 MPa), Roy et al. (1972) achieved a cementitious paste with almost zero porosity and a compressive strength of approximately 510 MPa. On the other hand, Yudenfreund et al. (1972) obtained a cement paste having a compressive strength of about 240 MPa with normal curing (25 °C) for 180 days. To do this, Yudenfreund et al. (1972) provided a special treatment on-ground clinker, used the low water-to-cement ratio of 0.2, and Blaine surface areas ranging from 6000 to 9000 cm^2/g. After nearly 10 years, Birchall et al. (1981) and Bache (1981) could develop two types of ultra-high-strength paste (or concrete) with very low porosity, such as densified with small particles (DSPs) concrete and macro-defect free (MDF) paste, by developing a pozzolanic admixture and a high-range water-reducing agent. Birchall et al. (1981) achieved the development of the cement pastes with compressive strength over 200 MPa and flexural strengths of 60–70 MPa, by removing macroscopic flaws during material preparation without using fibers or high-pressure compaction. Bache (1981) also successfully developed the concrete that was DSPs and had a compressive strength of 120–270 MPa. The key technique to densely pack the spaces between the cement particles was to use ultra-fine particles and an extremely low water content, with a large quantity of high-range water-reducing agent. In the mid-1990s, Richard and Cheyrezy (1995) first introduced the concept of and mixing sequence for reactive powder concrete (RPC), which was the forerunner of UHPFRC. To obtain a very high strength, the granular size was optimized by the packing density theory, by excluding coarse aggregate and by providing heat (90 and 400 °C) and pressure treatments. In addition, 1.5–3 % (by volume) of straight steel microfibers, with a diameter of 0.15 mm and a length of 13 mm, were added to achieve high ductility; consequently, the RPC developed by Richard and Cheyrezy (1995) showed compressive strengths of 200–800 MPa and fracture energies up to 40 kJ/m^2.

3. Performance of Structural UHPFRC Elements

3.1 Bond Behavior Between UHPFRC and Reinforcements

In order to practically apply a newly developed UHPFRC in the structures, bond performance with reinforcements should be examined. Many researchers (Jungwirth and Muttoni 2004; Ahmad Firas et al. 2011; Yoo et al. 2014a, b, 2015a) have investigated the bond behavior of internal steel and fiber-reinforced polymer (FRP) reinforcements with UHPFRC. Jungwirth and Muttoni (2004) carried out pullout test of deformed steel reinforcing bar using a 160 mm cube. Various bond lengths ranging from 20 to 50 mm and two different bar diameters of 12 and 20 mm were adopted. In their study, the average bond strength of steel bars embedded in UHPFRC was found to be 59 MPa, approximately 10 times higher than the bond strength of steel bars embedded in ordinary concrete, and the theoretical development length of deformed steel bars in UHPFRC was suggested by $l_b = f_y d_b / 4\tau_{max}$, where f_y is the yield strength of steel bar, d_b is the nominal diameter of steel bar, and τ_{max} is the bond strength. Yoo et al. (2014c) examined the effects of fiber content and embedment length on the bond behavior of deformed steel bars embedded in UHPFRC. For this, Yoo et al. (2014c) performed a number of pullout tests by modifying the test method, proposed by RILEM recommendations (RILEM 1994); the 150 mm cubic specimens with a single bar embedded vertically along the central axis were fabricated and used for testing. The embedment lengths were determined by 1 and 2 times the bar diameter, instead of using $5d_b$, as suggested by the RILEM recommendations. The bond strength was insignificantly affected by the fiber content and embedment length, but it clearly correlated with the compressive strength. The CEB-FIP Model Code (MC90) (CEB-FIP 1993), which defined τ_{max} as $2.0 f_c'^{0.5}$, substantially underestimated the bond strength of steel bars in UHPFRC because the parameters were suggested based on test data from previous concretes. Thus, Yoo et al. (2014c) proposed modified coefficients for the bond strength of steel bars in UHPFRC, based on a number of test data, as follows (Fig. 1):

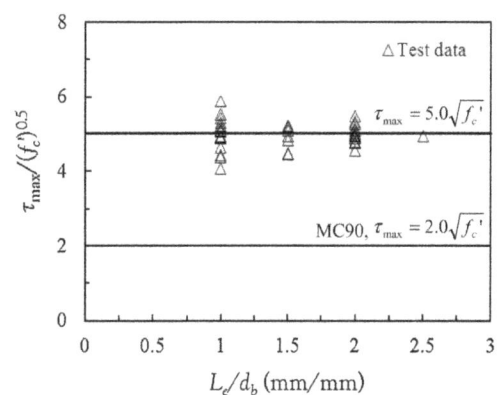

Fig. 1 Normalized bond strengths of deformed steel bars embedded in UHPFRC.

$$\tau_{\max} = 5.0\sqrt{f_c'} \tag{1}$$

In addition, CMR model (Cosenza et al. 1995), which sets $\tau = \tau_{\max} \times (1 - e^{-s/sr})^{\beta}$, was found to be appropriate for simulating the ascending bond stress versus slip behavior of steel bars embedded in UHPFRC, and the parameters were proposed as $\tau_{\max} = 5.0f_c'^{0.5}$, $s_r = 0.07$, and $\beta = 0.8$, where s_r and β are coefficients based on the curve fitting of test data.

Ahmad Firas et al. (2011) experimentally investigated the bond performance between carbon-fiber-reinforced polymer (CFRP) bars and UHPFRC according to the surface treatment, embedment length, bar diameter, and concrete age. Based on the test data, it was noted that the bond strength was insignificantly affected by the surface treatment of the glass-fiber-reinforced polymer (GFRP) bar; similar bond strengths for smooth bars and sand-coated bars were obtained. On the other hand, a decrease in bond strength was obtained by increasing both bar diameter and embedment length. The ultimate bond strength of CFRP bars in UHPFRC was insignificantly changed by age after 3 days, because it was primarily affected by the shear strength of the connection between the core and the outer layer of the CFRP bars. Ahmad Firas et al. (2011) suggested a development length for sand-coated bars of approximately $40d_b$, and a development length for a smooth bar of longer than $40d_b$. Yoo et al. (2015b) also examined the local bond behavior of GFRP bars embedded in UHPFRC. The average bond strengths of GFRP bars in UHPFRC were found to be from 16.7 to 22.8 MPa for a d_b of 12.7 mm, and from 19.3 to 27.5 MPa for a d_b of 15.9 mm, which are approximately 73 and 66 % less, respectively, than the bond strengths of deformed steel bars. Similar to the case of CFRP bars in UHPFRC (Ahmad Firas et al. 2011), bond failure was generated by the delamination of the resin and fiber in the bar. Based on a database of 68 pullout test results for GFRP bars in UHPFRC, Yoo et al. (2015b) suggested an equation for the relationship between normalized bond strength and development length by using regression analysis and by assuming no influence of the normalized cover parameter on bond strength, as follows (Fig. 2b):

$$\frac{u}{\sqrt{f_c'}} = 1.05 + 0.85\frac{d_b}{L_e} \tag{2}$$

where u is the bond strength ($=\tau_{\max}$), d_b is the bar diameter, and L_e is the embedment length.

The American Concrete Institute (ACI) 440.1R model (ACI 2006) was inappropriate for UHPFRC; it significantly overestimated the test data (normalized bond strength), as shown in Fig. 2a.

Yoo et al. (2015b) also pointed out that the previous model for development length of FRP bar in concrete, suggested by Wambeke and Shield (2006), was not appropriate for UHPFRC; thus, they proposed an expression for the development length of GFRP bars in UHPFRC, which is only valid for the case of pullout failure, as follows:

$$L_{d,pullout} = \frac{d_b f_{fu}}{3.4\sqrt{f_c'}} \tag{3}$$

where $L_{d,pullout}$ is the development length and f_{fu} is the ultimate strength of rebar.

Schäfers and Seim (2011) performed experimental and numerical investigations on the bond performance between timber and UHPFRC. The glued-laminated timber was bonded to sandblasted and ground UHPFRC with the "Sikadur 330" epoxy resin. Regardless of the bond length and surface treatment, most of specimens showed failure of the bond in the timber close to the bond-line. Based on the Volkersens theory, Schäfers and Seim (2011) suggested a bond length of 400 mm for standard test method to evaluate the bond strength of timber-concrete composites and noted that the effect of tensile stresses, orthogonal to the bond-line, can be neglected when the bond length is beyond 300 mm.

3.2 Flexural Dominated Reinforced UHPFRC Beams, Girders, and Composite Structures

Due to its excellent post-cracking tensile performance with multiple micro-cracks occurred, UHPFRC has attracted attention from engineers for application in structural elements subjected to bending. Several international recommendations

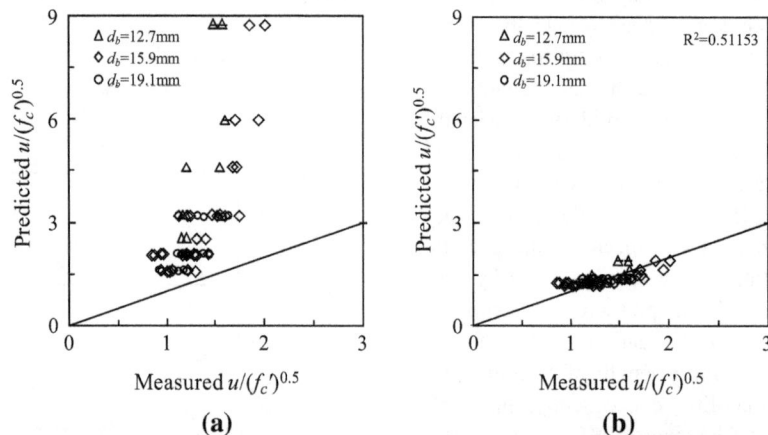

Fig. 2 Comparison of measured and predicted values of normalized bond strength of GFRP bars in UHPFRC; **a** ACI 440.1R, **b** proposed equation (Yoo et al. 2015b).

(AFGC-SETRA, JSCE, and KCI) from France, Japan, and South Korea (AFGC-SETRA 2002; JSCE 2004; KCI 2012) thus provide stress–strain models for compressive and tensile stress blocks in the cross-section, as well as the detailed process of predicting the ultimate capacity of UHPFRC elements under flexure. Since strain (and stress) distribution in the cross-section varies according to the curvature of a beam, multilayer sectional analysis (Yoo and Yoon 2015) is required to calculate an appropriate neutral axis depth and moment capacity at a certain curvature level.

Yoo and Yoon (2015) first reported test results of a number of reinforced UHPFRC beams to investigate the effects of steel fiber aspect ratio and type on flexural performance. Since a portion of the tensile stress after cracking was resisted by the steel fibers, low reinforcement ratios (percent) of 0.94 and 1.50 % were selected. In order to prevent brittle shear failure, stirrups were conservatively designed based on the specimens made of ultra-high-performance concrete (UHPC) without fibers. From the test results (Fig. 3), the beams made by UHPFRC with 2 % by volume of steel fibers exhibited much higher post-cracking stiffness and ultimate load capacity, compared to those made by UHPC without fiber, called 'NF'. In addition, the use of long straight or twisted steel fibers (S19.5, S30, and T30) led to a higher ductility than the use of short straight steel fibers (S13), which are applied for commercial UHPFRC available in North America (Graybeal 2008), at the identical fiber volume fraction. However, it is very interesting to note that much lower ductility indices were obtained by including steel fibers. This is caused by the fact that due to the very high bond strength between UHPFRC and steel rebar and its crack localization behavior, the steel rebar ruptured at a relatively smaller mid-span deflection, as compared with UHPC beams without fiber. Thus, Yoo and Yoon (2015) concluded that the strain-hardening behavior of UHPFRC was unfavorable to the ductility of reinforced beams.

In order to establish reasonable design codes for UHPFRC, Yang et al. (2010) carried out several four-point flexural tests for UHPFRC beams having reinforcement ratios less than 0.02. Test variables were the amount of steel rebar and the placement method. From their test results, placing concrete at the ends of the beams yielded better performance than when concrete was placed at the mid-length because of better fiber orientation to the direction of beam length at the maximum moment zone. In addition, they reported that all test beams showed a ductile response with the ductility index ranging from 1.60 to 3.75 and were effective in controlling cracks. However, the meaning of 'ductile response' could be incorrectly delivered to readers because no test results of reinforced UHPC beams without fiber were reported. In accordance with the test results by Yoo and Yoon (2015), reinforced UHPFRC beams exhibited lower ductility indices compared to beams without fiber due to the crack localization behavior, and Dancygier and Berkover (2016) also reported that the inclusion of steel fibers resulted in a decrease of flexural ductility of beams with low conventional reinforcement ratios.

Yang et al. (2011) examined the flexural behavior of large-scale prestressed UHPFRC I-beams. They indicated that the high volume content of steel fibers in UHPFRC effectively controlled the increase in crack widths, and led to multiple micro-cracks due to the fiber bridging at crack surfaces. The flexural strength of prestressed UHPFRC I-beams was insignificantly affected by the presence of stirrups. Graybeal (2008) also investigated the flexural behavior of a full-scale prestressed UHPFRC I-girder (AASHTO Type II girder) containing 26 prestressing strands. Based on the experimentally observed behavior, he reported that a UHPFRC I-girder shows larger flexural capacities than that of a conventional concrete girder with similar cross-sectional geometry. In addition, an inversely proportional relationship between crack spacing and maximum tensile strain was experimentally observed, as shown in Fig. 4, and the following equation was suggested:

$$\varepsilon = 450 + \frac{2520}{\sqrt{s_{cr}}} + \frac{25,800}{\sqrt{s_{cr}^2}} \quad \text{(in mm)} \tag{4}$$

where ε is the tensile strain and s_{cr} is the crack spacing.

In recent years, several studies (Ferrier et al. 2015; Yoo et al. 2016) have been carried out to develop a new type of

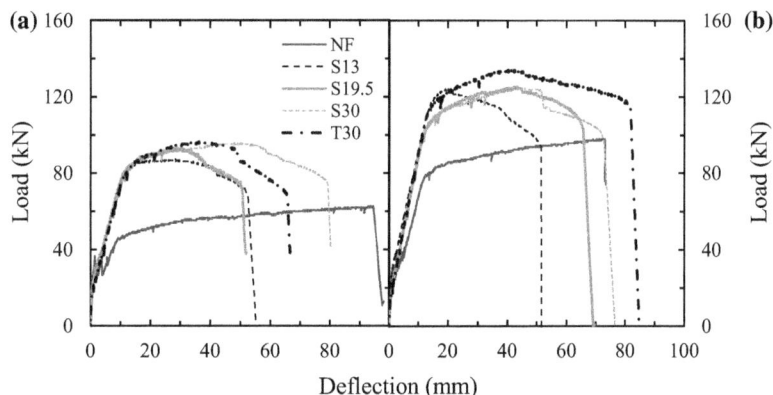

Fig. 3 Load-deflection curves of steel bar-reinforced UHPFRC/UHPC beams; a $\rho = 0.94$ %, b $\rho = 1.50$ % [NF = UHPC w/o fiber, S13 = UHPFRC w/ straight steel fibers ($L_f/d_f = 13/0.2$ mm/mm), S19.5 = UHPFRC w/ straight steel fibers ($L_f/d_f = 19.5/0.2$ mm/mm), S30 = UHPFRC w/ straight steel fibers ($L_f/d_f = 30/0.3$ mm/mm), T30 = UHPFRC w/ twisted steel fibers ($L_f/d_f = 30/0.3$ mm/mm)] (Yoo and Yoon 2015).

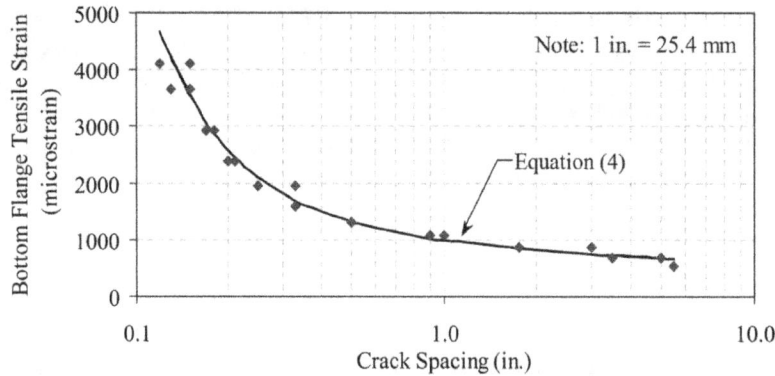

Fig. 4 Tensile strain related to flexural crack spacing of UHPFRC I-girder (Graybeal 2008).

high-performance lightweight beams by applying UHPFRC and FRP rebar. Ferrier et al. (2015) investigated the structural behavior of I-shaped UHPFRC beams reinforced with CFRP and GFRP rebar, according to the rebar axial stiffness ranging from 9 MN to 30 MN. Experimental results indicated that the CFRP rebar was effective in increasing the bending stiffness, which results in a lower mid-span deflection, as compared with the case of the GFRP rebar due to the higher elastic modulus of the former. Thus, they concluded that the axial stiffness of the FRP reinforcement is the most influential parameter of bending stiffness of beams. Yoo et al. (2016) also examined the flexural behavior of UHPFRC beams reinforced with GFRP rebar and hybrid reinforcements (steel + GFRP rebar), according to the axial stiffness ranging from 13 to 95.5 MN. Hybrid reinforcements were considered in their study because it has been considered as one of the most promising methods to overcome the large service deflection problems of conventional FRP-reinforced concrete beams reported by several researchers (Lau and Pam 2010; Yoon et al. 2011). Due to the strain-hardening characteristics of UHPFRC, all tested beams provided very stiff load versus deflection response even after the formation of cracks (Fig. 5), which is distinctive response with conventional FRP-reinforced concrete beams, and satisfied the service crack width criteria of the

Canadian Standards Association (CAN/CSA) S806 (CAN 2002). Furthermore, the deformability factors suggested by Jaeger et al. (1995) were higher than the lower limit ($D_f = 4$) of CAN/CSA-S6 (CAN 2006) for all test beams. Therefore, it was noted that the use of UHPFRC could be a new solution for solving the major drawbacks limiting the practical application of FRP rebar instead of steel rebar. An increase in the GFRP reinforcement ratio led to an improvement in the flexural performance, such as higher post-cracking stiffness, load carrying capacity, and ductility. However, the application of hybrid reinforcements to UHPFRC nullified the main advantage of using FRP to solve the corrosion problem and showed insignificant improvement in the structural performance. Synthetically, Yoo et al. (2016) recommended the use of GFRP rebar with UHPFRC, rather than the use of hybrid reinforcements.

Ferrier et al. (2009) also examined the flexural behavior of a new type of hybrid beam, made of glued-laminated wood and UHPFRC planks, including steel and FRP rebar. They mention that structural efficiency was obtained by using the hybrid beams, as a consequence of the increased bending stiffness due to the high elastic modulus of UHPFRC planks. In addition, the inclusion of steel and FRP rebar in the lower UHPFRC plank significantly increased the ultimate load capacity of the hybrid beams, as compared with when only

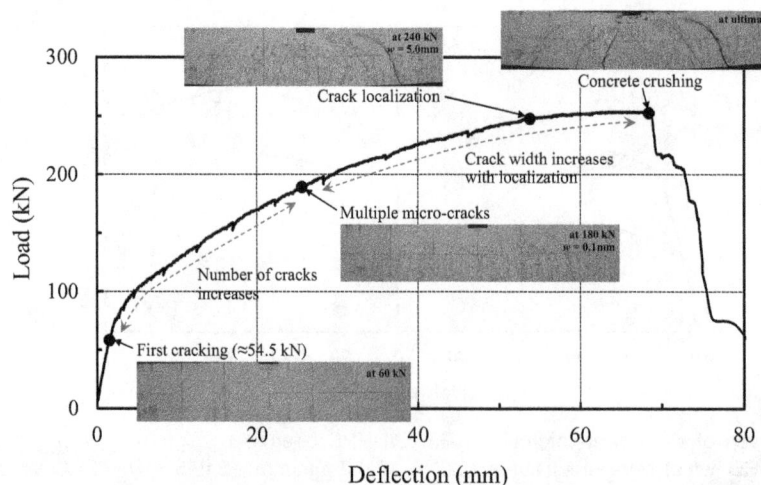

Fig. 5 Typical load–deflection response of GFRP bar-reinforced UHPFRC beam (Yoo et al. 2016).

pure wood elements were used. These advantages of using hybrid beams lead to the potential for reducing the beam depth or increasing the span length of the beam, compared with conventional timber structures.

To practically apply UHPFRC in real architectural and civil structures, appropriate design technique should be suggested based on the material models. Several international recommendations (AFGC-SETRA 2002; JSCE 2004; KCI 2012) thus provide material models for designing flexural members made of UHPFRC. Based these recommendations, many researchers have already precisely predicted the flexural behaviors of reinforced UHPFRC beams (Yang et al. 2010; Yang et al. 2011; Ferrier et al. 2015; Yoo and Yoon 2015; Yoo et al. 2016). In particular, the UHPFRC beams without stirrups were well predicted by AFGC-SETRA recommendations without consideration of fiber orientation coefficient ($K = 1$) (Yang et al. 2011; Yoo et al. 2016) because the fiber alignment in the direction of beam length was insignificantly disturbed by the internal rebars. However, Yoo and Yoon (2015) recently reported that the fiber orientation coefficient that is proposed by the AFGC-SETRA recommendations (i.e., $K = 1.25$), should be considered for simulating the flexural behavior of reinforced UHPFRC beams with stirrups, since the fiber orientation was clearly disturbed by the stirrups (Fig. 6). In the case of FRP-reinforced concrete elements, it is well known that the service deflection prediction is the most important parameter for designing such structures, because of the larger service deflection than that of beams reinforced with steel rebar. Ferrier et al. (2015) and Yoo et al. (2016) successfully predicted the load versus deflection curves of FRP-bar-reinforced UHPFRC beams by sectional analysis, in which they considered compressive and tensile stress blocks in the cross-section, similar to the method used for the above steel-bar-reinforced beams. Yoo and Banthia (2015) also accurately predicted the service deflection of UHPFRC beams reinforced with GFRP rebar and hybrid reinforcements (steel + GFRP rebar), based on a micromechanics-based finite element (FE) analysis; the average ratios of the serviceability deflections from predictions and experiments were found to be 0.91 with a standard deviation of 0.07.

3.3 Shear Resistance of Structural UHPFRC Beams, Girders, and Bridge Decks

Baby et al. (2013b) carried out shear tests of eleven 3-m long UHPFRC I-shaped girders with various shear reinforcements (stirrups and/or steel fibers, or neither) combined with longitudinal prestressing or passive steel bars. To examine the actual fiber orientation effect on the shear performance, the three-point flexural tests were performed by using notched prism specimens extracted from both of the undamaged ends of I-girders at different inclination angles. Test results, as shown in Fig. 7, clearly indicated that the fiber orientation significantly influenced the mechanical (flexural) performance; thus, they noted that the actual fiber orientation needs to be taken into account for shear design, as recommended by AFGC-SETRA recommendations (AFGC-SETRA 2002). By including 2.5 % steel fibers, an almost 250 % increase in shear strength was observed (Baby et al. 2013c). The stirrups yielded first, while localization of the shear crack took place significantly later, as shown in Fig. 8. Thus, crack localization is primarily influenced by the strain capacity of the UHPFRC, and the contributions of the fiber bridging and the stirrups up to their yield strength seem to be effective only when the tensile strain capacity of the UHPFRC is much higher than the yield strain of the stirrups. In their study (Baby et al. 2013c), the AFGC-SETRA recommendations were conservative for the shear-cracking strength, but reasonable for the ultimate shear strength prediction of UHPFRC I-girders. Baby et al. (2013a) also examined the feasibility of applying the modified compression field theory for the shear capacity of reinforced or prestressed UHPFRC beams. Based on their analytical results, the modified compression field theory was determined to be applicable for predicting the shear behavior with an effective estimation of the reorientation of the compressive struts with an increase in the load.

Voo et al. (2010) investigated the shear strength of prestressed UHPFRC I-beams without stirrups, according to the shear span-to-depth ratio (a/d) and the type of steel fibers. They indicated that a higher shear strength was obtained by using a higher fiber volume content and a lower a/d. The theory of the plastic shear variable engagement model presented a good basis for their shear design and a good

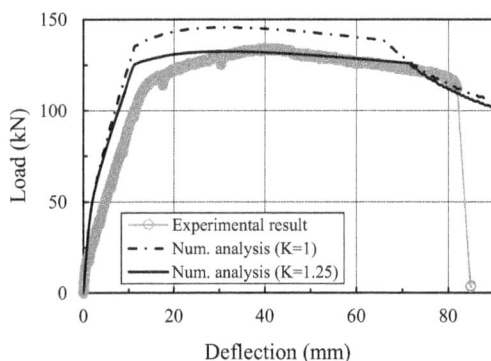

Fig. 6 Comparison of experimental and numerical results for steel bar-reinforced UHPFRC beams with stirrups according to fiber orientation coefficient (Yoo and Yoon 2015).

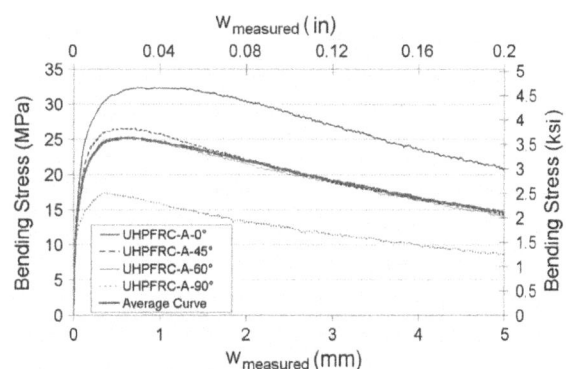

Fig. 7 Three-point bending test results of prisms extracted from I-girders at different inclinations (Baby et al. 2013b).

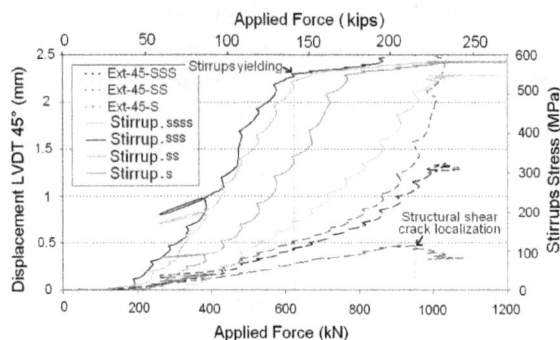

Fig. 8 Evolution during tests of stresses inside stirrups and displacements measured by LVDTs attached at 45° in the web (Baby et al. 2013b).

relationship to the experimental results; the ratio of shear strengths obtained from experiment and theory was found to be 0.92, with a coefficient of variation of 0.12. In addition, Bertram and Hegger (2012), Yang et al. (2012), and Tadepalli et al. (2015) mentioned that the shear strength increased with an increase in the fiber content and a decrease in the a/d ratio. For instance, the inclusion of 2.5 % steel fibers led to a 177 % higher ultimate load than that without fiber, and by changing the a/d ratio from 3.5 to 4.4, the shear capacity was reduced by 10 % (Bertram and Hegger 2012). Bertram and Hegger (2012) also noted that the size effect on shear strength was more substantially affected by the beam height as compared with the web thickness, and that about 12–14 % higher shear capacity was obtained when the effective prestressing force increased by 20 %. By comparing the test results with computed values, Yang et al. (2012) noticed that the predictions using the AFGC-SETRA and JSCE recommendations provided accurate estimates of the shear strength of UHPFRC I-beams (Fig. 9).

In order to replace the open-grid steel decks from moveable bridges, which have several drawbacks, such as poor rideability, high noise levels, susceptibility to fatigue damage, and high maintenance costs, Saleem et al. (2011) examined the structural performance of lightweight UHPFRC bridge decks reinforced with high-strength steel rebar. They properly designed and proposed UHPFRC waffle decks to satisfy the strength, serviceability, and self-weight requirements for moveable bridges. The governing

failure mode was shear, and in the multi-unit decks, shear failure was followed by punching shear failure at close to the ultimate state. However, the shear failure was less abrupt and catastrophic as compared with the commonly seen shear failure mode. Thus, Xia et al. (2011) recommended the ductile shear failure with higher post-cracking shear resistance of UHPFRC beams containing high-strength steel rebar as an acceptable failure mode, rather than including transverse reinforcements, because of their economic problems. The use of 180° hooks at both ends of the steel rebar, recommended by ACI 318 (ACI 2014), was also effective in avoiding bond failure, compared with specimens without end anchorage. Based on a thorough analysis of the experimental results, Saleem et al. (2011) noted that although the proposed UHPFRC waffle deck system exhibits shear failure mode, it has great potential to serve as an alternative to open-grid steel decks, which are conventionally used for lightweight or moveable bridges.

3.4 Torsional Behavior of Structural UHPFRC Beams and Girders

Empelmann and Oettel (2012) examined the effect of adding steel fibers (v_f of 1.5 and 2.5 %) on the torsional behavior of UHPFRC box girders. They experimentally observed that the inclusion of steel fibers led to a better cracking performance such as smaller crack widths and multitudinous cracks, higher ultimate and cracking torque, and improved torsional stiffness. Interestingly, the angle of the diagonal cracks was found to be approximately 45° for all test series, regardless of the steel fiber contents. Yang et al. (2013) also investigated the torsional behavior of UHPFRC beams reinforced with mild steel rebars. In order to estimate the effects of steel fiber content and transverse and longitudinal rebar ratios, thirteen UHPFRC beams were fabricated and tested. Based on their test results (Yang et al. 2013), an improvement in the initial cracking and ultimate torque were obtained by increasing the fiber volume fractions (Fig. 10a), which is consistent with the findings from Empelmann and Oettel (2012). Moreover, higher ultimate torque was found with increases in the ratio of stirrups with longitudinal rebar (Figs. 10b and 11c). In addition, the torsional stiffness after initial cracking was also improved by increasing the ratio of stirrups, as shown in Fig. 10b. In contrast to Empelmann and Oettel's findings, Yang et al. (2013) reported that the angle of the diagonal compressive stress ranged from 27° to 53°, and was affected by the number of stirrups and longitudinal rebar. For example, the angle of localized diagonal cracks increased with an increase in the number of stirrups, as illustrated in Fig. 11. Fehling and Ismail (2012) also reported similar test results for the torsional behavior of UHPFRC elements. They specifically said that the inclusion of steel fibers was effective in improving the torsional performance, such as cracking and ultimate torsional capacities, torsional ductility, post-cracking stiffness, and toughness. The use of longitudinal reinforcements and stirrups also obviously improved the torsional performance.

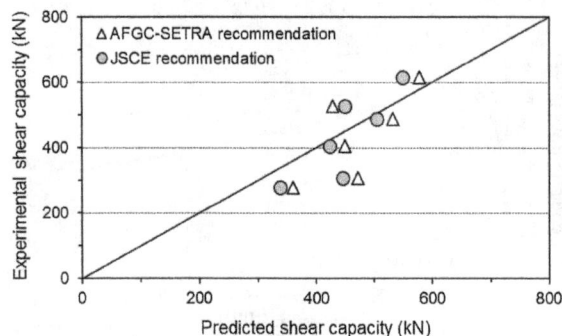

Fig. 9 Comparison of experimental and predicted shear capacities (Yang et al. 2012).

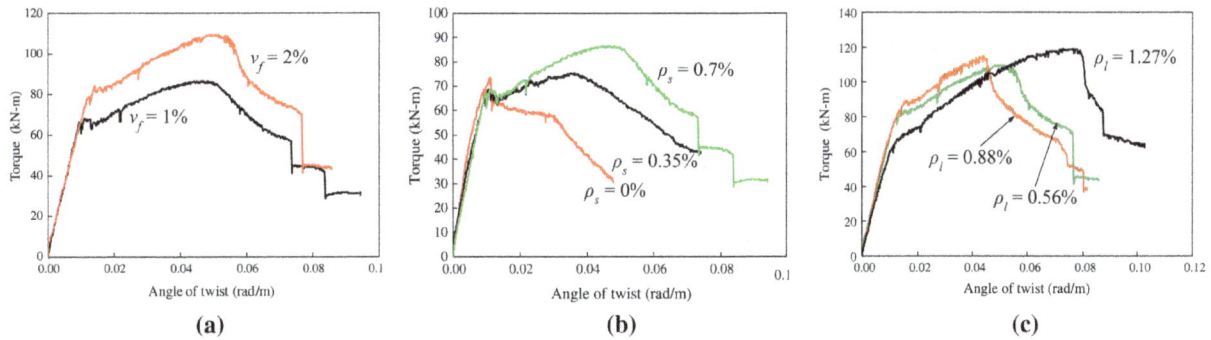

Fig. 10 Torque-twist curves of UHPFRC beams (v_f = volume fraction of steel fiber, ρ_s = transverse reinforcement ratio, ρ_l = longitudinal reinforcement ratio); **a** effect of steel fiber content, **b** effect of transverse reinforcement ratio, **c** effect of longitudinal reinforcement ratio (Yang et al. 2013).

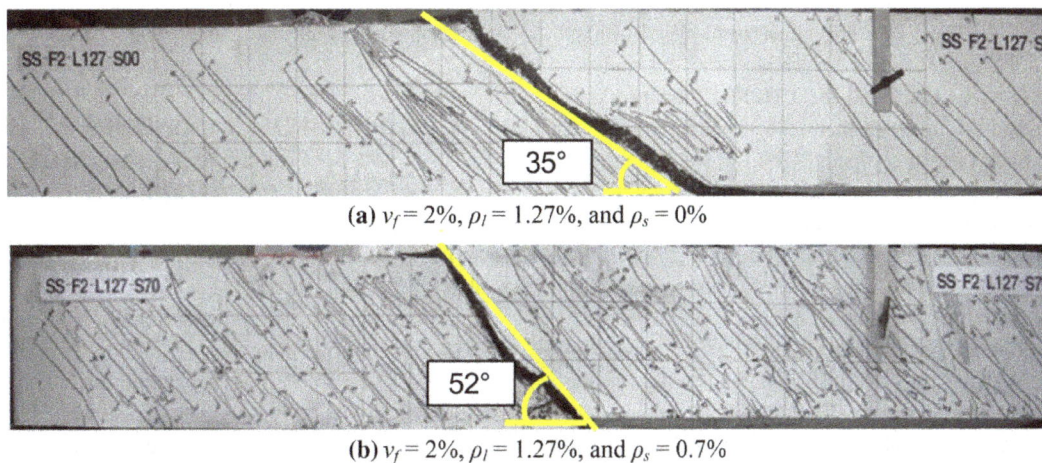

Fig. 11 Angle of localized diagonal cracks (v_f = volume fraction of steel fiber, ρ_s = transverse reinforcement ratio, ρ_l = longitudinal reinforcement ratio) (Yang et al. 2013).

3.5 Performance of Structural UHPFRC Beams, Slabs, and Columns Under Extreme Loadings

Fujikake et al. (2006a) and Yoo et al. (2015a, c) examined the impact resistance of reinforced or prestressed UHPFRC beams by testing a number of specimens using a drop-weight impact test machine. In their studies (Fujikake et al. 2006a), an increase in the maximum deflection of UHPFRC beams was observed by increasing the drop height while maintaining the weight of the hammer, owing to the increase of kinetic energy. The initial stiffness in the UHPFRC beams was insignificantly affected by the impact damage because of the excellent fiber bridging capacities after matrix cracking, and the residual load–deflection (or moment–curvature) curves, shifted based on the maximum deflection by impact, exhibited quite similar behaviors with those of the virgin specimens without impact damage. Hence, Fujikake et al. (2006a) mentioned that the maximum deflection response can be used as the most rational index for estimating the overall flexural damage of reinforced UHPFRC beams. Yoo et al. (2015a) reported that better impact resistance, i.e., lower maximum and residual deflections and higher deflection recovery, was obtained by increasing the amount of longitudinal steel rebars, and the maximum and

residual deflections of reinforced UHPFRC beams decreased significantly by adding 2 % (by volume) of steel fibers, leading to a change in the damage level from severe to moderate, whereas slight decreases in the maximum and residual deflections were found by increasing the fiber length at identical volume fractions (Yoo et al. 2015c). A higher ultimate load capacity was also obtained for the beams under impact loading, compared to those under quasi-static loading, and the residual load capacity after impact damage improved by including 2 % steel fibers and using the longer steel fibers. Fujikake et al. (2006a) and Yoo et al. (2015a) successfully predicted the mid-span deflection versus the time response of structural UHPFRC beams by using the sectional analysis and single- (or multi-) degree-of-freedom model. Improved mechanical compressive and tensile strengths according to the strain-rate were considered in the analysis by using the equations for the dynamic increase factor (DIF) of the UHPFRC, as suggested by Fujikake et al. (2006b, 2008).

Aoude et al. (2015) investigated the blast resistance of full-scale self-consolidating concrete (SCC) and UHPFRC columns under various blast-impulse combinations based on a shock-tube instrumentation. They verified that the steel bar-reinforced UHPFRC columns showed substantially higher blast resistance than the reinforced SCC columns in terms of reducing the maximum and residual deflections, enhancing damage tolerance, and eliminating secondary blast fragments. Based on the single-degree-of-freedom (SDOF) model and lumped inelasticity approach, Aoude et al. (2015) predicted the inelastic deflection-time histories. From the numerical results, several important findings were obtained as follows; (1) since the numerical predictions are sensitive to the choice of DIF, as given in Fig. 12a, further study needs to be done to develop the strain-rate models for using in the blast analysis of UHPFRC columns, and (2) the plastic hinge length (L_p) seems to be reduced in UHPFRC columns from $L_p = d$ (column effective depth), which has been used for the analysis in conventional reinforced concrete columns, as shown in Fig. 12b. Astarlioglu and Krauthammer (2014) numerically simulated the response of normal-strength concrete (NC) and UHPFRC columns subjected to blast loadings based on SDOF models using the dynamic structural analysis suite (DSAS) and reported that the UHPFRC columns presented lower mid-span displacement and sustained more than four times the impulse as compared with the NC columns.

Wu et al. (2009) carried out a series of blast tests of NC and UHPFRC slabs w/ and w/o reinforcements to examine their blast resistance. When the similar blast loads were applied, the UHPFRC slabs without reinforcement exhibited less damage than the NC slabs with reinforcements, and thus, they noticed that the application of UHPFRC is effective in blast design. The UHPFRC slab with passive reinforcements was superior to all other slab specimens, and the strengthening of NC slabs with external FRP plates in the compressive zone was efficient in improving the blast resistance. Yi et al. (2012) examined the blast resistance of the reinforced slabs made of NC, ultra-high-strength concrete (UHSC), and RPC, which is identical to UHPFRC. By analyzing the crack patterns and maximum and residual deflections, they indicated that RPC has the best blast-resistant capacity, followed by UHSC and then NC. For example, the maximum deflections of NC, UHSC, and RPC slabs from 15.88 ANFO charge were found to be 18.57,

15.14, and 13.09 mm, respectively. Mao et al. (2014) investigated the capability of modeling the impact behavior of UHPFRC slabs using the commercial explicit FE program, LS-DYNA (2007). Through FE analysis, they also studied the effects of steel fibers and rebar on the blast resistance of UHPFRC slabs. Importantly, they observed that the K&C model (mostly used for simulating the blast behavior of concrete structures) with automatically generated parameters provided a much better ductile response than the actual behavior, and thus, a modified parameter b_2 from 1.35 to -2 should be applied for UHPFRC. After verifying the numerical modeling with test data, a parametric study was carried out, and some useful results were obtained: (1) the additional use of steel fibers and rebar provide similar influence in the form of extra resistance to the UHPFRC panel under far field blast loading, and (2) under near field blast loading, the resistance of the UHPFRC panels increased substantially with steel rebar, as shown in Fig. 13.

4. Structural Design of UHPFRC Based on AFGC-SETRA and JSCE Recommendations

4.1 AFGC-SETRA Recommendations
4.1.1 Material Models

In AFGC-SETRA recommendations, UHPFRC is referred to as a cementitious material with a compressive strength in excess of 150 MPa, possibly obtaining 250 MPa, and including steel (or polymer) fibers to provide a ductile tensile behavior. The parameters of the design strength were suggested based on the mechanical test results of Ductal® (Orange et al. 1999), as follows: $f_{ck} = 150$–250 MPa, $f_{tj} = 8$ MPa, and $E_c = 55$ GPa, where f_{ck} is the compressive strength, f_{tj} is the post-cracking direct tensile strength, and E_c is the elastic modulus. A partial safety factor γ_{bf} is also introduced, with $\gamma_{bf} = 1.3$ in the case of fundamental combinations and $\gamma_{bf} = 1.05$ in the case of accident combinations. To consider the fiber orientation effect on the tensile behavior, three different fiber orientation coefficients were

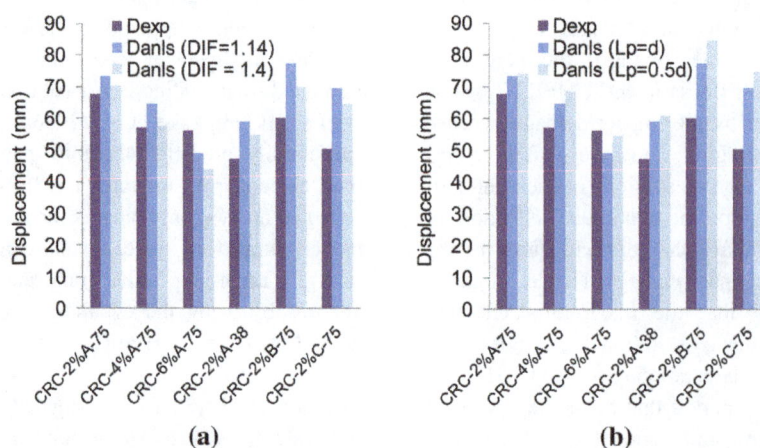

Fig. 12 Displacement predictions from SDOF analysis for UHPFRC columns (*Note* CRC means UHPFRC); a DIF = 1.14 vs. 1.4, b hinge length, $L_p = d$ vs. 0.5d (Aoude et al. 2015).

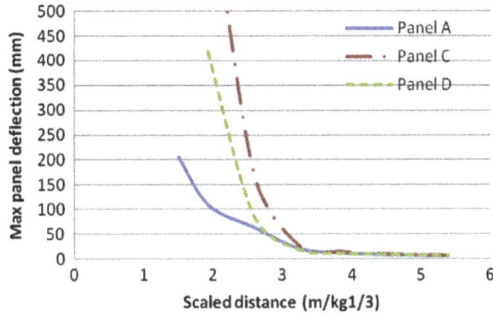

Fig. 13 Variation of maximum panel deflection with scaled distance from three UHPFRC panels (*Panel A* has 2 % of 13 mm long steel fibers with 3.4 % steel rebars, *Panel C* has 2 % of 13 mm long steel fibers with 0.3 % steel rebars, and *Panel D* has 2 % of 13 mm and 2 % of 25 mm long steel fibers with 0.3 % steel rebars) (Mao et al. 2014).

suggested, as follows: $K = 1$ for placement methods validated from test results of a representative model of actual structure, $K = 1.25$ for all loading other than local effects, and $K = 1.75$ for local effects.

In the case of compressive model, a bilinear stress–strain model can be used, as shown in Fig. 14, with the design strength and strain parameters—$\sigma_{bcu} = 0.85f_{ck}/\theta\gamma_b$ and $\varepsilon_u = 0.003$.

The tensile stress–crack opening displacement (σ–w) model is recommended to be first derived based on an inverse analysis. To apply the σ–w model into the tensile stress block in the cross-section, it needs to be transformed to the tensile stress–strain (σ–ε) model by using the characteristic length, l_c, which is $l_c = 2/3 \times h$ for the case of rectangular or T-beams, where h is the beam height.

To obtain the tensile σ–ε model, the elastic tensile strain and strains at crack widths of 0.3 mm and 1 % of beam height need to be calculated based on the following equations:

$$\varepsilon_e = \frac{f_{tj}}{E_c} \tag{5}$$

$$\varepsilon_{0.3} = \frac{w_{0.3}}{l_c} + \frac{f_{tj}}{\gamma_{bf}E_c} \tag{6}$$

$$\varepsilon_{1\%} = \frac{w_{1\%}}{l_c} + \frac{f_{tj}}{\gamma_{bf}E_c} \tag{7}$$

where ε_e is the elastic strain, $w_{0.3}$ is the crack width of 0.3 mm, $\varepsilon_{0.3}$ is the strain at crack width of 0.3 mm, $w_{1\%}$ is the crack width corresponding to $0.01H$ (H is the specimen height), and $\varepsilon_{1\%}$ is the strain at crack width corresponding to $0.01H$.

The ultimate tensile strain is expressed by $\varepsilon_{\lim} = L_f/4l_c$, where $\varepsilon_{\lim}$ is the ultimate tensile strain and L_f is the fiber length.

The stresses at two different characteristic points (at $w_{0.3}$ and $w_{1\%}$) are expressed as follows:

$$f_{bt} = \frac{f(w_{0.3})}{K\gamma_{bf}} \tag{8}$$

$$f_{1\%} = \frac{f(w_{1\%})}{K\gamma_{bf}} \tag{9}$$

where f_{bt} is the stress at a crack width of 0.3 mm and $f_{1\%}$ is the stress at a crack width corresponding to $0.01H$.

The completed material models under compression and tension are given in Fig. 14. Based on the capacity of tensile resistance with crack opening displacement, the AFGC-SETRA recommendations classify the tensile response by two different laws: (1) the strain-softening law ($f_{tj} > f_{bt}$) and (2) the strain-hardening law ($f_{tj} < f_{bt}$), as shown in Fig. 14.

Other important material properties of UHPFRC with heat treatment are given by AFGC-SETRA recommendations, as follows:

– Poisson's ratio: 0.2.
– Thermal expansion coefficient ($\times 10^{-6}/°C$): 11.0.
– Long-term creep coefficient: 0.2.
– Total (autogenous) shrinkage: 550×10^{-6}.

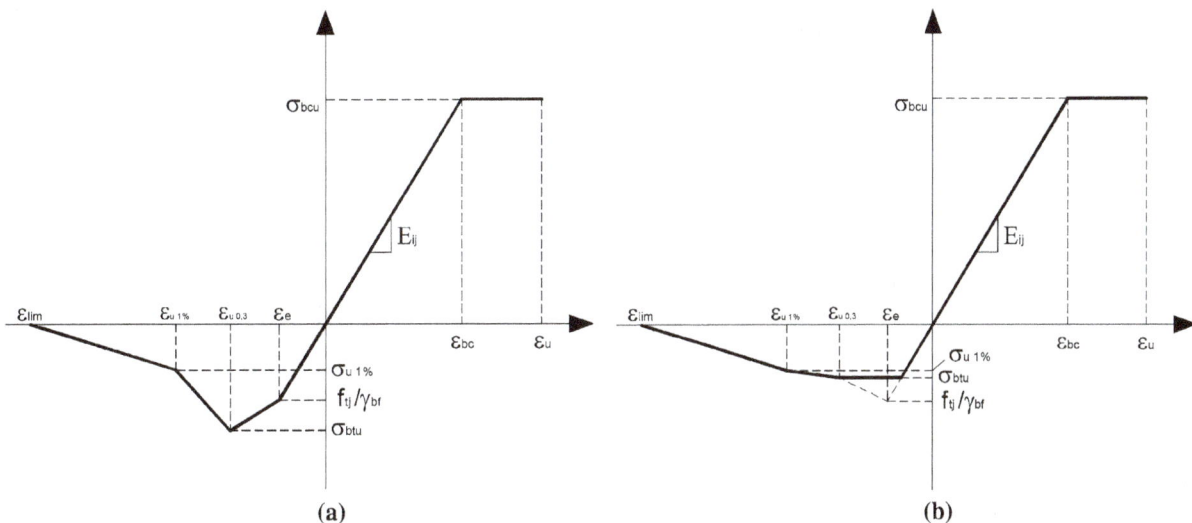

Fig. 14 Material models [AFGC-SETRA recommendations (AFGC-SETRA 2002)]; **a** strain-hardening, **b** strain-softening.

4.1.2 Flexural Design

A chapter for the flexural design of UHPFRC structures is not included in the AFGC-SETRA recommendations. However, based on the suggested material models in Fig. 14, the design of a UHPFRC element subjected to bending can be performed by using the sectional analysis (Fujikake et al. 2006a; Yang et al. 2011; Ferrier et al. 2015; Yoo and Yoon 2015; Yoo et al. 2016). A schematic description of a multi-layered cross-section with strain and stress distributions and an algorithm for sectional analysis are shown in Fig. 15 (Yoo and Yoon 2015). The cross-section of the elements is first divided into a number of layers along the height. After that, the compressive and tensile stresses at each layer can be calculated by assuming that the plane section remains a plane at a given curvature. Then, the neutral axis depth can be calculated from the force equilibrium condition. Lastly, the moment is calculated. The calculation was repeated until the ultimate strain of the steel rebar was reached.

4.1.3 Shear Design

Since the steel fibers can resist a portion of the stress at shear cracks, they mentioned that stirrups may be used, but the shear strength given by the fibers may make it possible to dispense with the stirrups. The ultimate shear strength V_u is given by:

$$V_u = V_{Rb} + V_a + V_f \tag{10}$$

where V_{Rb} is the term for the participation of the concrete, V_a is the term for the participation of the reinforcement, and V_f is the term for the participation of the fibers.

The shear strength by reinforcement V_a is given by:

$$V_a = 0.9d \frac{A_t f_e}{s_t \gamma_s}(\sin\alpha + \cos\alpha) \tag{11}$$

where d is the effective depth, A_t is the area of the stirrups, s_t is the spacing of the stirrups, f_e is the stress in the stirrups, γ_s is the partial safety factor, and α is the inclination angle of the stirrups.

V_{Rb} is expressed by two equations for reinforced and prestressed concrete, as follow:

$$V_{Rb} = \frac{1}{\gamma_E}\frac{0.21}{\gamma_b}k\sqrt{f_{ck}}b_0d \quad \text{(for reinforced concrete)} \tag{12}$$

$$V_{Rb} = \frac{1}{\gamma_E}\frac{0.24}{\gamma_b}\sqrt{f_{ck}}b_0z \quad \text{(for prestressed concrete)} \tag{13}$$

where γ_E is the safety coefficient such that: $\gamma_E \times \gamma_b = 1.5$, b_0 is the element width, σ_m is the mean stress in the total section of concrete under the normal design force, $k = 1 + 3\sigma_{cm}/f_{tj}$ for compression, $k = 1 - 0.7\sigma_{tm}/f_{tj}$ for tension, and z is the distance from the top fiber to the center of prestressing strand.

Lastly, the shear strength of the fibers is calculated by using the following equation:

$$V_f = \frac{S\sigma_p}{\gamma_{bf}\tan\beta_u} \tag{14}$$

$$\sigma_p = \frac{1}{K}\frac{1}{w_{\lim}}\int_0^{w_{\lim}}\sigma(w)dw \tag{15}$$

with $w_{lim} = \max(w_u; 0.3 \text{ mm})$ and $w_u = l_c\varepsilon_u$

where S is the area of the fiber effect ($S = 0.9b_0d$ or b_0d for rectangular or T-sections, and $S = 0.8(0.9d)^2$ or $0.8z^2$ for circular sections), β_u is the inclination angle between a diagonal crack and the longitudinal direction of the beam, K is the orientation coefficient for general effects, $\sigma(w)$ is the experimental characteristic post-cracking stress for a crack width of w, and w_u is the ultimate crack width.

4.2 JSCE Recommendations
4.2.1 Material Models

In JSCE recommendations, UHPFRC is defined as the concrete with $f_c' \geq 150$ MPa, $f_{crk} \geq 4$ MPa, and $f_{tk} \geq 5$ MPa, where f_c' is the compressive strength, f_{crk} is the cracking strength, and f_{tk} is the tensile stress at crack width of 0.5 mm. JSCE recommendations were proposed based on Ductal® (Orange et al. 1999), which is commercially available UHPFRC with heat treatment and 2 vol.% of steel fibers having $d_f = 0.2$ mm and $L_f = 15$ mm, and provided the strength properties used for structural design as follows; $f_c' = 180$ MPa, $f_{crk} = 8$ MPa, $f_{tk} = 8.8$ MPa, and $E_c = 50$ GPa. Importantly, they suggested a bilinear stress–strain curve for compression (Fig. 16a) and a bilinear tension-softening curve (TSC) with $f_{tk} = 8.8$ MPa, $w_{1k} = 0.5$ mm, and $w_{2k} = 4.3$ mm for tension (Fig. 16b). In order to obtain the structural safety, $\gamma_c = 1.3$ was proposed for the partial safety factor. To take into account the suggested TSC for tensile stress block in the cross-section, the crack opening displacement should be transformed to a strain by using the equivalent specific length, L_{eq}, as follows

$$L_{eq} = 0.8h\left[1 - \frac{1}{\left(1.05 + 6h/l_{ch}\right)^4}\right] \tag{16}$$

where h is the overall depth of beam, l_{ch} is the characteristic length ($=G_F E_c/f_{tk}^2$), and G_F is the fracture energy.

By using the equivalent specific length, the tensile stress–strain model is obtained by the following equations, based on the TSC:

$$\varepsilon_{cr} = \frac{f_{tk}}{\gamma_c E_c} \tag{17}$$

$$\varepsilon_1 = \varepsilon_{cr} + \frac{w_{1k}}{L_{eq}} \tag{18}$$

$$\varepsilon_2 = \frac{w_{2k}}{L_{eq}} \tag{19}$$

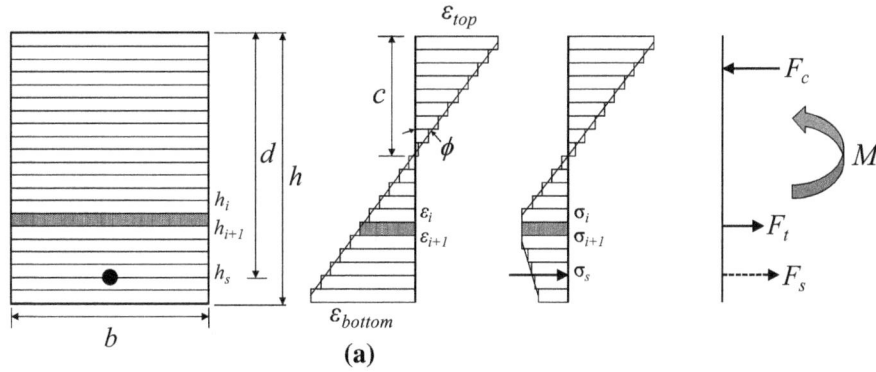

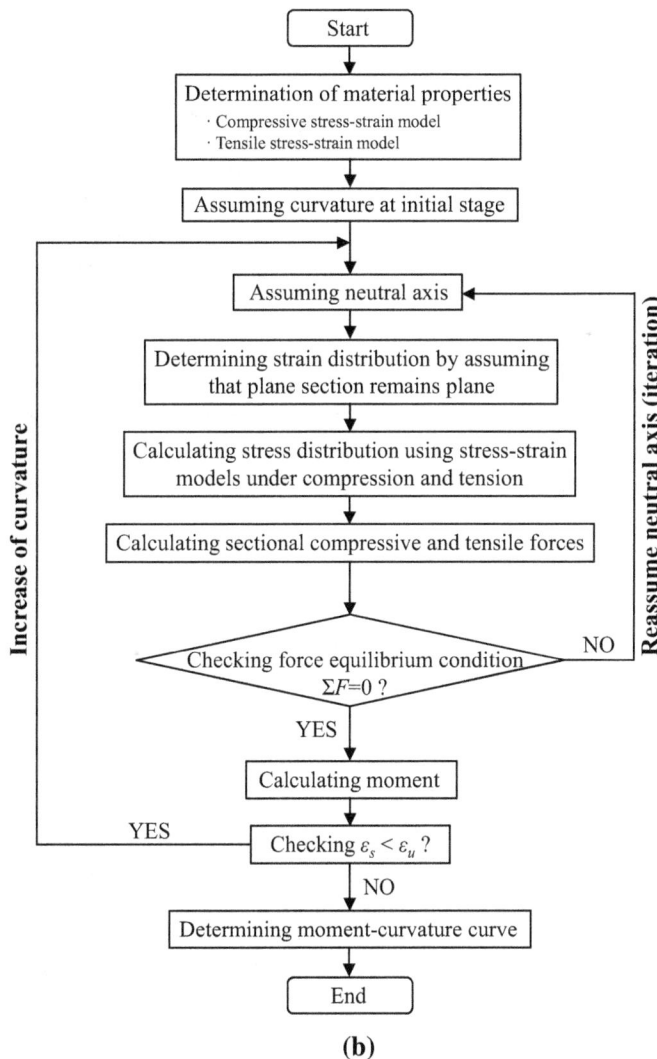

Fig. 15 Sectional analysis; **a** schematic description of stress and strain distributions in cross-section, **b** algorithm for sectional analysis (Yoo and Yoon 2015).

where ε_{cr} is the factored elastic strain, ε_1 is the strain at the end of the initial plateau, w_{1k} is the crack opening displacement for which a certain stress level is retained after the first crack, ε_2 is the strain at zero tensile stress, and w_{2k} is the crack opening displacement at zero tensile stress.

Other material properties of the UHPFRC, proposed by the JSCE recommendations, are as follows:

 – Poisson's ratio: 0.2.

 – Coefficient of thermal expansion ($\times 10^{-6}/°C$): 13.5.
 – Thermal conductivity (kJ/mh °C): 8.3.
 – Thermal diffusivity ($\times 10^{-3} m^2/h$): 3.53.
 – Specific heat (kJ/kg °C): 0.92.
 – Total shrinkage: 550×10^{-6}.
 – Creep coefficient: 0.4.
 – Density (kN/m^3) for calculating dead load: 25.5.

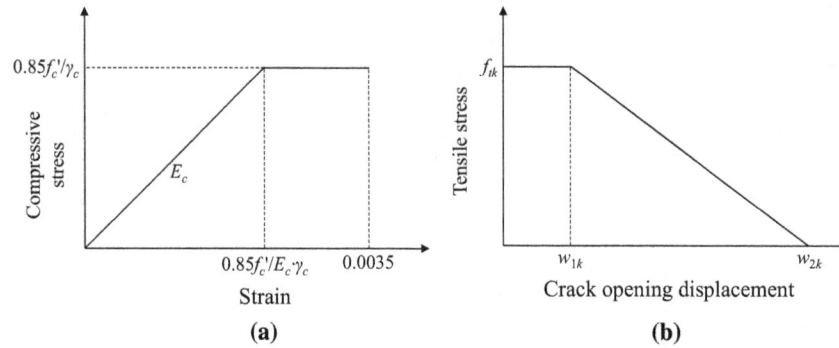

Fig. 16 Material models [JSCE recommendations (JSCE 2004)]; **a** compressive stress–strain curve, **b** tension-softening curve.

4.2.2 Flexural Design

In the JSCE recommendations, for the structural design of UHPFRC elements under bending, two simple assumptions are required to be satisfied: (1) the linear strain distribution and (2) the use of the proposed material models, as given in Fig. 16. The steel fiber contribution in the tensile zone after cracking needs to be considered in the structural design, and the compressive and tensile stress blocks in the cross-section should be considered based on the proposed material stress–strain models (Fig. 16) by considering the equivalent specific length for tension. Although a detailed procedure for calculating the ultimate moment capacity is not introduced in the JSCE recommendations, sectional analysis can be adopted for calculating the moment–curvature behavior of the UHPFRC elements (Yang et al. 2011; Yoo et al. 2016), similar to the case of the AFGC-SETRA recommendations.

4.2.3 Shear Design

For the shear design of the UHPFRC elements, the shear resistance from the matrix and the included steel fibers is required to be calculated. In accordance with the JSCE recommendations, the total shear resistance can be calculated by:

$$V_{yd} = V_{rpcd} + V_{fd} + V_{ped} \tag{20}$$

where V_{yd} is the total shear resistance of the reinforced UHPFRC beams, V_{rpcd} is the shear resistance of the matrix without fiber, V_{fd} is the shear resistance of the steel fibers, and V_{ped} is the shear resistance of the stirrups.

Herein, V_{rpcd} and V_{ped} are obtained by using Eqs. (21) and (22), respectively.

$$V_{rpcd} = 0.18\sqrt{f_c'}b_w d/\gamma_b \tag{21}$$

$$V_{fd} = (f_{vd}/\tan\beta_u)b_w z/\gamma_b \tag{22}$$

where b_w is the web width of the beam, d is the effective depth of the beam, γ_b is the strength reduction factor ($\gamma_b = 1.3$), f_{vd} is the design tensile strength perpendicular to the diagonal crack ($= f_{tk}/\gamma_b$), z is the distance between the point-acting compressive force and the center tensile reinforcement ($=d/1.15$), and β_u is the inclination angle between

the diagonal crack and the longitudinal direction of the beam. The inclination angle β_u should be larger than 30° and is calculated by $\beta_u = 1/2\tan^{-1}[2\tau/(\sigma_{xu} - \sigma_{yu})] - \beta_0$, where τ is the average shear stress, σ_{xu} and σ_{yu} are the average compressive stress in longitudinal and transverse directions, and β_0 is the inclination angle without axial force.

In addition, V_{ped} can be calculated as follows

$$V_{ped} = P_{ed}\sin\alpha_p/\gamma_b \tag{23}$$

where P_{ed} is the effective tensile force of the prestressing strand, α_p is the inclination angle between the prestressing strand and longitudinal axis of beam, and γ_b is the strength reduction factor ($\gamma_b = 1.1$).

As was reported by Yang et al. (2012), the shear design of UHPFRC elements can be carried out by using both AFGC-SETRA and JSCE recommendations, which provide good estimates with test data of I-shaped UHPFRC beams, as shown in Fig. 9.

5. Field Applications of UHPFRC

Due to its excellent mechanical performance, UHPFRC can lead to a reduction in the number of sections, the elimination of passive reinforcements, and the possibility for the design of structures that is not possible with ordinary concrete (NPCA 2011). For this reason, UHPFRC has attracted much attention from engineers for field applications from 1995 to 2010 (Voo et al. 2012). The representative application examples in North America, Europe, and Asia are briefly explained herein. The first structural application of UHPFRC was the prestressed hybrid pedestrian bridge at Sherbrooke in Canada, constructed in 1997 (Resplendino 2004) (Fig. 17a). This precast and prestressed pedestrian bridge includes a post-tensioned open-web space UHPFRC truss with 4 access spans made by conventional high-performance concrete. A ribbed slab with a 30-mm thickness was adopted, and a transverse prestressing was applied with sheathed monostrands. The UHPFRC truss webs were confined by steel tubes, and the structure was longitudinally prestressed by both internal and external prestressing strands. The total span length of the bridge was 60 m, and the main span was assembled from six 10-m prefabricated match-cast segments. The Bourg-lès-Valence bridge was the first UHPC road bridge in the world, built in France in 2001 (Hajar et al. 2004), as shown in Fig. 17b. The bridge was built from five assembled π-shaped precast UHPFRC beams, and the joints were made by in situ UHPFRC with internal reinforcements. The bridge consisted of two isolated spans with a length of approximately 20 m, and all the π-shaped beams were prestressed without transverse passive reinforcement. The Seonyu Footbridge, completed in 2002 in Seoul, Korea, is currently the longest footbridge made by UHPFRC (Ductal®) with a single span of 120 m and no central support (Fig. 17c). It consists of a π-shaped arch supporting a ribbed UHPFRC slab with a thickness of 30 mm, and transverse prestressing was provided by sheathed monostrands. With equivalent load carrying capacity and strength properties, the bridge needed only half of the amount of materials required for conventional concrete construction (Voo et al. 2014). The first UHPFRC highway bridge, built in Iowa, USA, is the Mars Hill bridge, as shown in Fig. 17d. This is a simple single-span bridge consisting of three precast and prestressed concrete beams with a length of 33.5 m (modified 1.14-m-deep Iowa bulb-tee beams), and the cast-in-place concrete bridge deck was topped. No stirrup was applied, and each beam included 47 low-relaxation prestressing strands with a diameter of 15.2 mm (Russell and Graybeal 2013).

Curved UHPFRC panels were applied to the building named the Atrium, which was built in Victoria, BC, Canada in 2013 (Fig. 18a) (NPCA 2011). In this project, UHPFRC was selected as an appropriate material for the curved panel system, due to its ability to form monolithically tight radial curves, and consequently, it could improve the energy efficiency of the building by eliminating unattractive seams and openings. The Community Center in Sedan, France was built in 2008 with a double skin facade to protect the glass fascia behind and to provide privacy using light blue UHPFRC perforated panels (Fig. 18b). They designed the UHPFRC panels to be 2 m by 4 m by 45 mm thick, to permit sunlight to stream through to the interior spaces. The main reason for choosing UHPFRC panels instead of traditional perforated panels, made of metal, painted steel, cast aluminum, cast iron, and stainless steel, was that UHPFRC is durable, and requires less energy consumption for fabrication and for maintenance over time (Henry et al. 2011).

Mega-architectural projects were carried out to build the Stade Jean Bouin and the MuCEM in France (Fig. 19) (NPCA 2011). In the case of the Stade Jean Bouin, the first application of a precast UHPFRC lattice-style facade system in the world was made. The 23,000 m² envelope, which contains a 12,000 m² roof, was built from 3600 self-supporting UHPFRC panels that are 8–9 m long by 2.5 m wide, and 45 mm thick. The MuCEM consists of several types of precast UHPFRC structural elements, such as 78-m-long footbridge, main cubic structures with a 15,000 m² surface area, flooring, and lattice-style envelope with a series of slender N- and Y-shaped columns. The architect designed the MuCEM to carry the entire external load by columns, and to satisfy the maximum requirements regarding both seismic and fire resistance, as specified by CSTB.

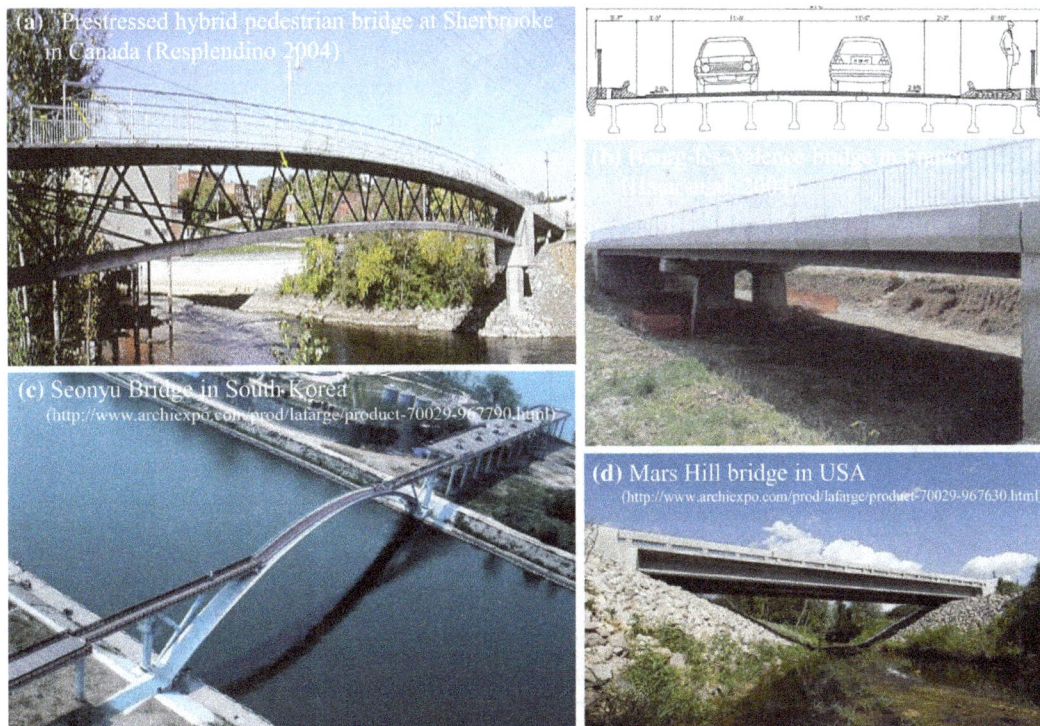

Fig. 17 Examples of UHPFRC applications in bridges.

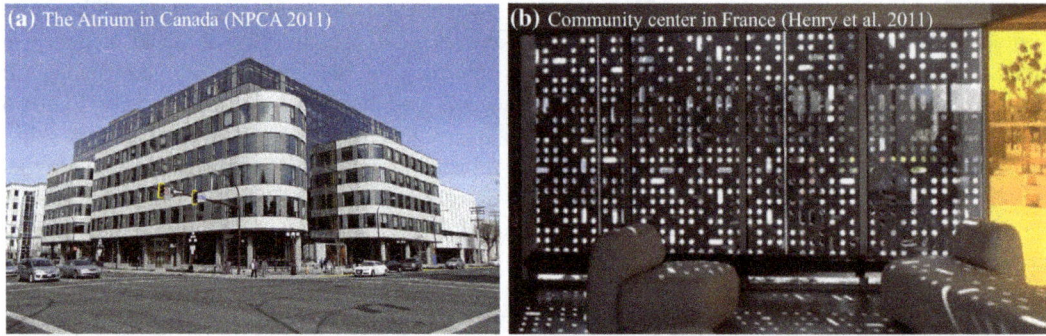

Fig. 18 Examples of UHPFRC applications in buildings.

Fig. 19 Mega architectural projects in France (NPCA 2011).

6. Conclusion

This paper reviewed the current state-of-the-art for structural performance, design recommendations, and applications of UHPFRC. From the above literature review and discussions, the following conclusions are drawn:

(1) The use of 2 % steel fibers resulted in a higher post-cracking stiffness and ultimate load capacity of steel bar-reinforced UHPFRC beams, but it decreased their ductility because of the superb bond strength with steel bars and crack localization characteristics. Importantly, the use of UHPFRC was effective in overcoming the major drawbacks of conventional FRP-reinforced concrete structures (large service deflection) due to its strain-hardening response, and the use of CFRP bars was efficient in improving the flexural stiffness of reinforced UHPFRC beams, compared to that of GFRP bars. The hybrid reinforcements (steel + FRP bars), which have been adopted to reduce the service deflection of conventional FRP-reinforced beams, were ineffective in UHPFRC beams, and thus, the use of single FRP bars, instead of hybrid reinforcements, was recommended for the case of UHPFRC.

(2) With the inclusion of 2.5 % steel fibers, approximately 250 % higher shear strength was obtained, compared to that without fibers. The shear strength also increased with an increase in the fiber contents and a decrease in the shear span-to-depth ratio. Due to the excellent fiber bridging at crack surfaces, the shear failure of UHPFRC beams was less abrupt than the commonly seen shear failure mode in conventional concrete. In addition, both AFGC-SETRA and JSCE recommendations provided accurate predictions of shear strength of UHPFRC beams.

(3) The inclusion of steel fibers provided a better cracking performance, higher ultimate and cracking torque, and improved torsional stiffness. The higher ultimate torque was also found by increasing the ratios of stirrups and longitudinal rebars. The diagonal crack angle was influenced by the amount of stirrups and longitudinal rebars, whereas it was not affected by the amount of steel fibers.

(4) Better impact resistance of reinforced UHPFRC beams was obtained by including 2 % steel fibers and a larger number of longitudinal steel bars. The residual capacity after impact damage was also improved by adding 2 % steel fibers and using the longer steel fibers. UHPFRC columns exhibited significantly higher blast resistance such as lower maximum deflection, improved damage tolerance, and higher resistance, compared with the reinforced SCC and NC columns. Thus, it was noted that the application of UHPFRC in the structures subjected to blast loads is effective.

(5) Finally, the international design recommendations on UHPFRC were discussed minutely, and examples of practical applications of UHPFRC in architectural and civil structures were investigated.

Acknowledgments

This research was supported by a grant from a Construction Technology Research Project 13SCIPS02 (Development of impact/blast resistant HPFRCC and evaluation technique thereof) funded by the Ministry of Land, Infrastructure and Transport.

References

ACI Committee 440. (2006). *Guide for the design and construction of concrete reinforced with FRP bars (ACI 440.1R-06)*. Farmington Hills, MI: American Concrete Institute.

AFGC-SETRA. (2002). *Ultra high performance fibre-reinforced concretes. Interim recommendations*. Bagneux, France: SETRA.

Ahmad Firas, S., Foret, G., & Le Roy, R. (2011). Bond between carbon fibre-reinforced polymer (CFRP) bars and ultra high performance fibre reinforced concrete (UHPFRC): Experimental study. *Construction and Building Materials, 25*(2), 479–485.

American Concrete Institute (ACI). (2014). *Building code requirements for structural concrete and commentary. ACI 318-14 and ACI 318R-14*. Farmington Hills, MI: American Concrete Institute (ACI).

Aoude, H., Dagenais, F. P., Burrell, R. P., & Saatcioglu, M. (2015). Behavior of ultra-high performance fiber reinforced concrete columns under blast loading. *International Journal of Impact Engineering, 80*, 185–202.

Astarlioglu, S., & Krauthammer, T. (2014). Response of normal-strength and ultra-high-performance fiber-reinforced concrete columns to idealized blast loads. *Engineering Structures, 61*, 1–12.

Baby, F., Marchand, P., Atrach, M., & Toutlemonde, F. (2013a). Analysis of flexure-shear behavior of UHPFRC beams based on stress field approach. *Engineering Structures, 56*, 194–206.

Baby, F., Marchand, P., & Toutlemonde, F. (2013b). Shear behavior of ultrahigh performance fiber-reinforced concrete beams. I: Experimental investigation. *Journal of Structural Engineering, 140*(5), 04013111.

Baby, F., Marchand, P., & Toutlemonde, F. (2013c). Shear behavior of ultrahigh performance fiber-reinforced concrete beams. II: Analysis and design provisions. *Journal of Structural Engineering, 140*(5), 04013112.

Bache, H. H. (1981). Densified cement ultra-fine particle-based materials. In *Proceedings of the 2nd international conference on superplasticizers in concrete*, Ottawa, Canada, p. 33.

Bertram, G., & Hegger, J. (2012). Shear behavior of pretensioned UHPC beams—Tests and design. In *Proceedings of the third international symposium on UHPC and nanotechnology for high performance construction materials*, Kassel, pp. 493–500.

Birchall, J. D., Howard, A. J., & Kendall, K. (1981). Flexural strength and porosity of cements. *Nature, 289*(5796), 388–390.

CAN/CSA S806. (2002). *Design and construction of building components with fibre reinforced polymers*, Rexdale, Canada.

CAN/CSA-S6. (2006). *Canadian highway bridge design code*, Toronto, ON.

CEB-FIP. (1993). *Model code for concrete structures*. CEB Bulletin d'Information. Comite Euro international du Beton, Lausanne, Switzerland.

Choi, W. C., Yun, H. D., Cho, C. G., & Feo, L. (2014). Attempts to apply high performance fiber-reinforced cement composite (HPFRCC) to infrastructures in South Korea. *Composite Structures, 109*, 211–223.

Cosenza, E., Manfredi, G., & Realfonzo, R. (1995). Analytical modelling of bond between FRP reinforcing bars and concrete. In L. Taerwe (Ed.), *Proceedings of second international RILEM symposium (FRPRCS-2)* (pp. 164–171). London, UK: E and FN Spon.

DAfStB UHPC. (2003). *State-of-the-art report on ultra high performance concrete—Concrete technology and design*. Deutscher Ausschuss für Stahlbeton/German Association for Reinforced Concrete, Berlin, Germany, draft 3.

Dancygier, A. N., & Berkover, E. (2016). Cracking localization and reduced ductility in fiber-reinforced concrete beams with low reinforcement ratios. *Engineering Structures, 111*, 411–424.

Empelmann, M., & Oettel, V. (2012). UHPFRC box girders under torsion. In *Proceedings of the third international symposium on UHPC and nanotechnology for high performance construction materials*, Kassel, Germany, pp. 517–524.

Farhat, F. A., Nicolaides, D., Kanellopoulos, A., & Karihaloo, B. L. (2007). High performance fibre-reinforced cementitious composite (CARDIFRC)—Performance and application to retrofitting. *Engineering Fracture Mechanics, 74*(1–2), 151–167.

Fehling, E., & Ismail, M. (2012). Experimental investigations on UHPC structural elements subjected to pure torsion. In *Proceedings of the third international symposium on UHPC and nanotechnology for high performance construction materials*, Kassel, Germany, pp. 501–508.

Ferrier, E., Labossiere, P., & Neale, K. W. (2009). Mechanical behavior of an innovative hybrid beam made of glulam and ultrahigh-performance concrete reinforced with FRP or steel. *Journal of Composites for Construction, 14*(2), 217–223.

Ferrier, E., Michel, L., Zuber, B., & Chanvillard, G. (2015). Mechanical behaviour of ultra-high-performance short-fibre-reinforced concrete beams with internal fibre reinforced polymer bars. *Composites Part B: Engineering, 68,* 246–258.

Fujikake, K., Senga, T., Ueda, N., Ohno, T., & Katagiri, M. (2006a). Study on impact response of reactive powder concrete beam and its analytical model. *Journal of Advanced Concrete Technology, 4*(1), 99–108.

Fujikake, K., Senga, T., Ueda, N., Ohno, T., & Katagiri, M. (2006b). Effects of strain rate on tensile behavior of reactive powder concrete. *Journal of Advanced Concrete Technology, 4*(1), 79–84.

Fujikake, K., Uebayashi, K., Ohno, T., Shimoyama, Y., & Katagiri, M. (2008). Dynamic properties of steel fiber reinforced mortar under high-rates of loadings and triaxial stress states. In *Proceedings of the 7th international conference on structures under shock and impact* (pp. 437–446). Montreal, Canada: WIT Press.

Graybeal, B. A. (2008). Flexural behavior of an ultrahigh-performance concrete I-girder. *Journal of Bridge Engineering, 13*(6), 602–610.

Graybeal, B., & Tanesi, J. (2007). Durability of an ultrahigh-performance concrete. *Journal of Materials in Civil Engineering, 19*(10), 848–854.

Hajar, Z., Lecointre, D., Simon, A., & Petitjean, J. (2004). Design and construction of the world first ultra-high performance concrete road bridges. In *Proceeding of the international symposium on ultra high performance concrete*, University of Kassel, Kassel, Germany, pp. 39–48.

Henry, K. A., Seibert, P. J., & America, L. N. (2011). Manufacturing UHPC architectural products. http://www.ductal. fr/CPI_-_OCTOBER_2011.pdf.

Jaeger, G. L., Tadros, G., & Mufti, A. A. (1995). *Balanced section, ductility and deformability in concrete with FRP reinforcement*. Technical report no. 2-1995, Nova Scotia computer aided design/computer aided manufacturing centre, Technical University of Nova Scotia, Halifax, Canada.

JSCE. (2004). *Recommendations for design and construction of ultra-high strength fiber reinforced concrete structures (Draft)*. Tokyo, Japan: Japan Society of Civil Engineers.

Jungwirth, J., & Muttoni, A. (2004). Structural behavior of tension members in UHPC. In *Proceedings of the international symposium on ultra-high-performance concrete*, Kassel, Germany, pp. 533–544.

KCI. (2012). *Design recommendations for ultra-high performance concrete K-UHPC. KCI-M-12-003*. Seoul: Korea Concrete Institute.

Kim, S. W., Park, J. J., Kang, S. T., Ryo, G. S., & Koh, K. T. (2008). Development of ultra high performance cementitious composites (UHPCC) in Korea. In *Proceedings of the fourth international IABMAS conference*, Seoul, Korea, p. 110.

Lau, D., & Pam, H. J. (2010). Experimental study of hybrid FRP reinforced concrete beams. *Engineering Structures, 32*(12), 3857–3865.

Li, H., & Liu, G. (2016). Tensile properties of hybrid fiber-reinforced reactive powder concrete after exposure to

elevated temperatures. *International Journal of Concrete Structures and Materials, 10*(1), 29–37.

Livermore Software Technology Corporation. (2007). *LS-DYNA user's manual—Version 971*. Livermore, CA: Livermore Software Technology Corporation.

Mao, L., Barnett, S., Begg, D., Schleyer, G., & Wight, G. (2014). Numerical simulation of ultra high performance fibre reinforced concrete panel subjected to blast loading. *International Journal of Impact Engineering, 64*, 91–100.

NPCA White Paper. (2011). Ultra high performance concrete (UHPC), guide to manufacturing architectural precast UHPC elements. http://precast.org/wp-content/uploads/2011/05/NPCA-ultra-high-performance-concrete.pdf

Orange, G., Acker, P., & Vernet, C. (1999). A new generation of UHP concrete: Ductal® damage resistance and micromechanical analysis. In *Proceedings of the third international workshop on high performance fiber reinforced cement composites (HPFRCC3)*, Mainz, Germany, pp. 101–111.

Resplendino, J. (2004). First recommendations for ultra-high-performance concretes and examples of application. In *Proceeding of the international symposium on ultra high performance concrete*, University of Kassel, Kassel, Germany, pp. 79–90.

Richard, P., & Cheyrezy, M. (1995). Composition of reactive powder concretes. *Cement and Concrete Research, 25*(7), 1501–1511.

RILEM TC. (1994). *RILEM recommendations for the testing and use of constructions materials. RC 6 bond test for reinforcement steel. 2. Pull-out test, 1983* (pp. 218–220). London, UK: E & FN SPON.

Roy, D. M., Gouda, G. R., & Bobrowsky, A. (1972). Very high strength cement pastes prepared by hot pressing and other high pressure techniques. *Cement and Concrete Research, 2*(3), 349–366.

Russell, H. G., & Graybeal, B. A. (2013). *Ultra-high performance concrete: A state-of-the-art report for the bridge community, FHWA-HRT-13-060*.

Saleem, M. A., Mirmiran, A., Xia, J., & Mackie, K. (2011). Ultra-high-performance concrete bridge deck reinforced with high-strength steel. *ACI Structural Journal, 108*(5), 601–609.

Schäfers, M., & Seim, W. (2011). Investigation on bonding between timber and ultra-high performance concrete (UHPC). *Construction and Building Materials, 25*(7), 3078–3088.

Tadepalli, P. R., Dhonde, H. B., Mo, Y. L., & Hsu, T. T. (2015). Shear strength of prestressed steel fiber concrete I-beams. *International Journal of Concrete Structures and Materials, 9*(3), 267–281.

Tam, C. M., Tam, V. W., & Ng, K. M. (2012). Assessing drying shrinkage and water permeability of reactive powder concrete produced in Hong Kong. *Construction and Building Materials, 26*(1), 79–89.

Voo, Y. L., Foster, S. J., & Voo, C. C. (2014). Ultrahigh-performance concrete segmental bridge technology: Toward sustainable bridge construction. *Journal of Bridge Engineering, 20*(8), B5014001.

Voo, Y. L., Nematollahi, B., Said, A., Gopal, A., & Yee, T. Y. (2012). Application of ultra high performance fiber reinforced concrete—The Malaysia perspective. *International Journal of Sustainable Construction Engineering and Technology, 3*(1), 26–44.

Voo, Y. L., Poon, W. K., & Foster, S. J. (2010). Shear strength of steel fiber-reinforced ultrahigh-performance concrete beams without stirrups. *Journal of Structural Engineering, 136*(11), 1393–1400.

Wambeke, B. W., & Shield, C. K. (2006). Development length of glass fiber-reinforced polymer bars in concrete. *ACI Structural Journal, 103*(1), 11–17.

Wu, C., Oehlers, D. J., Rebentrost, M., Leach, J., & Whittaker, A. S. (2009). Blast testing of ultra-high performance fibre and FRP-retrofitted concrete slabs. *Engineering Structures, 31*(9), 2060–2069.

Xia, J., Mackie, K. R., Saleem, M. A., & Mirmiran, A. (2011). Shear failure analysis on ultra-high performance concrete beams reinforced with high strength steel. *Engineering Structures, 33*(12), 3597–3609.

Yang, I. H., Joh, C., & Kim, B. S. (2010). Structural behavior of ultra high performance concrete beams subjected to bending. *Engineering Structures, 32*(11), 3478–3487.

Yang, I. H., Joh, C., & Kim, B. S. (2011). Flexural strength of large-scale ultra high performance concrete prestressed T-beams. *Canadian Journal of Civil Engineering, 38*(11), 1185–1195.

Yang, I. H., Joh, C., Lee, J. W., & Kim, B. S. (2012). An experimental study on shear behavior of steel fiber-reinforced ultra high performance concrete beams. *Journal of The Korean Society of Civil Engineers, 32*(1A), 55–64.

Yang, I. H., Joh, C., Lee, J. W., & Kim, B. S. (2013). Torsional behavior of ultra-high performance concrete squared beams. *Engineering Structures, 56*, 372–383.

Yi, N. H., Kim, J. H. J., Han, T. S., Cho, Y. G., & Lee, J. H. (2012). Blast-resistant characteristics of ultra-high strength concrete and reactive powder concrete. *Construction and Building Materials, 28*(1), 694–707.

Yoo, D. Y., & Banthia, N. (2015). Numerical simulation on structural behavior of UHPFRC beams with steel and GFRP bars. *Computers and Concrete, 16*(5), 759–774.

Yoo, D. Y., Banthia, N., Kim, S. W., & Yoon, Y. S. (2015a). Response of ultra-high-performance fiber-reinforced concrete beams with continuous steel reinforcement subjected to low-velocity impact loading. *Composite Structures, 126*, 233–245.

Yoo, D. Y., Banthia, N., & Yoon, Y. S. (2016). Flexural behavior of ultra-high-performance fiber-reinforced concrete beams reinforced with GFRP and steel rebars. *Engineering Structures, 111*, 246–262.

Yoo, D. Y., Kang, S. T., & Yoon, Y. S. (2014a). Effect of fiber length and placement method on flexural behavior, tension-softening curve, and fiber distribution characteristics of UHPFRC. *Construction and Building Materials, 64*, 67–81.

Yoo, D. Y., Kwon, K. Y., Park, J. J., & Yoon, Y. S. (2015b). Local bond-slip response of GFRP rebar in ultra-high-performance fiber-reinforced concrete. *Composite Structures, 120*, 53–64.

Yoo, D. Y., Park, J. J., Kim, S. W., & Yoon, Y. S. (2014b). Influence of reinforcing bar type on autogenous shrinkage stress and bond behavior of ultra high performance fiber reinforced concrete. *Cement and Concrete Composites, 48*, 150–161.

Yoo, D. Y., Shin, H. O., Yang, J. M., & Yoon, Y. S. (2014c). Material and bond properties of ultra high performance fiber reinforced concrete with micro steel fibers. *Composites Part B: Engineering, 58*, 122–133.

Yoo, D. Y., & Yoon, Y. S. (2015). Structural performance of ultra-high-performance concrete beams with different steel fibers. *Engineering Structures, 102*, 409–423.

Yoo, D. Y., Yoon, Y. S., & Banthia, N. (2015c). Impact and residual capacities of ultra-high-performance concrete beams with steel rebars. In *Proceedings of the fifth international workshop on performance, protection & strengthening of structures under extreme loading*, East Lansing, MI.

Yoon, Y. S., Yang, J. M., Min, K. H., & Shin, H. O. (2011). Flexural strength and deflection characteristics of high-strength concrete beams with hybrid FRP and steel rebar reinforcement. In *Proceedings of the 10th symposium on fiber reinforced polymer reinforcement for concrete structures (FRPRCS-10), SP-275-04*, American Concrete Institute, Farmington Hills, MI, pp. 1–22.

Yudenfreund, M., Odler, I., & Brunauer, S. (1972). Hardened portland cement pastes of low porosity I. Materials and experimental methods. *Cement and Concrete Research, 2*(3), 313–330.

15

Review of Design Flexural Strengths of Steel–Concrete Composite Beams for Building Structures

Lan Chung[1], Jong-Jin Lim[1], Hyeon-Jong Hwang[2], and Tae-Sung Eom[1],*

Abstract: Recently, as the use of high-performance materials and complex composite methods has increased, the need for advanced design specifications for steel–concrete composite structures has grown. In this study, various design provisions for ultimate flexural strengths of composite beams were reviewed. Design provisions reviewed included the load and resistance factor design method of AISC 360-10 and the partial factor methods of KSSC–KCI, Eurocode 4 and JSCE 2009. The design moment strengths of composite beams were calculated according to each design specification and the variation of the calculated strengths with design variables was investigated. Furthermore, the relationships between the deformation capacity and resistance factor for flexure were examined quantitatively. Results showed that the design strength and resistance factor for flexure of composite beams were substantially affected by the design formats and variables.

Keywords: composite beam, flexural strength, partial factor method, load and resistance factor method, composite structure.

1. Introduction

In Korea, the design of steel–concrete composite members for building structures has been conventionally addressed in a section of the design code for steel structures, KBC 2014 Sec. 0709 (Architectural Institute of Korea 2014). However, as the use of high performance materials and complex composite methods has increased, the need for a more advanced design code for composite members and structures is growing. For this, joint research to develop an independent design code for composite structures was performed by the Korean Society of Steel Construction (KSSC)–Korean Concrete Institute (KCI) joint composite structure committee (KSSC–KCI Joint Composite Structure Committee 2014). By reviewing existing design standards and recent studies, the KSSC–KCI joint research was aimed at developing a performance-based design code to accommodate high-strength materials and new composite systems.

According to the review of existing design standards, such as AISC 360-10 (American Institute of Steel Construction 2010), KBC 2014 (Architectural Institute of Korea 2014), Eurocode 4 (European Committee for Standardization 2004a), and JSCE 2009 (Japan Society of Civil Engineers 2009), the calculation methods for design strengths of steel–concrete composite members can be divided into the load and resistance factor design method (LRFD) and the partial factor method (PFM). For AISC 360-10 and KBC 2014, using LRFD, the design strength of a composite member is determined by multiplying the nominal member strength and the resistance factor ϕ, which is not greater than 1.0. For Eurocode 4 and JSCE 2009, adopting PFM, in contrast, the partial safety factor γ of not less than 1.0 is applied directly to material characteristic strengths rather than to the member strength. This difference in the calculation format between the LRFD and PFM can result in significant differences in the design strength of composite members, even though the material and section properties are the same. Furthermore, by using a safety factor at the material level, rather than at the member level, PFM may be more flexible in accommodating high-strength materials and new composite methods.

In this study, the provisions for flexural design of composite beams specified in AISC 360-10, KBC 2014, Eurocode 4, and JSCE 2009 were reviewed in terms of design format, resistance and safety factors, and the method of section analysis. For a quantitative comparison, the design moment strength of fully composite beams was calculated according to the provisions specified in each design code. Then, the variation of the calculated strengths with design variables (steel yield strength, concrete strength, and effective width of concrete slab) was investigated. Furthermore, the relationships between the deformation capacity and resistance factor for flexure were analyzed quantitatively. Particular attention was given to the applicability of 800 MPa grade high-strength steel to composite beams, which was included in KBC 2014 (Architectural Institute of Korea 2014).

[1]Department of Architectural Engineering, Dankook University, Yongin-si 448-701, Gyeonggi-do, Korea.
*Corresponding Author; E-mail: tseom@dankook.ac.kr
[2]College of Civil Engineering, Hunan University, Yuelu Mountain, Changsha 410082, Hunan, China.

2. Provisions for Flexural Design

2.1 Design Format and Material Strength

Table 1 compares the design formats and material strengths specified in AISC 360-10, KBC 2014, Eurocode 4, and JSCE 2009. In the table, the characteristic strength, design strength, and safety factor for materials are denoted as f_k, f_d ($=f_k/\gamma$), and γ, respectively. For example, f_{ck}, f_{cd}, and γ_c are the characteristic compressive strength, design compressive strength, and safety factor for concrete, F_{yk}, F_{yd}, and γ_s are the characteristic yield strength, design yield strength, and safety factor for structural steel, and f_{yrk}, f_{yrd}, and γ_r are the values for reinforcing steel bars. Additionally, $M(f_k)$ and $M(f_d)$ denote the ultimate moment strengths of composite beams calculated by using the characteristic and design material strengths, f_k and f_d, respectively. In table 1 M_d is the design moment strength including a safety margin against the nominal strength, and ϕ is the resistance factor used for LRFD.

For AISC 360-10 and KBC 2014, which use LRFD as the design format, the design moment strength of composite beams is calculated by multiplying the nominal strength $M(f_k)$ and resistance factor ϕ ($=0.9$): $M_d = \phi M(f_k)$. Thus, the LRFD can ensure a constant safety margin for bending, regardless of the behavior of composite beams. For Eurocode 4, which uses PFM as its design format, in contrast, the design moment strength is directly calculated from the reduced material strengths f_d divided by the partial safety factors for concrete, steel, and reinforcing bar (i.e., $\gamma_c = 1.5$, $\gamma_s = 1.0$, and $\gamma_r = 1.15$): $M_d = M(f_d)$. Thus, the safety margin for bending of composite beams designed by PFM is affected substantially by their behavior (this will be discussed in detail later, in the Sect. 5). In JSCE 2009, on the other hand, a safety factor for member γ_b ($=1.1$) addressing the effects of accuracy in section analysis/design, variations in member size, and the importance of the role of the member, is used along with the safety factors for materials (i.e. $\gamma_c = 1.3$, $\gamma_s = 1.0$, and $\gamma_r = 1.0$; see Table 1). The design moment strength is determined as $M_d = M(f_d)/\gamma_b$. Because $1/\gamma_b$ can be considered as a resistance factor for

bending, the design format of JSCE 2009 can be seen as a mixed form of PFM and LRFD.

Table 1 also shows the upper and lower limits on characteristic strengths of materials specified in each design code. The upper limit of the compressive strength of concrete is generally similar to $f_{ck} = 60$–80 MPa in all design codes. On the other hand, AISC 360-10 and KBC 2014 allow the use of relatively high-strength steels ($F_{yk} = 525$ and 650 MPa). In particular, KBC 2014 has increased the upper limit of steel yield strength as $F_{yk} = 650$ MPa, based on recent studies (Kim et al. 2012a, b, 2014; Lee et al. 2012, 2013a, b, c, 2014; Youn 2013a).

2.2 Design Moment Strength

The ultimate moment strength of composite beam sections can be calculated using the plastic stress distribution method (PSDM) and strain-compatibility method (SCM). Table 2 compares stress distributions of concrete, steel, and reinforcing bars over a composite section required for the PSDM prescribed in each design code. The stress distributions illustrated in Table 2 are for positive bending, where the concrete flange is subjected to compression. For AISC 360-10 and KBC 2014 that use LRFD as their design format, the plastic stresses of concrete, steel, and reinforcing bars are defined as $0.85f_{ck}$, F_{yk}, and f_{yrk}, respectively. The plastic moment M_{pl} and the depth D_p of plastic neutral axis are then calculated from the force equilibrium between internal resultant forces produced by the plastic stresses $0.85f_{ck}$, F_{yk}, and f_{yrk}. On the other hand, Eurocode 4 and JSCE 2009 adopting PFM as the design format define the design plastic stresses of concrete, steel, and reinforcing bars as $0.85f_{cd}$, F_{yd}, and f_{yrd}, respectively. Because the design plastic stresses are decreased by dividing by the safety factors γ_c, γ_s, and γ_r (≥ 1.0), the values of M_{pl} and D_p determined from PSDM specified in Eurocode 4 and JSCE 2009 are not equivalent to those of AISC 360-10 and KBC 2014.

In fact, the plastic stress distributions shown in Table 2 are different from the actual stress distributions at the ultimate limit state. Furthermore, a composite beam may suffer a premature failure due to crushing failure in the concrete slab

Table 1 Design format and material strength of existing specifications.

	AISC 360-10	KBC 2014	Eurocode 4	JSCE 2009
Design format	Load and resistance factor design	Load and resistance factor design	Partial factor method	Partial factor method
Design moment strength M_d	$M_d = \phi M(f_k)$	$M_d = \phi M(f_k)$	$f_d = f_k/\gamma$ and $M_d = M(f_d)$	$f_d = f_k/\gamma$ and $M_d = M(f_d)/\gamma_b^{(1)}$
Resistance factor ϕ or Safety factor γ for materials	$\phi = 0.9$	$\phi = 0.9$	Concrete $\gamma_c = 1.5$ Steel $\gamma_s = 1.0^{(2)}$ Reinforcing bar $\gamma_r = 1.15$	Concrete $\gamma_c = 1.3$ Steel $\gamma_s = 1.0$ Reinforcing bar $\gamma_r = 1.0$
Characteristic material strength f_k (MPa)	Concrete $21 \leq f_{ck} \leq 70$ Steel $F_{yk} \leq 525$	Concrete $21 \leq f_{ck} \leq 70$ Steel $F_{yk} \leq 650$	Concrete $20 \leq f_{ck} \leq 60$ Steel $F_{yk} \leq 460$	Concrete $18 \leq f_{ck} \leq 80$ Steel $-^{(3)}$

[1] Partial safety factor for members $\gamma_b = 1.1$.

[2] $\gamma_s = 1.0$ is used for yielding.

[3] Although steel strength limitation is not specified, $F_{yk} \leq 450$ MPa is generally accepted.

Table 2 Design moment strengths by plastic stress distribution method and strain compatibility method.

	AISC 360-10 and KBC 2014	Eurocode 4	JSCE 2009
	PSDM[1]		
Plastic stress	Conc. $0.85f_{ck}$, steel F_{yk}, and reinforcing bar f_{yrk}	Conc. $0.85f_{cd}$, steel F_{yd}, and reinforcing bar f_{yrd}	Conc. $0.85f_{cd}$, steel F_{yd}, and reinforcing bar f_{yrd}
Stress distribution			
Design strength M_d	$M_d = \phi M_{pl}$ and $\phi = 0.9$	$M_d = M_{pl}$ or $\beta M_{pl}^{(2)}$	$M_d = M_{pl}/\gamma_b$ and $\gamma_b = 1.1$
	SCM[1]		
Conc. σ–ε curve	Not specified[3] Maximum compressive strain = 0.003	Parabola-rectangle[4]	Parabola-rectangle[5]
Steel σ–ε curve	Not specified[3]	Elastic-perfectly plastic	Bilinear with $0.01E_s$ hardening
Stress and strain distributions (positive bending)			
Design strength M_d	$M_d = \phi M_{nl}$ and $\phi = 0.9$	$M_d = M_{nl}$	$M_d = M_{nl}/\gamma_b$ and $\gamma_b = 1.1$

[1] *PSDM* plastic stress distribution method, *SCM* strain compatibility method.
[2] β is the reduction factor for high strength steels of 420 and 460 MPa. β is used for positive moment only.
[3] The σ–ε relationships of concrete and steel shall be obtained from tests or from published results for similar materials.
[4] The σ–ε relationship is given in Eurocode 2 (European Committee for Standardization 2004b). The bilinear relationship is also available.
[5] The σ–ε relationship is given in JSCE 2007 (Japan Society of Civil Engineers 2007) ($k_1 = 1 - 0.003f_{ck}$).

even before the plastic stress is fully developed in the steel section. This is more likely to occur when high-strength steel is used. Thus, to secure a relatively greater margin of safety, Eurocode 4 requires the design moment strength M_d of the composite beam under positive bending be modified as $M_d = \beta M_{pl}$ by multiplying by a reduction factor β (≤ 1.0) (see Table 2). The reduction factor β is applied only when high-strength steels of $F_{yk} = 420$ and 460 MPa are used. Figure 1 shows the reduction factor β specified in Eurocode 4. If $D_p/D_t \leq 0.15$, $\beta = 1.0$ is applied and thus the moment strength calculated by the PSDM is not reduced (D_t = overall depth of composite section); if $0.15 < D_p/$

$D_t \leq 0.4$, then β decreases linearly, from 1.0 to 0.85. The PSDM should not be used for $D_p/D_t > 0.4$ because a brittle failure of the composite beam can occur as a result of early crushing in the concrete slab.

The reduction factor β is also used in the AASHTO LRFD bridge design specification (American Association of State Highway and Transportation Officials 2012; Wittry 1993). As shown in Fig. 1, β specified in AASHTO LRFD 2012 decreases from 1.0 to 0.78 as the ratio of D_p/D_t increases from 0.1 to 0.42. The reduction factor β specified in AASHTO LRFD 2012 is applied to steels of all strength grades of $F_{yk} = 485$ MPa or less, while the β specified in

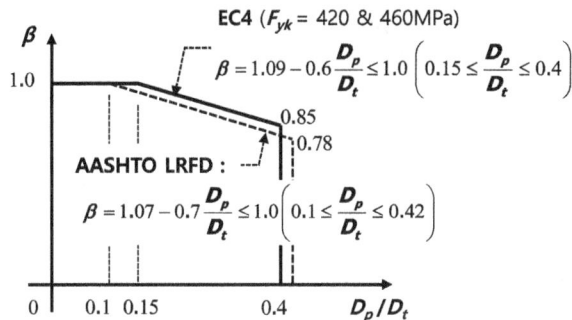

Fig. 1 Reduction factor β for plastic moment strength under positive bending: Eurocode 4 and AASHTO LRFD 2005.

Eurocode 4 is applied only to high-strength steels of $F_{yk} = 420$ and 460 MPa. AISC 360-10, KBC 2014, and JSCE 2009 do not define a reduction factor β for the plastic moment determined from the PSDM.

The stress and strain distributions at the ultimate limit state for the design of a cross section by the SCM are also shown in Table 2 (European Committee for Standardization 2004b; Japan Society of Civil Engineers 2007). Linear strain distribution along the height of the composite section is assumed in all design codes. However, the maximum compressive strain ε_{cu} of concrete varies: AISC 360-10 and KBC 2014 use a constant value of $\varepsilon_{cu} = 0.003$, while Eurocode 4 and JSCE 2009 define a varying $\varepsilon_{cu} = 0.0025{-}0.0035$, according to the characteristic compressive strength f_{ck} of concrete [refer to notes (4) and (5) of Table 2]. In the SCM, the stresses of concrete, steel, and reinforcing bars corresponding to the linear stain distribution are basically determined by the stress–strain relationship of each material. Eurocode 4 prescribes the bilinear, parabolic-rectangle, and rectangular stress distributions for concrete specified in Eurocode 2 (European Committee for Standardization 2004b). JSCE 2009 is similar to Eurocode 4. In contrast, AISC 360-10 and KBC 2014 allow the use of the relationship obtained from tests or from published results for similar materials, without providing a specific stress–strain relationship for concrete. For steel, bilinear relationships without hardening and with hardening are allowed for Eurocode 4 and JSCE 2009, respectively. For AISC 360-10 and KBC 2014, however, any stress–strain relationship obtained from tests or from published results for similar materials can be used.

In the SCM, the nonlinear moment strength Mnl is obtained by integrating the stresses and forces of concrete, steel, and reinforcing bars over the cross-section. For AISC 360-10 and KBC 2014 that use LRFD as their design format, the design moment strength of the cross-section is determined as $M_d = \phi M_{nl}$, by multiplying by the resistance factor ϕ (=0.9). For Eurocode 4 that uses PFM as its design format, in contrast, the design moment strength is determined as $M_d = M_{nl}$ because a safety margin is already addressed in the design strength of materials. JSCE 2009 defines the design moment strength as $M_d = M_{nl}/\gamma_b$ by dividing M_{nl} by the safety factor for the member.

3. KSSC–KCI Provisions for Flexural Design

The KSSC–KCI joint composite structure committee developed a draft version of a performance-based design specification for composite structures, KSSC–KCI (KSSC–KCI Joint Composite Structure Committee 2014). KSSC–KCI adopted PFM as a design format. Thus, the design strengths of concrete, steel, and reinforcing bars (f_{cd}, F_{yd}, and f_{yrd}, respectively) are defined using the resistance factors for materials, as follows.

$$f_{cd} = \phi_c f_{ck} \tag{1a}$$

$$F_{yd} = \phi_s F_{yk} \tag{1b}$$

$$f_{yrd} = \phi_r f_{yrk} \tag{1c}$$

where, ϕ_c, ϕ_s, and ϕ_r, respectively, are the resistance factors for concrete, steel, and reinforcing bars. In KSSC–KCI, the resistance factors were defined as $\phi_c = 0.65$, $\phi_s = 1.0$, and $\phi_r = 0.9$.

Basically, the moment strength for the design of cross sections can be calculated from PSDM and SCM. In the case of the PSDM, first, the plastic moment M_{pl} and the depth D_p of plastic neutral axis are calculated using the plastic stresses at the ultimate limit state, such as $0.85f_{cd}$ (=$0.85\phi_c f_{ck}$) for concrete, F_{yd} (=$\phi_s F_{yk}$) for steel, and f_{yrd} (=$\phi_r f_{yrk}$) for reinforcing bars. The design moment strength of the cross section is determined as $M_d = M_{pl}$ and βM_{pl} for positive and negative bending, respectively. In KSSC–KCI, the reduction factor β (≤ 1.0) is defined as follows on the basis of Youn's study (Youn 2013b) (see Fig. 2).

$$\beta = 1.045 - 0.375\left(\frac{D_p}{D_t}\right) \quad \text{for} \quad F_{yk} = 450 \text{ MPa or less} \tag{2a}$$

$$\beta = 1.066 - 0.550\left(\frac{D_p}{D_t}\right) \quad \text{for} \quad F_{yk} = 650 \text{ MPa} \tag{2b}$$

In Eqs. (2a) and (2b), D_p/D_t should not be greater than 0.42. If $D_p/D_t > 0.42$, the PSDM cannot be used for the design of cross sections. Similar to AASHTO LRFD 2012,

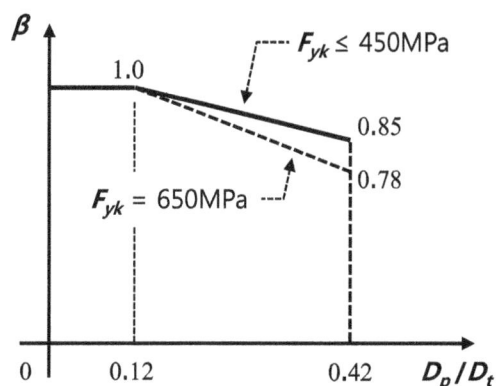

Fig. 2 Reduction factor β for plastic moment strength under positive bending: KSSC–KCI 2014.

KSSC–KCI requires the reduction factor β be applied to steels of all strength grades of $F_{yk} = 650$ MPa or less. However, to secure a greater margin of safety for high-strength steels of $F_{yk} = 650$ MPa, a relatively smaller value of β is defined, as shown in Eq. (2b) and Fig. 2. The values of β corresponding to each strength grade of steel are given in detail in Youn's study (2013).

KSSC–KCI also allows the use of SCM for the design of composite cross sections. For the strain-compatible section analysis, a linear strain distribution over the cross section is assumed, as illustrated in Table 2. The stress–strain relationship of concrete including the ultimate compressive strain ε_{cu} is defined as follows (see Fig. 3).

$$\sigma_c = f_{cd}\left[1 - \left(1 - \frac{\varepsilon_c}{\varepsilon_{co}}\right)^n\right] \quad \text{for} \quad 0 \le \varepsilon_c \le \varepsilon_{co} \tag{3a}$$

$$\sigma_c = f_{cd} \quad \text{for} \quad \varepsilon_{co} \le \varepsilon_c \le \varepsilon_{cu} \tag{3b}$$

where, σ_c and ε_c = compressive stress and strain of concrete, respectively, n = an exponent determining the shape of ascending parabola, ε_{co} = strain at the peak stress ($=f_{cd}$), and ε_{cu} = ultimate compressive strain at failure.

$$n = 2.0 - \left(\frac{f_{ck} - 40}{100}\right) \le 2.0 \tag{4}$$

$$\varepsilon_{co} = 0.002 + \left(\frac{f_{ck} - 40}{100000}\right) \ge 0.002 \tag{5}$$

$$\varepsilon_{cu} = 0.0033 - \left(\frac{f_{ck} - 40}{100000}\right) \le 0.0033 \tag{6}$$

The stress distribution and ultimate strain (ε_{cu}) of concrete can have substantial effects on design results, particularly when high-strength steel is used and the cross section is under positive bending. Thus, the σ_c–ε_c relationship of concrete is specified in KSSC–KCI so that engineers can use the SCM with convenience for the design of cross sections.

For steel sections, KSSC–KCI allows the use of a bilinear stress–strain relationship, representing the elastic-perfectly plastic or strain-hardening behavior. For reinforcing bars, in contrast, only an elastic-perfectly plastic model is allowed. Such bilinear models of steel and reinforcing bars are used for the strain-compatible section analysis of composite sections in conjunction with the linear strain distribution.

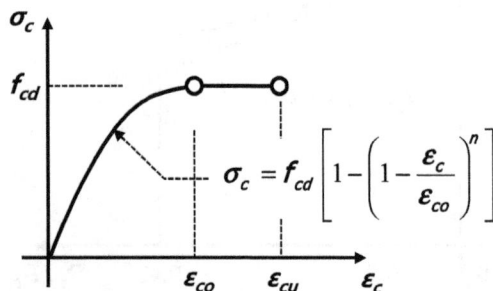

Fig. 3 Stress–strain relationship of concrete for section design.

4. Design Resistance by Plastic Stress Distribution Method

In this section, the design moment strength of cross sections calculated by the PSDM specified in KSSC–KCI, AISC 360-10 (or KBC 2014), Eurocode 4, and JSCE 2009 were compared. For KBC 2014, the design format, resistance factor, and plastic stresses of the materials are the same as those of AISC 360-10 (refer to Tables 1 and 2). Figure 4 shows the cross sections of interior and exterior composite beams used for the study. The sectional properties of interior and exterior composite beams were equivalent except for the effective widths b_{eff} of the concrete slabs. The overall height and flange width of the steel section were 600 and 200 mm, respectively, and the thicknesses of the web and flange were 11 and 17 mm, respectively. The overall and net thicknesses of the concrete slabs were 150 and 95 mm, respectively. The effective widths of the concrete slabs were $b_{eff} = 2400$ and 1000 mm for the interior and exterior beams, respectively. In the calculation of design moment strengths, the reinforcement of concrete slabs ($A_s = 1980$ and 824 mm^2) was ignored.

In this study, $f_{ck} = 21$ and 30 MPa were considered as the characteristic compressive strength of concrete. For the steel section, $F_{yk} = 235, 315, 355, 450$, and 650 MPa were considered as the characteristic yield strengths. Such yield strengths are the same as those of weldable structural steel specified in KSSC–KCI and KBC 2014.

4.1 Positive Bending

Tables 3 and 4, respectively, show the design moment strengths M_d of the interior and exterior composite beams under positive bending, calculated by the PSDM specified in each design code. M_d includes the effects of the resistance factor (ϕ) or the safety factor for materials (γ_c, γ_s, and γ_r or ϕ_c, ϕ_s, and ϕ_r) [refer to Table 2 and note (1) of Tables 3 and 4]. Because Eurocode 4 and JSCE 2009 do not allow the use of a high-strength steel of $F_{yk} = 650$ MPa, M_d corresponding to $F_{yk} = 650$ MPa was not calculated in the tables [refer to note (3) of Tables 3 and 4]. Although AISC 360-10 is also not applicable to $F_{yk} = 650$ MPa, M_d calculated according to AISC 360-10 is given for a comparison to KSSC–KCI. Additionally, the PSDM specified in KSSC–KCI and Eurocode 4 was applied only for $D_p/D_t \le 0.4$ and $D_p/D_t \le 0.42$, respectively [refer to note (4) of Table 4].

Figures 5 and 6 show the variation of the design strengths M_d by KSSC–KCI, Eurocode 4, and JSCE 2009 according to steel yield strengths F_{yk} (=235, 315, 355, 450, and 650 MPa). The vertical and horizontal axes indicate the ratio of design strengths (i.e., $M_d/M_{pl,AISC}$) and the characteristic yield strength F_{yk} of steel, respectively. The variation of $M_d/M_{pl,AISC}$ for the interior and exterior beams are presented in Figs. 5 and 6, respectively. It is noted that, for comparisons between comparable design codes, the design strengths M_d of KSSC–KCI, Eurocode 4, and JSCE 2009 were divided by the nominal strength $M_{pl,AISC}$ of AISC 360-10 (see M_{pl} of Table 3). If all safety and resistance factors for materials are

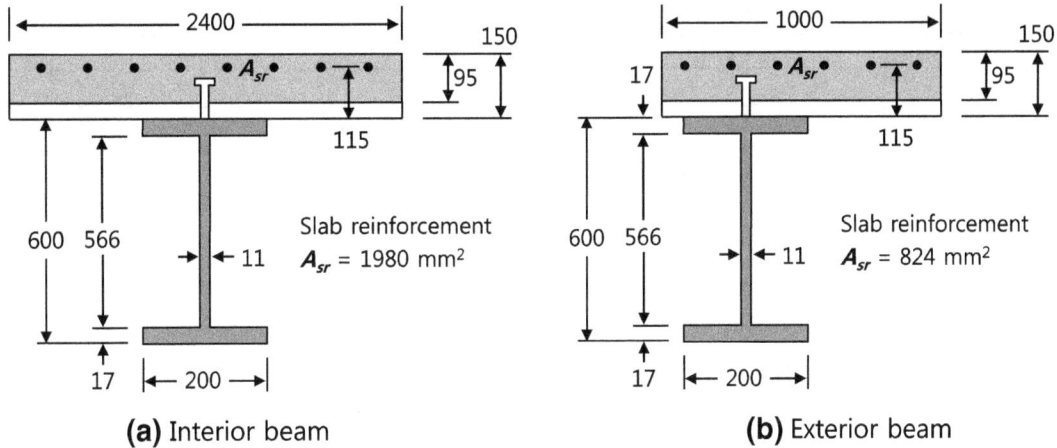

(a) Interior beam   **(b)** Exterior beam

Fig. 4 Cross sections of composite beams (mm).

Table 3 Design results calculated by PSDM: interior beams under positive bending (kN m).

f_{ck} (MPa)	F_{yk} (MPa)	AISC 360-10		KSSC–KCI		Eurocode 4		JSCE 2009
		M_{pl}	$M_d^{(1)}$	β	$M_d^{(1)}$	β	$M_d^{(1)}$	$M_d^{(1)}$
21	235	1268	1141	0.957	1138	_(2)	1196	1123
	315	1648	1483	0.952	1422	_(2)	1501	1407
	355	1803	1623	0.951	1564	_(2)	1653	1545
	450	2167	1950	0.934	1868	0.949	1907	1871
	650	2920	2628	0.782	2097	_(3)	_(3)	_(3)
30	235	1301	1171	1.000	1287	_(2)	1293	1203
	315	1709	1538	0.958	1550	_(2)	1628	1516
	355	1907	1716	0.956	1694	_(2)	1783	1677
	450	2354	2119	0.949	2025	0.961	2062	2010
	650	3122	2810	0.881	2539	_(3)	_(3)	_(3)

(1) $M_d = \phi M_{pl}$ for AISC 360-10 (or KBC 2014), M_{pl} or βM_{pl} for Eurocode 4, βM_{pl} for KSSC–KCI 2014, and M_{pl}/γ_b for JSCE 2009.
(2) For Eurocode 4, β shall be applied for high strength steel of $F_{yk} = 420$ and 460 MPa.
(3) For Eurocode 4 and JSCE 2009, the plastic stress distribution method shall not be applied for $F_{yk} = 650$ MPa.

ignored (i.e., assumed to be 1.0), the nominal plastic moment strengths calculated from KSSC–KCI, Eurocode 4, and JSCE 2009 are the same as that of AISC 360-10, $M_{pl,AISC}$. In this regard, $M_d/M_{pl,AISC}$, shown in Figs. 5 and 6, reflects not only the difference in design moment strengths between design codes but also the variation of the resistance factor for bending (=ϕ) of each design code, depending on the design variables, such as F_{yk}, f_{ck}, and b_{eff}.

As shown in Figs. 5 and 6, $M_d/M_{pl,AISC}$ of KSSC–KCI and Eurocode 4 showed decreasing trends as F_{yk} was increased from 235 to 650 MPa. Additionally, $M_d/M_{pl,AISC}$ for a lower concrete strength of $f_{ck} = 21$ MPa was mostly less than that for a higher concrete strength of $f_{ck} = 30$ MPa. $M_d/M_{pl,AISC}$ of the exterior beam with a narrower concrete flange ($b_{eff} = 1000$ mm) was mostly less than those of the interior beam with a wider concrete flange ($b_{eff} = 2400$ mm). Such trends of $M_d/M_{pl,AISC}$ with respect to F_{yk}, f_{ck}, and b_{eff} show that the design strength of the cross sections and the resistance factor for bending under positive bending are affected

substantially by the compression resistance of the concrete flange. That is, the greater F_{yk} of the steel section and the smaller f_{ck} and b_{eff} of the concrete flange made the depth D_p of the plastic neutral axis greater, which, in turn, resulted in increasing the contribution of the concrete flange to M_d. Because KSSC–KCI and Eurocode 4 define relatively higher resistance and safety factors for concrete (i.e. $\phi_c = 0.65$ and $\gamma_c = 1.5$, respectively), the safety margin for bending of the design strength increased along with the increased contribution of the concrete flange. In contrast, AISC 360-10 defines a constant resistance factor ϕ (=0.9) for bending regardless of material and section properties. As a result, $M_d/M_{pl,AISC}$ (or the resistance factor ϕ for bending) of KSSC–KCI and Eurocode 4 showed decreasing trends with respect to F_{yk}, f_{ck}, and b_{eff} in Figs. 5 and 6.

For KSSC–KCI and Eurocode 4, the reduction factor β, defined as a function of D_p/D_t, also affected the decreasing trends of $M_d/M_{pl,AISC}$ (or the resistance factor ϕ for bending). In particular, KSSC–KCI requires β be applied to steels of

Table 4 Design results calculated from PSDM: exterior beams under positive bending (kN m).

f_{ck} (MPa)	F_{yk} (MPa)	AISC 360-10		KSSC–KCI		Eurocode 4		JSCE 2009
		M_{pl}	$M_d^{(1)}$	β	$M_d^{(1)}$	β	$M_d^{(1)}$	$M_d^{(1)}$
21	235	1082	974	0.901	907	$-^{(2)}$	1004	939
	315	1377	1239	0.866	1089	$-^{(2)}$	1265	1185
	355	1514	1363	0.854	1181	$-^{(4)}$	$-^{(4)}$	1302
	450	1826	1643	$-^{(4)}$	$-^{(4)}$	$-^{(4)}$	$-^{(4)}$	1570
	650	2443	2199	$-^{(4)}$	$-^{(4)}$	$-^{(3)}$	$-^{(3)}$	$-^{(3)}$
30	235	1164	1048	0.949	1014	$-^{(2)}$	1073	1001
	315	1468	1321	0.911	1237	$-^{(2)}$	1364	1274
	355	1619	1457	0.894	1334	$-^{(2)}$	1499	1403
	450	1967	1770	0.866	1556	0.861	1555	1693
	650	2631	2368	$-^{(4)}$	$-^{(4)}$	$-^{(3)}$	$-^{(3)}$	$-^{(3)}$

$^{(1)}$ $M_d = \phi M_{pl}$ for AISC 360-10 (or KBC 2014), M_{pl} or βM_{pl} for Eurocode 4, βM_{pl} for KSSC–KCI 2014, and M_{pl}/γ_b for JSCE 2009.
$^{(2)}$ For Eurocode 4, β shall be applied for high strength steel of $F_{yk} = 420$ and 460 MPa.
$^{(3)}$ For Eurocode 4 and JSCE 2009, the plastic stress distribution method shall not be applied for $F_{yk} = 650$ MPa.
$^{(4)}$ The use of the plastic stress distribution method is restrained for $D_p/D_t \leq 0.4$ for Eurocode 4 and $D_p/D_t \leq 0.42$ for KSSC–KCI.

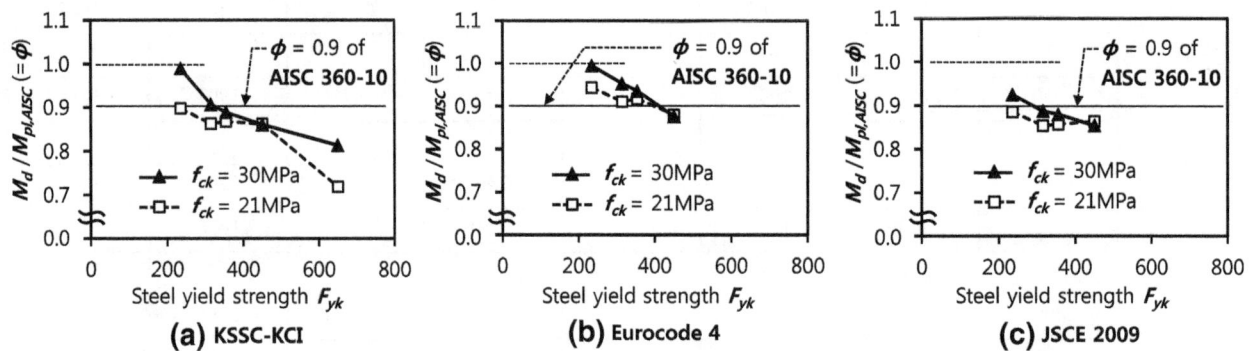

Fig. 5 Comparison of design strengths calculated from PSDM: interior beam under positive moment.

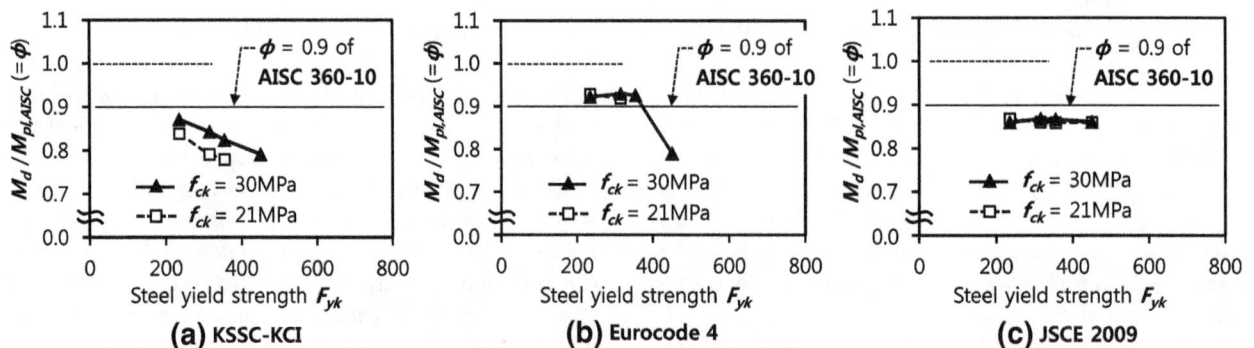

Fig. 6 Comparison of design strengths calculated from PSDM: exterior beam under positive moment.

all strength grades between $F_{yk} = 235$ and 650 MPa, while Eurocode 4 does not apply β to normal strength steels of $F_{yk} = 235$–355 MPa (see Fig. 2; Table 2). Thus, the decreasing trend of $M_d/M_{pl,AISC}$ was steeper in KSSC–KCI than in Eurocode 4.

For JSCE 2009 where the design format is a mixed form of PFM and LRFD, as shown in Figs. 5c and 6c, the variation of $M_d/M_{pl,AISC}$ (or the resistance factor ϕ for bending) according to the design variables, such as F_{yk}, f_{ck}, and b_{eff}, was not as significant as those of KSSC–KCI and Eurocode 4. Because the safety factor for member ($\gamma_b = 1.1$) acting as a resistance factor for bending had a significant impact on the design strengths, $M_d/M_{pl,AISC}$ (or ϕ) was almost constant regardless of F_{yk}, f_{ck}, and b_{eff}.

Figures 5 and 6 also compare the design strengths M_d of KSSC–KCI, Eurocode 4, and JSCE 2009 (PFM) with those of AISC 360-10 (LRFD). For AISC 360-10, the ratio of $M_d/M_{pl,AISC}$ is constant at 0.9, regardless of design variables. Therefore, if $M_d/M_{pl,AISC}$ of a design code is greater than 0.9, M_d of the design code is greater than that of AISC 360-10. The values of M_d calculated from Eurocode 4 were mostly greater than those of AISC 360-10, except for the cases of $F_{yk} = 420$ and 460 MPa. For KSSC–KCI, on the other hand, the values of M_d were mostly less than those of AISC 360-10, except for the interior beam with $F_{yk} = 235$ and 315 MPa. The averages of $M_d/M_{pl,AISC}$ were only 0.87 and 0.82 for the interior and exterior beams, respectively. Although the safety margins for materials specified in KSSC–KCI and Eurocode 4 were almost equivalent in magnitude, M_d of KSSC–KCI was reduced further even in $F_{yk} = 235$, 315, and 355 MPa as a result of applying the reduction factor β to all strength grades of steel. JSCE 2009 also showed the values of M_d less than those of AISC 360-10.

4.2 Negative Bending

Table 5 compares the design moment strengths M_d of the interior and exterior composite beams under negative bending, calculated by the PSDM specified in each design code. Properties of the cross sections are shown in Fig. 4. In the calculation of M_d, the tensile stress of concrete was ignored but the effect of slab reinforcement ($f_{yrk} = 400$ MPa and $A_{sr} = 1980$ and 824 mm^2) was included. Because of the effects of slab reinforcements, the values of M_d for the interior and exterior beams were slightly different (see Table 5). Figure 7 shows the variation of the design strengths M_d of KSSC–KCI, Eurocode 4, and JSCE 2009 according to steel yield strengths, F_{yk}. For comparisons between comparable design codes, the design strengths M_d of KSSC–KCI, Eurocode 4, and JSCE 2009 were divided by the nominal strength $M_{pl,AISC}$ of AISC 360-10 (see M_{pl} of Table 5). As discussed in the previous section, the ratio of

$M_d/M_{pl,AISC}$ is equivalent to the resistance factor for bending ($=\phi$) of each design code.

For KSSC–KCI and Eurocode 4, $M_d/M_{pl,AISC}$ was 1.0 regardless of material and section properties, such as F_{yk}, f_{ck}, and b_{eff}. Thus, KSSC–KCI and Eurocode 4 had a constant resistance factor for bending of $\phi = 1.0$. This is because M_d under negative bending was governed by the steel section, rather than the concrete flange. KSSC–KCI and Eurocode 4 that use PFM as their design format do not define any safety margin for steel (i.e., $\phi_s = 1.0$ and $\gamma_s = 1.0$, respectively). In contrast, AISC 360-10 uses the resistance factor ϕ ($=0.9$) for bending. For JSCE 2009, $M_d/M_{pl,AISC}$ ($=\phi$) was slightly less than 0.9 as the result of dividing by the member safety factor γ_b ($=1.1$), though the safety factor for steel was $\gamma_s = 1.0$.

As shown in Fig. 7, the design strengths M_d of KSSC–KCI and Eurocode 4 were about 10 % greater than those of AISC 360-10. For moment-resisting frame structures, the negative moment of composite beams at both ends are generally greater than the positive moment at the mid-span because lateral and gravity load effects are combined. Thus, from a practical view point, the greater M_d under negative bending of KSSC–KCI and Eurocode 4 can lead to a more economical design.

5. Design Resistance by Strain-Compatibility Method

5.1 Rotational Capacity and Resistance Factor for Bending

As discussed in the previous sections, AISC 360-10 that uses LRFD as its design format can secure a constant resistance factor for bending (i.e., $\phi = 0.9$), regardless of the rotational capacity of cross sections. For KSSC–KCI, Eurocode 4, and JSCE 2009 that use PFM as the design format, however, the resistance factor for bending may vary significantly according to design variables, such as the

Table 5 Design results calculated from PSDM: interior and exterior beams under negative bending (kN m).

| Type | F_{yk} (MPa) | AISC 360-10 | | KSSC–KCI | Eurocode 4 | JSCE 2009 |
		M_{pl}	$M_d^{(1)}$	$M_d^{(1)}$	$M_d^{(1)}$	$M_d^{(1)}$
Exterior beam	235	799	719	787	784	726
	315	1031	928	1019	1015	937
	355	1147	1032	1134	1130	1041
	450	1420	1278	1407	1403	1290
	650	1994	1795	1981	1977	1812
Interior beam	235	940	846	919	912	855
	315	1184	1066	1161	1153	1077
	355	1304	1174	1279	1272	1185
	450	1586	1427	1558	1550	1440
	650	2168	1951	2139	2130	1970

$^{(1)}$ $M_d = \phi M_{pl}$ for AISC 360-10 (or KBC 2014), M_{pl} for Eurocode 4 and KSSC–KCI, and M_{pl}/γ_b for JSCE 2009.

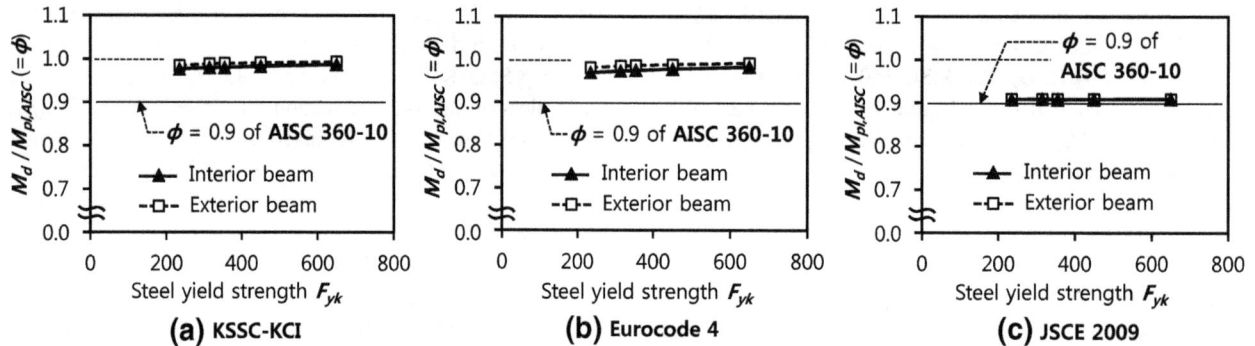

Fig. 7 Comparison of design strengths calculated from PSDM: interior and exterior beams under negative bending.

strength of materials and the geometry and rotational capacity of cross sections, because the margin of safety for bending is indirectly determined from the resistance or safety factor for each material. In this section, the quantitative relationship between the resistance factor for bending and the rotational capacity was investigated.

The investigation requires a strain-compatible section analysis addressing the stress–strain relationships of materials. For this, a fiber section analysis program to calculate the moment–curvature relationship of the cross section of composite members was developed. In the fiber section analysis, the cross section of a composite member is divided into a number of fiber elements with infinitesimal area and then internal forces of the steel section, concrete slab, and reinforcements are determined by integrating the infinitesimal stress and moment of each fiber element corresponding to strain. Figure 8 shows a typical moment–curvature relationship of the cross section of composite beams. For composite beams under positive bending, the ultimate limit state is defined as when the compressive strain of the extreme fiber of concrete slab reaches the ultimate strain ε_{cu}. The moment strength and curvature at the ultimate limit state are denoted as M_{nl} and κ_u, respectively (see Fig. 8). For KSSC–KCI, Eurocode 3, and JSCE 2009 that use PFM as their design format, the design moment strength M_d is determined as M_{nl} (KSSC–KCI and Eurocode 4) and M_{nl}/γ_b ($\gamma_b = 1.1$, JSCE 2009) (refer to Table 2).

For the cross section of a composite beam, the resistance factor for bending can be defined as M_d/M_k, where M_k is the nonlinear moment strength M_{nl} calculated from the fiber

section analysis using the characteristic strengths for materials f_k. Additionally, the rotational capacity can be quantified as the curvature ductility μ_d, determined by dividing the ultimate curvature κ_u by the yield curvature κ_y: $\mu_d = \kappa_u/\kappa_y$ (see Fig. 8). The yield curvature κ_y is defined from an idealized bilinear moment–curvature relationship constructed to pass through the point where the tensile flange of steel section reaches its yield stress first. In Fig. 8, the strain energy using the idealized bilinear curve until κ_u is the same as that using the actual moment–curvature curve.

The rotational capacity (i.e., μ_d) and the resistance factor for bending (i.e., ϕ) for the cross sections of interior and exterior beams, shown in Fig. 4, were evaluated. The characteristic yield strength of steel and the characteristic compressive strength of concrete varied between $F_{yk} = 235$–650 MPa and between $f_{ck} = 21$ and 30 MPa, respectively. Although not allowed in Eurocode 4 and JSCE 2009, high-strength steel of $F_{yk} = 650$ MPa was included in this investigation for a comparison between comparable design codes. For Eurocode 4 and JSCE 2009, the stress–strain relationships of concrete and steel presented in Table 2 were used for the fiber section analysis. For KSSC–KCI, the stress–strain relationships of concrete and steel proposed in the Sect. 3 were used. Reinforcements under compression in the concrete slab were ignored in the fiber section analysis.

Tables 6 and 7 show the values of M_k, M_d, ϕ_f, and μ_d for each design code, calculated from the fiber section analysis. Tables 6 and 7 are the results for the interior and exterior beams, respectively. For all design codes, as F_{yk} was increased from 235 to 650 MPa, M_k and M_d were increased, but μ_d was decreased. In particular, the value of μ_d of the exterior beam for $F_{yk} = 650$ MPa was 1.0, indicating brittle failure due to concrete crushing of the slab before the tensile yielding of the steel flange could occur. Thus, the rotational capacities of the composite beams were inversely proportional to F_{yk}. Additionally, when f_{ck} was increased from 21 to 30 MPa or b_{eff} was increased from 1000 to 2400 mm, M_k and M_d did not vary significantly but μ_d was increased. This indicates that to enhance the rotational capacity of composite beams under positive bending, the compression resistance of the concrete slab (e.g., concrete strength and effective flange width) need to be secured.

Figures 9 and 10 show the resistance factor for bending (ϕ)-curvature ductility (μ_d) relationships of interior and

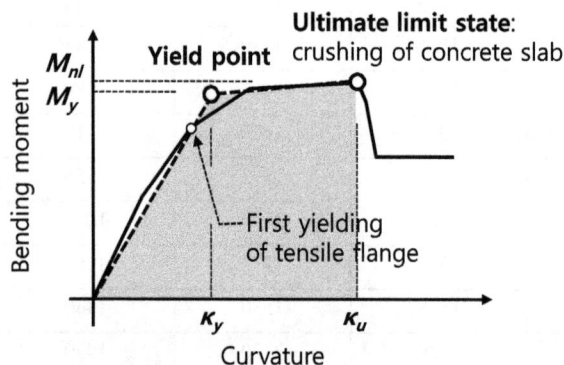

Fig. 8 Definition of ultimate limit state and yield point.

Table 6 Design results calculated from SCM: interior beams under positive bending.

f_{ck} (MPa)	F_{yk} (MPa)	KSSC–KCI				Eurocode 4				JSCE 2009			
		$M_k^{(1)}$	$M_d^{(1)}$	ϕ	μ_d	$M_k^{(1)}$	$M_d^{(1)}$	ϕ	μ_d	$M_k^{(1)}$	$M_d^{(1)}$	ϕ	μ_d
21	235	1282	1221	0.952	9.36	1282	1232	0.961	10.7	1461	1245	0.852	11.0
	315	1674	1534	0.916	5.73	1675	1544	0.921	6.17	1778	1504	0.845	6.70
	355	1841	1684	0.914	4.51	1847	1695	0.917	4.93	1916	1629	0.850	5.44
	450	2223	2008	0.903	2.99	2226	2023	0.908	3.30	2265	1920	0.847	3.58
	650	2914	2497	0.856	1.68	2915	2537	0.870	1.78	2874	2398	0.834	1.88
30	235	1310	1274	0.972	14.9	1311	1277	0.974	15.4	1575	1352	0.858	15.1
	315	1726	1655	0.958	8.30	1726	1667	0.965	8.76	1915	1652	0.862	8.80
	355	1928	1814	0.940	6.54	1928	1827	0.947	7.37	2085	1792	0.859	7.38
	450	2377	2189	0.920	4.51	2378	2204	0.926	4.81	2460	2107	0.856	4.82
	650	3194	2844	0.890	2.19	3200	2873	0.897	2.43	3195	2692	0.842	2.77

[1] M_k and M_d are the moment strengths for characteristic and design strengths of materials, respectively (Unit: kN m).

Table 7 Design results calculated from SCM: exterior beams under positive bending.

f_{ck} (MPa)	F_{yk} (MPa)	KSSC–KCI				Eurocode 4				JSCE 2009			
		$M_k^{(1)}$	$M_d^{(1)}$	ϕ	μ_d	$M_k^{(1)}$	$M_d^{(1)}$	ϕ	μ_d	$M_k^{(1)}$	$M_d^{(1)}$	ϕ	μ_d
21	235	1109	1020	0.919	4.37	1110	1024	0.922	4.38	1145	986	0.861	4.73
	315	1392	1271	0.913	2.79	1394	1281	0.918	2.96	1405	1208	0.859	3.08
	355	1519	1377	0.906	2.31	1524	1393	0.914	2.52	1524	1309	0.858	2.59
	450	1782	1581	0.887	1.67	1782	1604	0.900	1.76	1769	1506	0.851	1.86
	650	2078	1789	0.860	1.00[2]	2134	1821	0.853	1.00[2]	2099	1703	0.811	1.00[2]
30	235	1198	1092	0.911	5.51	1201	1098	0.914	5.86	1252	1065	0.850	6.18
	315	1506	1369	0.909	3.54	1506	1379	0.915	3.55	1525	1305	0.855	3.86
	355	1650	1493	0.904	2.93	1651	1507	0.912	2.93	1657	1419	0.856	3.15
	450	1960	1743	0.889	2.04	1964	1763	0.897	2.04	1938	1640	0.846	2.20
	650	2402	2032	0.846	1.00[2]	2434	2107	0.865	1.00[2]	2341	1953	0.834	1.00[2]

[1] M_k and M_d are the moment strengths for characteristic and design strengths of materials, respectively (Unit: kN m).
[2] Crushing failure of the extreme fiber of concrete slab occurs before the tensile yielding of bottom flanges.

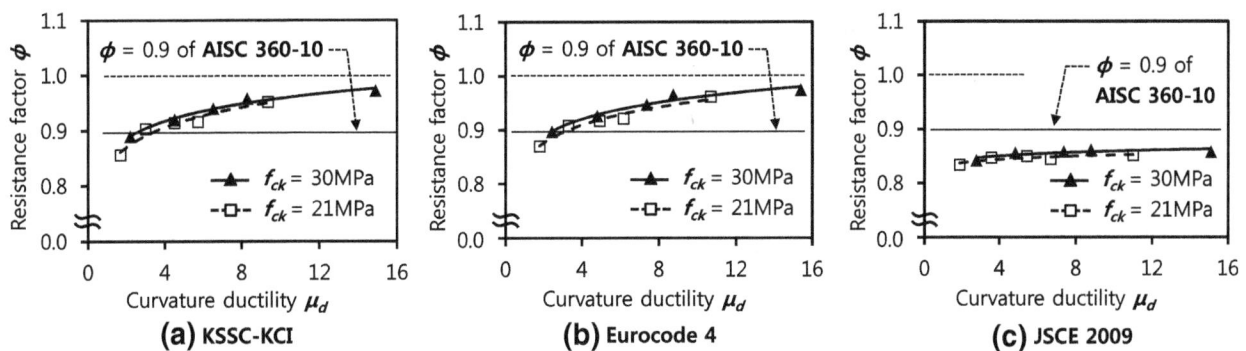

Fig. 9 Resistance factor for bending versus rotational capacity: interior beam.

exterior beams, respectively. In the figures, the values corresponding to f_{ck} = 21 and 30 MPa are marked as rectangles and triangles, respectively. For KSSC–KCI and Eurocode 4 that use PFM as their design format, ϕ ($=M_d/M_k$) was

increased, close to 1.0, as μ_d was increased. The trend in the ϕ–μ_d relationships of the interior and exterior beams was very similar (compare Figs. 9 and 10). The reason for this trend in the ϕ–μ_d relationships can be explained as follows.

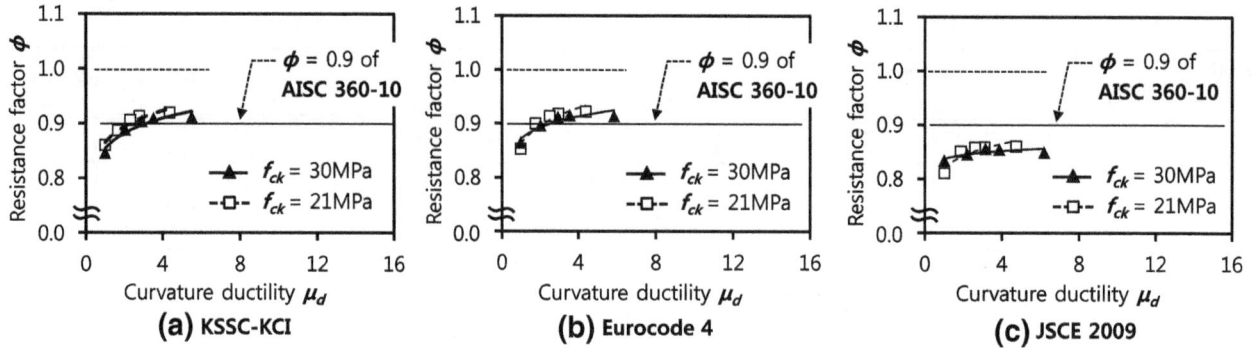

Fig. 10 Resistance factor for bending versus rotational capacity: exterior beam.

When the rotational capacity is small (e.g., $\mu_d \leq 3.0$), ϕ is primarily determined by the resistance and safety factors for concrete (i.e., $\phi_c = 0.65$ and $\gamma_c = 1.5$, respectively) because the moment strength of the cross section is dominated by the concrete flange, rather than by the steel section. On the other hand, when the rotational capacity is large (e.g., $\mu_d \geq 4.0$), ϕ is determined primarily by the resistance and safety factors for the steel (i.e., $\phi_s = 1.0$ and $\gamma_s = 1.0$, respectively) because the moment strength of the cross section is dominated by the steel section.

In contrast, for AISC 360-10 that uses LRFD as the design format, the resistance factor for bending ϕ is constant at 0.9, regardless of the rotational capacity (see the dashed lines in Figs. 9 and 10). Furthermore, ϕ of JSCE 2009 did not vary much according to μ_d because the member safety factor γ_b (=1.1) acted as a constant safety factor for bending (see Figs. 9c and 10c).

5.2 Comparison Between Design Strengths of PSDM and SCM

Figure 11 compares the design strengths of the interior beam under positive bending, calculated by the PSDM and SCM, $M_{d,PSDM}$ and $M_{d,SCM}$, respectively. The values of $M_{d,PSDM}$ and $M_{d,SCM}$ for each design codes are shown in Tables 3 and 6, respectively. The results for the exterior beam under positive bending are presented in Tables 4 and 7 and Fig. 12. For KSSC–KCI and Eurocode 4, the ratios of $M_{d,SCM}/M_{d,PSDM}$ were mostly greater than 1.0, and increased as the yield strength of steel was increased from $F_{yk} = 235$ to 650 MPa. This indicates that by using the SCM, an

economical structural design for composite beams may be possible, especially if high-strength steel is used.

$M_{d,SCM}$ greater than $M_{d,PSDM}$ shown in Figs. 11 and 12 were attributed to two reasons. First, the reduction factor β specified in KSSC–KCI and Eurocode 4 did decrease the design strengths of cross sections calculated from the PSDM. Additionally, because β decreases as D_p/D_t increases, $M_{d,PSDM}$ decreased further especially when high-strength steels of $F_{yk} = 450$ and 650 MPa were used. Second, the compressive stress distribution of concrete flange did increase the design strengths calculated by the SCM. Figure 13 illustrates the stress and strain distributions of the interior beam for KSSC–KCI ($f_{ck} = 21$ MPa and $F_{yk} = 235$ MPa), calculated from the fiber section analysis. The neutral axis at the ultimate limit state was located in between the concrete slab and the compression flange of steel section (i.e., 124 mm deep from the top surface of the concrete slab). The calculated maximum and minimum compressive stresses in the concrete flange were $1.0f_{cd}$ and $0.608f_{cd}$, respectively, and the mean value was $0.93f_{cd}$. Clearly, the mean stress $0.93f_{cd}$ was 13 % greater than the plastic stress of concrete assumed for the PSDM, $0.85f_{cd}$. This, along with the reduction factor β (=0.957; see Table 3), resulted in the 7.0 % greater $M_{d,SCM}$ (=1221 kN m) than $M_{d,PSDM}$ (=1141 kN-m).

Figures 11c and 12c show the ratios of $M_{d,SCM}/M_{d,PSDM}$ of the interior and exterior beams, respectively, calculated from JSCE 2009. The ratios of $M_{d,SCM}/M_{d,PSDM}$ were mostly greater than 1.0 but, in contrast to KSSC–KCI and Eurocode 4, decreased as the yield strength of steel was increased from $F_{yk} = 235$ to 650 MPa. This difference between $M_{d,SCM}$ and

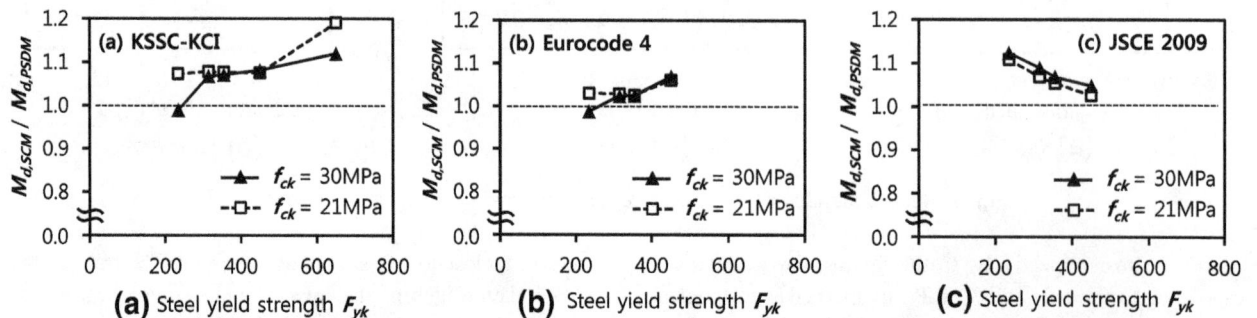

Fig. 11 Comparison of design strengths calculated from PSDM and SCM: interior beam.

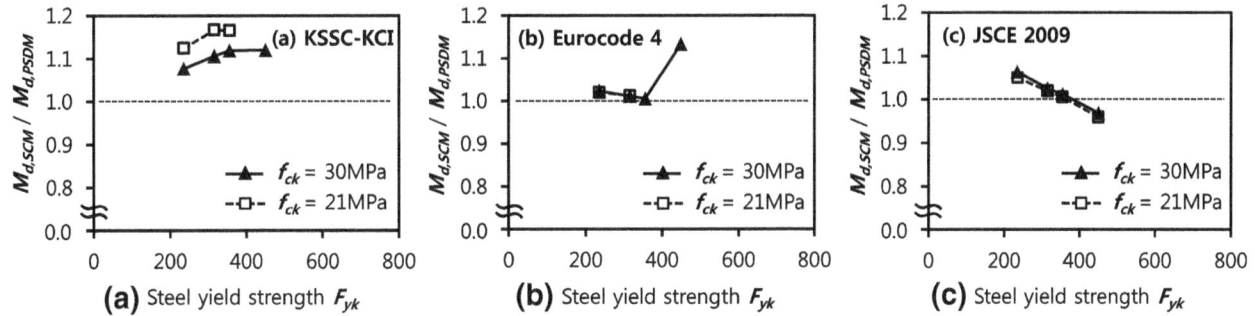

Fig. 12 Comparison of design strengths calculated from PSDM and SCM: exterior beam.

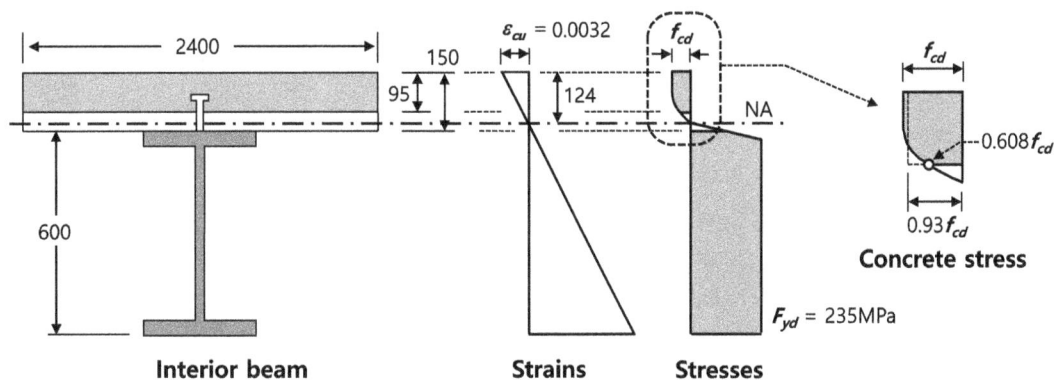

Fig. 13 Strain and stress distributions at ultimate limit state: interior beam for KSSC–KCI 2014 (f_{ck} = 21 MPa and F_{yk} = 235 MPa).

$M_{d,PSDM}$ was attributed to the strain-hardening behavior of steel addressed in the SCM (see Table 2), as follows. Because JSCE 2009 allows a tensile stress of steel greater than the yield strength due to the strain-hardening behavior, basically, $M_{d,SCM}$ can be greater than $M_{d,PSDM}$. However, when high-strength steel is used, the stress increase of steel is less significant because the rotational capacity of cross sections is poor. Thus, the difference between $M_{d,SCM}$ and $M_{d,PSDM}$ is greatly reduced, especially if high-strength steels of F_{yk} = 450 and 650 MPa are used.

6. Summary and Conclusion

In this study, provisions for the flexural design of composite beams specified in KSSC–KCI (i.e., a draft version prepared by the KSSC–KCI joint composite structure committee), Eurocode 4, and JSCE 2009, which use PFM as their design format, were compared with those of AISC 360-10 and KBC 2014 based on LRFD, in terms of design format, material strength, and resistance or safety factor. Additionally, the design moment strengths M_d of the cross sections, calculated by the plastic stress design method (PSDM) and strain-compatibility method (SCM) specified in each design code, were investigated quantitatively. The major findings of this study can be summarized as follows.

1. The design strength M_d and resistance factor for bending ϕ, calculated from the PSDM specified in KSSC–KCI, Eurocode 4, and JSCE 2009, varied significantly with

material and section properties. For positive bending, M_d and ϕ of KSSC–KCI and Eurocode 4 showed decreasing trends as the depth of the plastic neutral axis increased. In particular, the reduction factor β reduced the design values further for high-strength steel. M_d and ϕ of Eurocode 4 were mostly greater than the design values of AISC 360-10. However, the design values of KSSC–KCI and JSCE 2009 were less than those of AISC 360-10. For negative bending, the design strengths of KSSC–KCI and Eurocode 4 that define the safety or resistance factor for steel as 1.0 were about 10 % greater than those of AISC 360-10 that use the resistance factor for bending as ϕ = 0.9.

2. The resistance factor for bending ϕ calculated from the SCM specified in KSSC–KCI and Eurocode 4 was increased, close to 1.0 from 0.85, as the rotational capacity of the cross section was increased. This is because, in the case of the PFM that uses different resistance factors for concrete and steel, the overall resistance factor for bending of the cross sections was determined primarily by concrete or steel, whichever was dominant. For JSCE 2009, on the other hand, ϕ did not vary much according to the rotational capacity because the member safety factor γ_b (=1.1) acted as a constant safety factor for bending.

3. For KSSC–KCI and Eurocode 4, the design strengths M_d of the cross section under positive bending calculated from the SCM were greater than those by PSDM. The SCM was beneficial to prevent brittle failure of

composite beams due to early concrete crushing and to achieve economical designs, especially when high-strength steel of $F_{yk} = 420$–650 MPa is used. For JSCE 2009, the SCM was most economical for composite beams using normal-strength steel.

Acknowledgments

The present research was conducted by the research fund of Dankook university in 2014.

References

American Association of State Highway and Transportation Officials. (2012). *AASHTO LRFD specifications* (6th ed.). Washington DC.

American Institute of Steel Construction. (2010). Specifications for structural steel buildings, AISC 360-10, Chicago, IL.

Architectural Institute of Korea. (2014). Korea Building Codes (Draft), KBC 2014, Seoul (in Korean).

European Committee for Standardization. (2004a). Eurocode 4: Design of composite steel and concrete structures, Part 1-1: General rules and rules for buildings, EN 1994-1-1:2004.

European Committee for Standardization. (2004b). Eurocode 2: Design of concrete structures—Part 1-1: General rules and rules for buildings, EN 1992-1-1:2004.

Japan Society of Civil Engineers. (2007). Standard specifications for concrete structures—2007—"Design", JSCE Guidelines for Concrete No. 15

Japan Society of Civil Engineers. (2009). Standard specifications for steel and composite structures.

Kim, D. H., Kim, J. H., & Chang, S. K. (2014). Material performance evaluation and super-tall building applicability of the 800 MPa high-strength steel plates for building structures. *International Journal of Steel Structures, 14*(4), 889–900.

Kim, T. S., Lee, M. J., Suk, O. Y., Lee, K. M., & Kim, D. H. (2012a). A study on compressive strength of built-up H-shaped columns fabricated with HSA800 high performance steels. *Journal of the Korean Society of Steel Construction, KSSC, 24*(6), 627–636. (in Korean).

Kim, C. S., Park, H. G., Chung, K. S., & Choi, I. R. (2012b). Eccentric axial load testing for concrete-encased steel columns using 800 MPa steel and 100 MPa concrete. *Journal of Structural Engineering, 138*(8), 1019–1031.

KSSC–KCI Joint Composite Structure Committee. (2014). *Design codes of composite structures (Draft)*. Seoul, korea: Korean Society of Steel Construction. (in Korean).

Lee, C. H., Han, K. H., Kim, D. K., Park, C. H., Kim, J. H., Lee, S. E., & Ha, T. H. (2012). Local buckling and inelastic behavior of 800 MPa high-strength steel beams. *Journal of Korean Society of Steel Construction, KSSC, 24*(4), 479–490. (in Korean).

Lee, C. H., Han, K. H., Uang, C. H., Kim, D. K., Park, C. H., & Kim, J. H. (2013a). Flexural strength and rotation capacity of I-shaped beams fabricated from 800-MPa. *Journal of Structural Engineering, ASCE, 139*(6), 1043–1058.

Lee, C. H., Kang, K. Y., Kim, S. Y., & Koo, C. H. (2013b). Review of structural design provisions of rectangular concrete filled tubular columns. *Journal of the Korean Society of Steel Construction, KSSC, 25*(4), 389–398. (in Korean).

Lee, C. H., Kim, D. K., Han, K. H., Park, C. H., Kim, J. H., Lee, S. E., & Kim, D. H. (2013c). Tensile testing of groove welded joints joining thick-HSA800 plates. *Journal of the Korean Society of Steel Construction, KSSC, 25*(4), 431–440. (in Korean).

Lee, M. J., Kim, C. W., & Kim, H. D. (2014). The evaluation of the axial strength of composite column with HSA800 grade steel. *Journal of the Korean Society of Steel Construction, KSSC, 24*(6), 627–636. (in Korean).

Wittry, D.M. (1993). An analysis study of the ductility of steel-concrete composite sections, MS Thesis, University of Texas, Austin, TX.

Youn, S. G. (2013a). Nominal moment capacity of hybrid composite sections using HSB600 high-performance steel. *International Journal of Steel Structures, KSSC, 13*(2), 243–252.

Youn, S. K. (2013b). Reevaluation of nominal flexural strength of composite girders in positive bending region. *Journal of the Korean Society of Steel Construction, 25*(2), 165–178. (in Korean).

16

Comparison of Strength–Maturity Models Accounting for Hydration Heat in Massive Walls

Keun-Hyeok Yang[1),*], Jae-Sung Mun[2)], Do-Gyeum Kim[3)], and Myung-Sug Cho[4)]

Abstract: The objective of this study was to evaluate the capability of different strength–maturity models to account for the effect of the hydration heat on the in-place strength development of high-strength concrete specifically developed for nuclear facility structures under various ambient curing temperatures. To simulate the primary containment-vessel of a nuclear reactor, three 1200-mm-thick wall specimens were prepared and stored under isothermal conditions of approximately 5 °C (cold temperature), 20 °C (reference temperature), and 35 °C (hot temperature). The in situ compressive strengths of the mock-up walls were measured using cores drilled from the walls and compared with strengths estimated from various strength–maturity models considering the internal temperature rise owing to the hydration heat. The test results showed the initial apparent activation energies at the hardening phase were approximately 2 times higher than the apparent activation energies until the final setting. The differences between core strengths and field-cured cylinder strengths became more notable at early ages and with the decrease in the ambient curing temperature. The strength–maturity model proposed by Yang provides better reliability in estimating in situ strength of concrete than that of Kim et al. and Pinto and Schindler.

Keywords: high-strength concrete, in situ strength, mock-up, hydration heat, maturity, curing temperature.

List of Symbols

D	Diameter of core
$E_a(i)$	Apparent activation energy at time step i
E_i	Initial apparent activation energy during the hardening phase
E_s	Apparent activation energy until the final setting time
F_d	Correction factor to damage to the surface of the core samples
F_{dia}	Correction factor to account for the diameter of the core samples
$F_{H/D}$	Correction factor to account for the slenderness of core samples
F_{mc}	Correction factor to account for the core moisture content
f_{core}	Compressive strength of core samples
H	Height of core samples
k_r	Rate constant at the reference temperature
k_t	Rate constant
R	Universal gas constant (= 8.314 J/mol/K)
R_{sp}	High-range water-reducing agent–cementitious material ratio by mass
S	Compressive strenth of concrete at an age of t
S_u	Limiting strength of concrete
S_{28}	28-day compressive strenth of concrete
$(S_{28})_{Tr}$	28-day compressive strenth of concrete under reference curing temperature
T_{A3}	Curing temperature at an age of 3 days
$T_c(i)$	Curing temperature of concrete at time step i
T_r	Reference curing temperature
t	Concrete age
t_e	Equivalent age
t_s	Final setting time at a given temperature
t_{sr}	Final setting time at the reference temperature
t_0	Offset time at a given temperature
t_{0r}	Offset time at the reference temperature
w/cm	Water–cementitious material ratio by mass
γ	Ratio between estimated compressive strength of concrete and test results
γ_m	Mean of the ratios between estimates and test results
γ_s	Standard deviation of the ratios between estimates and test results
γ_v	Coefficient of variation of the ratios between estimates and test results

[1)]Department of Plant Architectural Engineering, Kyonggi University, Suwon 443-760, Korea.
*Corresponding Author; E-mail: yangkh@kgu.ac.kr
[2)]Department of Architectural Engineering, Graduate School, Kyonggi University, Suwon 443-760, Korea.
[3)]Structural Engineering & Bridges Research Division, Korea Institute of Construction Technology, Goyang 411-712, Korea.
[4)]KHNP-Central Research Institute, Korea Hydro & Nuclear Power Co., LTD, Daejeon 305-343, Korea.

1. Introduction

Construction schedules for concrete structures significantly depend on the minimum stripping time for the concrete form work and shoring and the minimum concrete strength for applying a prestressing force to a structural element (Sofi et al. 2012; Vázquez-Herrero et al. 2012). Thus obtaining a higher strength gain of concrete at an early age is one of the most critical concerns for shortening the construction time for concrete structures. For this reason, the use of high-strength concrete (HSC) has been gradually encouraged for the fast-track construction of nuclear facility structures, especially in South Korea. The strength development of HSC is more sensitive to the curing temperature than normal-strength concrete (NSC) because the hydration rate of cement is greater at a lower water–cementitious materials ratio and a higher curing temperature (Kim et al. 2002a; Pinto and Schindler 2010). As a result, the strength–maturity models established from the NSC test data using standard cylinders are frequently pointed out to overestimate the in situ strength of HSC, particularly under a cold temperature at an early age (Parsons and Naik 1985; Hulshizer 2001). However, the strength–maturity relationship for HSC is still an equivocal issue because of the limited available data, although the maturity method is used as an effective means of estimating strength development of in situ concrete (Sofi et al. 2012).

The strength–maturity relationship of concrete is traditionally determined using standard cubic or cylindrical specimens cured in a laboratory at standard temperature. However, as commonly recognized (Haug and Jakobsen 1990; Puciontti 2013; Uva et al. 2013), the curing histories of a standard-cured specimen and an in-place concrete member would not be identical. Thus, the strength development of a laboratory specimen would differ from that of a structure under construction. This difference may be primarily attributed to the heat of hydration, which produces a higher internal temperature in structural members than the surrounding environment (Kim et al. 2002b). Hence, a large difference between a cylinder's strength and the in situ strength is expected with an increase in the design strength of the concrete and/or member thickness. However, available data (Harris et al. 2000; Schrader 2007) dealing with the difference between standard-cured or field-cured cylinder strength and in-place strength for a mass concrete element are still scare, although the concept of equivalent age derived from the Arrhenius function has been primarily used as a maturity function to describe the temperature sensitivity of the reaction of cementitious materials. Furthermore, to reasonably estimate construction schedules for mass concrete structures under various temperatures, the reliability of the strength–maturity relationships needs to be ascertained for HSC used in massive concrete members.

The purpose of this study was to examine the differences in the compressive strength development between field-cured cylindrical specimens and a mass concrete element made using HSC with a design strength of 55 MPa. The mixture proportions of the HSC were specifically determined based on its use in nuclear facility structures under different ambient temperatures in South Korea, while considering the hydration heat generation, economic efficiency and durability of the concrete (Yang 2014). To simulate the primary containment-vessel of a nuclear reactor, three 1200-mm-thick mock-up wall specimens were prepared and cured under isothermal ambient temperatures of approximately 5, 20, and 35 °C. The in situ compressive strengths of the mock-up walls were measured using cores drilled from the wall structures, in accordance with ASTM C42/C42 M (2011). The setting times, rate constants, and apparent activation energies of the prepared mixtures were also measured in accordance with ASTM procedures (2011) in order to calculate the equivalent age based on the Arrhenius function. The measured core strength was compared with the companion cylinder strength and estimates obtained from different maturity approaches (Kim et al. 2002a; Pinto and Schindler 2010; Yang 2014).

2. Experimental Details

2.1 Concrete Mixtures

To improve the economic efficiency by shortening the construction time and extending the expected service life of nuclear facility structures, a target of 55 MPa was set for the 28-day compressive strength of concrete. Considering the workability and need to minimize the bleeding of concrete in the primary containment-vessel of a nuclear reactor with large diameter reinforcing bars, a value of 150 ± 15 mm was selected for the target initial slump of the fresh concrete. For the targeted concrete strength and initial slump, tests of numerous laboratory mixtures were previously conducted under different ambient curing temperatures of approximately 5, 20, and 35 °C. As a result, three mixture proportions were specifically determined for use at the three ambient temperatures, as listed in Table 1 (Yang 2014).

According to the ambient curing temperatures, different supplementary cementitious materials (SCMs) were added as partial replacements for the cement. The heat of hydration and the rate of heat evolution in concrete commonly increase with increasing C_3S and C_3A contents of cement. On the other hand, the pozzolanic reaction is slower than C_3S hydration and it produces less heat than does cement hydration (Nili and Salehi 2010). As a result, concrete containing SCMs normally experiences slow hydration, accompanied by a lower temperature rise. Bamforth (1980) reported that ground granulated blast-furnace slag (GGBFS) as a partial replacement for ordinary portland cement (OPC) generates a lower temperature rise and a slower rate of increase than OPC mass concrete. The weight ratios of the SCMs selected for the three mixtures were as follows: 5 % silica fume (SF) for the ambient curing temperature of 5 °C, 50 % GGBFS for a temperature of 20 °C, and a combination of 65 % GGBFS and 5 % SF for a temperature of 35 °C. The three mixtures were identified as S5, G50, and G65S5

Table 1 Concrete mixture proportions.

Mixtures	Ambient curing temperature (°C)	w/cm (%)	Quantity (kg/m^3)						
			Water	Modified cement	SF	GGBFS	Sand	Gravel	R_{sp}* (%)
S5	5	34	155	433	23	–	737	941	1.5
G50	20	34	155	228	–	228	705	900	0.7
G65S5	35	34	155	137	23	296	703	897	0.5

* R_{sp} = high-range water-reducing agent–to–cementitious material ratio by mass.

based on the SCM replacements. For all three mixtures, the water–cementitious material ratio (w/cm) and unit water content were fixed at 0.34 and 155 kg/m^3, respectively. To achieve the targeted initial slump, a high-range water-reducing admixture was also added, as given in Table 1. The main composition of this admixture was acrylic acid-acrylic ester copolymer, lignosulfonate, and sodium gluconate.

2.2 Materials

The chemical compositions of the cementitious materials are given in Table 2, which were obtained using an X-ray fluorescence (XRF) analysis. The cement that is commonly used for nuclear plant structures in South Korea was selected as the main binder. The chemical composition of the cement was modified to reduce the hydration heat generation. As a result, the chemical composition of the cement was close to that of a moderate heat cement. As compared with the common chemical composition of Type I portland cement specified in ASTM C150 (2011), the aluminum oxide (Al_2O_3) content in the modified portland cement was lower by approximately 2 %, whereas the silicon oxide (SiO_2) content was 1 % higher. From a potential Bogue composition (1955) of the mineral compounds based on the percentage of the given oxide in the total mass of the modified portland cement, the C_3S, C_2S, C_3A and C_4AF contents were calculated to be 43.9, 33.9, 3.7 and 11.6 %, respectively. This indicates that the C_3S and C_3A contents of the modified cement were 16 and 54 % lower than those conventionally determined from the Type I cement, respectively, whereas C_2S content of the former was 37.6 % higher than that of latter. The GGBFS, which conformed to ASTM C989 (2011), had a high calcium oxide (CaO) content and a SiO_2–to–Al_2O_3 mass ratio of 2.29. The basicity of the GGBFS calculated from the chemical composition was 1.94. The primary component of the SF was SiO_2. The specific gravity and specific surface area of the cementitious materials were

3.15 and 3466 cm^2/g, respectively, for the modified portland cement, 2.94 and 4497 cm^2/g for GGBFS, and 2.32 and 200,000 cm^2/g for SF.

Natural sand and locally available crushed granite with a maximum particle size of 25 mm were used for fine and coarse aggregates, respectively. The specific gravity, water absorption, and fineness modulus are given in Table 3.

2.3 Mock-up Wall Specimens and Curing

To simulate the primary containment-vessel of a nuclear reactor, wall specimens were prepared. The size of these wall specimens was 1200 × 1000 × 2000 mm, as shown in Fig. 1. For vertical and horizontal reinforcements, deformed bars with a diameter of 35 mm were arranged at a spacing of 300 mm as a minimum configuration of wall reinforcement. Both ends of the walls were insulated using 50-mm-thick expanded polystyrene. The bottom of the walls had no insulation materials. To cure the wall specimens under isothermal ambient temperatures, three chambers were manufactured using 75-mm-thick sandwich panels, as shown in Fig. 2. Each curing chamber was equipped with an automatic constant-temperature control system. The average temperatures in the chambers were set to 5, 20, and 35 °C to simulate cold weather (winter), reference (control laboratory) and hot weather (summer) conditions, respectively, because the three mixture proportions of concrete were determined considering different ambient temperatures in South Korea. After the wall formworks were set up in the chambers, the concrete was placed using a concrete pumping vehicle. The concretes were produced at a ready-mixed concrete plant using the mixture proportions given in Table 1. Because the mock-up tests were carried out during cold weather, the minimum concrete temperatures at time of placement and mixing were maintained above 7 and 10 °C, respectively, in accordance with the ACI 306 report (2010). Immediately after casting, the specimens were covered with

Table 2 Chemical composition of cementitious materials (% by mass).

Materials	SiO$_2$	Al$_2$O$_3$	Fe$_2$O$_3$	CaO	MgO	K$_2$O	Na$_2$O	TiO$_2$	SO$_3$	LOI*
Modified cement	23.30	3.85	3.83	63.4	1.24	1.47	0.15	0.33	2.01	0.42
SF	98.94	0.30	0.08	0.12	0.04	0.13	0.05	–	0.28	0.06
GGBFS	33.18	14.07	0.51	44.6	4.31	0.45	0.24	0.55	1.36	0.73

* Loss on ignition.

Table 3 Physical properties of aggregates.

Type	Maximum size (mm)	Bulk density (kg/m^3)	Specific gravity	Water absorption (%)	Fineness modulus
Coarse aggregates	25	1.557	2.59	0.78	6.64
Fine aggregates	5	1.668	2.56	1.34	2.80

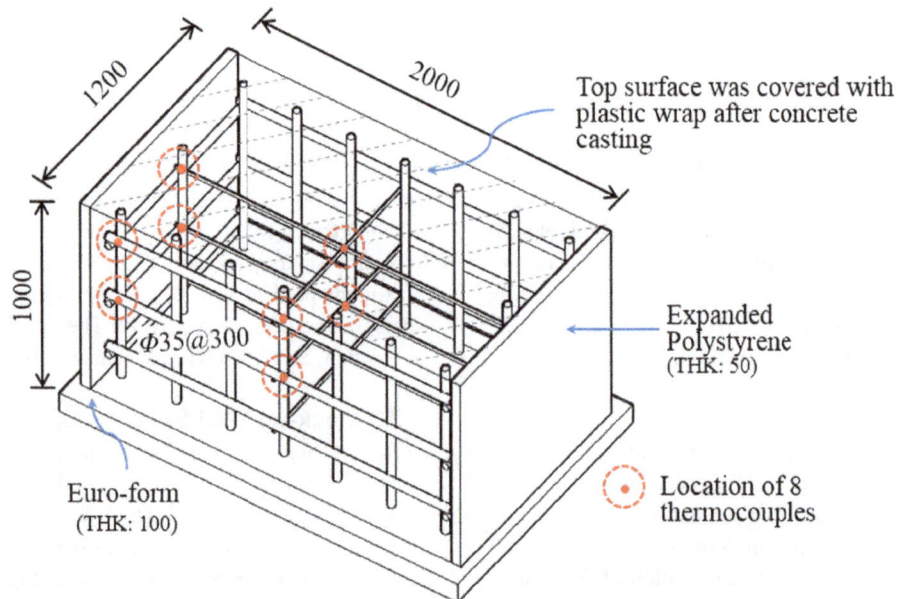

Fig. 1 Details of mock-up wall specimens and location of thermocouples (All dimensions are in mm).

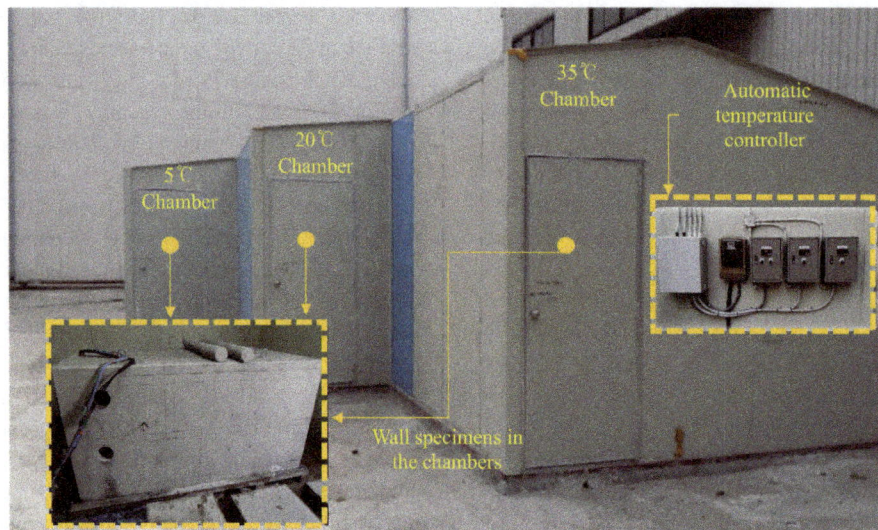

Fig. 2 Curing chambers equipped with constant-temperature control systems.

a vinyl chloride sheet to control evaporation. The wall specimens were continuously cured in the chambers under the previously set ambient temperature conditions. For comparison with the field-cured cylinder strength, 100 × 200 mm cylinders molded from each concrete mixture were simultaneously cured with the wall specimens at the same ambient temperatures. The cylinders and the wall specimen cured under the cold condition were stripped out at

an age of 3 days, whereas the others cured under the reference or hot condition were stripped out at 1 day.

2.4 Testing

Temperatures were monitored using thermocouples within the chambers and at eight different locations for the wall specimens, including the center and near surface regions, as shown in Fig. 1. Temperature data were recorded using a data logger at 20-min intervals to the end of the first day, and every 2 h thereafter. To evaluate setting times by penetration resistance and determine the apparent activation energy of the prepared concrete mixture, fresh mortar was extracted from each concrete mix using a 4.75-mm sieve. The mortars were simultaneously cured with each wall specimen at the same ambient temperatures. The penetration resistance testing to determine the setting time of the concrete was conducted in accordance with ASTM C403/C403 M (2011). The testing required to experimentally determine the apparent activation energy for the hardening phase was performed using 18 cubes with 50-mm dimensions and cured at three temperatures (5, 20, and 35 °C), in accordance with ASTM C1074 (2011). To measure the in situ strength of the concrete in the walls, cores with a diameter of 100 mm were drilled at different locations with different hydration heat histories, as shown in Fig. 3. All cores were drilled horizontally. Cores were not drilled at the bottom regions of the walls because it was difficult to install a core machine there. The extraction was carefully carried out by experienced operators to minimize drilling damage. The cores were classified into four groups according to their drilling locations as follows (Fig. 3): inner surface portion at top region [top surface (TS)], central portion at top region [top center (TC)], inner surface portion at middle region [middle surface (MS)], and central portion at middle region [middle center (MC)]. The cores, which did not contain reinforcement, were cut to a length of 200 mm and tested in the same manner as

standard cylinder specimens with the same dimension, in accordance with ASTM C42/C42 M (2011). The ends of the cores were ground to minimize an eccentricity. Immediately after the cores were drilled, the drilling water was wiped from their surface using dry towels. Measurements of the compressive strength of the cores were scheduled at the ages of 3, 7, 14, 28, 56, and 91 days. Only 2 cores per location and testing age were tested because of the size limitation of the mock-up walls.

The core strength is affected by aspect ratio and diameter of the cores, presence of embedded reinforcement, and disturbance owing to drilling (Uva et al. 2013). Thus, the compressive strength (f_{core}) of the cores was corrected using a single relationship specified in ACI 214.4R-10 (2010), as follows:

$$S = F_{H/D}F_{dia}F_{mc}F_d f_{core} \tag{1}$$

where S is the corrected strength of concrete, and $F_{H/D}$, F_{dia}, F_{mc}, and F_d are correction factors to account for the slenderness, diameter, moisture content, and damage to the surface of the core. The correction factors for the current cores, $F_{H/D}$, F_{dia}, and F_{mc} are calculated to be 1.0 and F_d is 1.06.

Because the mock-up walls were not subjected to any loads, no cracking owing to external loads existed in the cores. Furthermore, apparent damage or cracking owing to drilling was not observed in the surface of the cores, as shown in Fig. 4.

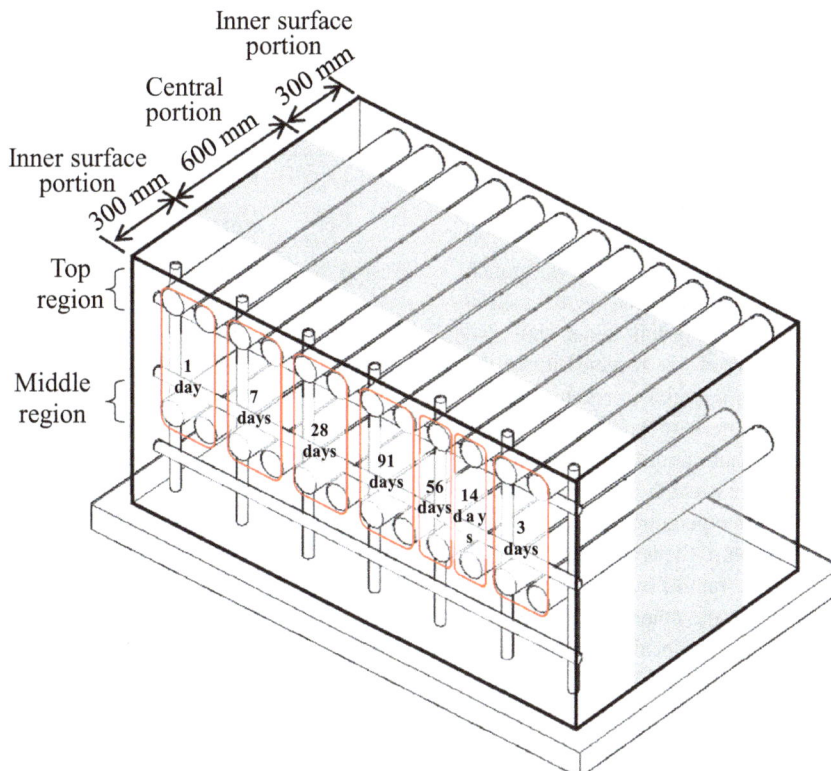

Fig. 3 Core drilling locations in wall specimens at different testing ages.

Fig. 4 Typical surface state of core samples.

3. Test Results and Discussion

3.1 Temperature Rise

The temperature rise profiles of the concrete resulting from the hydration of the cementitious materials are shown in Fig. 5. The ambient temperature profile of each chamber is also plotted on the same figure. As expected, peak temperature was higher at the central portion of the wall than at its inner surface. The highest temperature was recorded in the middle-center region. The differences between the peak temperatures at the center and inner surface regions tended to decrease with an increase in the ambient curing temperature. The peak temperature measured in the middle center was higher for S5 wall cured under 5 °C than for the other walls, whereas the peak temperature measured in the surface was higher for G65S5 wall cured under 35 °C than the other walls. A higher ambient curing temperature resulted in the hotter surface of the wall. This implies that the difference of the peak temperatures between the center and inner surface regions is more dependent on the ambient curing temperature than the mixture proportions used in the present study. It is also interesting that a greater slope at the ascending branch of the temperature rise curve was observed at the central portion compared to the surface portion under the cold curing condition, whereas this trend was reversed under the hot curing condition. The times required to reach the peak temperature at the central portion were 1.5, 2, and 2 days under the cold, reference, and hot conditions, respectively, showing the potential for a faster strength development with the S5 mixture under the cold temperature than the G65S5 mixture under the hot temperature. This was because the addition of GGBFS reduced the rate of hydration reactions of the cementitious materials. On the other hand, the rate of the temperature drop after reaching the peak increased with a decrease in the ambient curing temperature.

It is commonly known that the potential maximum heat production of SCMs is lower than that of portland cement (Nili and Salehi 2010). As a result, using SCMs as cement replacements in concrete reduces the temperature rise by reducing the cement content per a unit volume. Figure 5 also

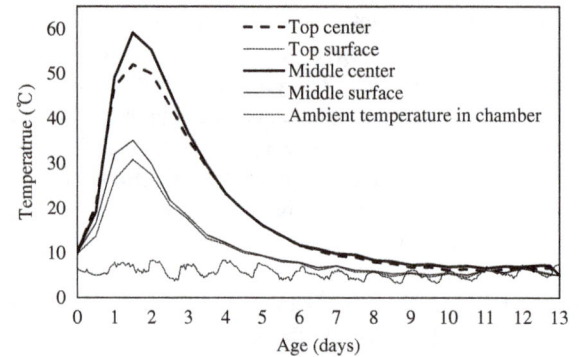

(a) S5 mixture under cold condition

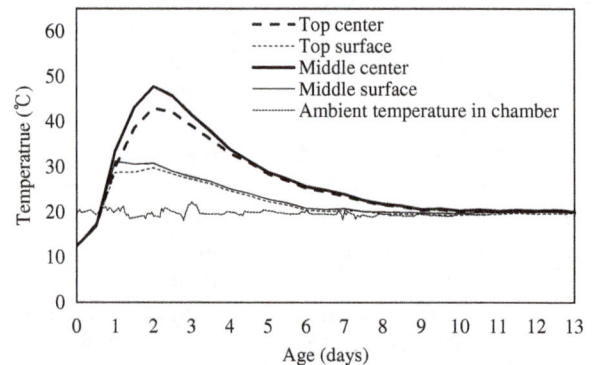

(b) G50 mixture under reference condition

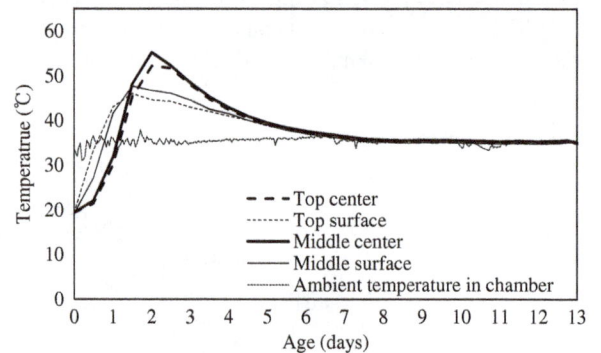

(c) G65S5 mixture under hot condition

Fig. 5 Temperature rise profile for each concrete wall.

clearly demonstrates that the peak temperature recorded in the middle-center region of the G50 wall was lower than that of the S5 wall, even though the G50 mixture was cured under a higher ambient temperature than the S5 mixture. The peak temperature of the G65S5 wall was higher than that of the G50 wall, even though a higher amount of SCM was used in the G65S5 concrete than in the G50 concrete. The heat production rate at an early age in cement paste commonly increases with an increase in the curing temperature (Zákoutský et al. 2012), which results in a higher cumulative heat production at an early age. The 35 °C ambient temperature increased the rate of hydration to compensate for the reduced portland cement content. This implies that a higher curing temperature produces a higher temperature rise in concrete but leads to a lower differential between the core and surface of concrete element.

3.2 Activation Energy Until Final Setting Time

Figure 6 shows the setting behavior of the tested mixtures and the fitting curves determined from each dataset. The penetration resistance versus time of each concrete mixture significantly depended on the ambient curing temperature and the addition of GGBFS. The differences in the setting times of the concrete mixtures decreased with an increase in the ambient curing temperature. As shown in Table 4, the G50 and G65S5 mixtures showed longer final setting times than the S5 mixture by 5.3 and 8.2 h under the cold temperature condition, 3.4 and 5.4 h under the reference temperature condition, and 1.0 and 1.5 h under the hot temperature condition, respectively. This indicates that the delay in final setting of concrete owing to the addition of slag cemnet is more notable with a decrease in the ambient curing temperature. The apparent activation energy (E_s) until the final setting time was calculated from the Arrhenius plot using the inverse of the final setting time as a rate constant. The obtained values for E_s were 22,700, 24,800, and 26,100 J/mol for the S5, G50 and G65S5 mixtures, respectively, as given in Table 4. Overall, E_s slightly increased with an increase in the addition of GGBFS.

3.3 Initial Activation Energy at Start of Hardening Phase

In general, the apparent activation energy is considered to be a key parameter in the maturity function using the equivalent age, because it describes the effect of the temperature on the rate of the strength development of the

concrete after final setting. Using a graph with the reciprocal of compressive strength of mortars cured under three different temperatures of 5, 20, and 35 °C and the reciprocal of age, the rate constant (k_t) of each concrete mixture was determined (see Table 4), in accordance with ASTM C1074 (2011). At the same curing temperature, the k_t values determined at the start of the hardening phase were lower in the G50 and G65S5 mixtures than in the S5 mixture, indicating that k_t decreases with an increase in the addition of GGBFS, as shown in Fig. 7. The temperature dependence of k_t showed a nonlinear variation rather than a linear relationship, regardless of the SCM addition. Consequently, the k_t values for each concrete were fitted using the Arrhenius function.

The initial apparent activation energy (E_i) values calculated from the Arrhenius plot of the natural logarithm of the k_t values versus the inverse of absolute temperature were 42,200, 54,700, and 68,900 J/mol for the S5, G50 and G65S5 mixtures, respectively, as given in Table 4. These values for such mixtures were higher by 1.8, 2.2, and 2.6 times, respectively, than the values of E_s. This implies that the temperature sensitivity of the strength development after final setting is more than that of setting time. The apparent activation energy was not a constant value, because it depends on the type of cement, the type and dosage of the SCMs, and the w/cm. In general, the values of E_i for normal-strength OPC concrete without SCMs have been reported to be between 40,000 and 45,000 J/mol (ASTM 2011). The values of E_i for the current concrete mixtures exceeded the general range for OPC, showing that the E_i value tended to increase with an increase in the dosage level of GGBFS. These higher values of E_i indicates that strength development of concrete with GGBFS would be more sensitive to temperature than OPC concrete.

3.4 28-Day Compressive Strength (S_{28})

Figure 8 shows comparisons of the field-cured cylinder strengths and core strengths of the concrete mixtures at an age of 28 days. The strength of the concrete mixtures measured using 100 × 200 mm cylinders were close to the design strength of 55 MPa. The strength of the S5 mixture at 5 °C was approximately 10 % higher than that of the G50 mixture cured at 20 °C. The G65S5 mixture cured at 35 °C had a 28-day strength similar to the G50 mixture. Compared with the field-cured cylinder strength, a higher 28-day core strength was commonly observed. The average ratios between core strengths and cylinder strengths were

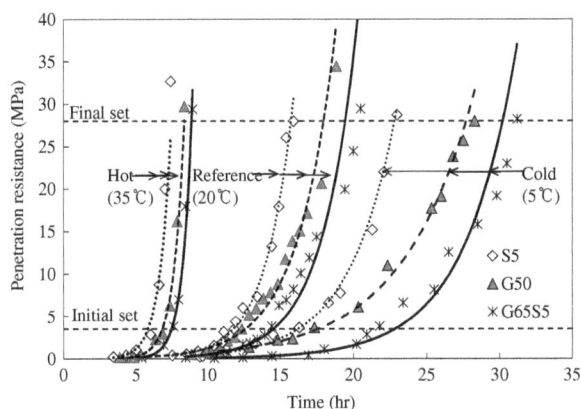

Fig. 6 Setting behavior of concrete mixtures according to curing temperature.

Table 4 Characteristics of concrete mixtures.

Mixtures	Setting time (hrs)						E_s (J/mol)	E_i (J/mol)	k_t (1/days)		
	5 °C		20 °C		35 °C				5 °C	20 °C	35 °C
	Initial	Final	Initial	Final	Initial	Final					
S5	16.30	23.00	10.80	14.92	6.50	7.70	22,700	42,200	0.16	0.30	0.95
G50	17.34	28.30	12.34	18.37	7.42	8.57	24,800	54,700	0.07	0.18	0.72
G65S5	21.80	31.20	15.20	20.30	8.00	8.90	26,100	68,900	0.05	0.15	0.70

1.15 and 1.19 at the top and middle regions, respectively, for the S5 wall, and 1.21 and 1.27 at those regions of the G50 wall. The 28-day core strength was slightly higher at the middle regions of the walls compared to the top regions, whereas the difference of the core strengths between the inner surface and center at each region was not significant. The ratios between the core strength and cylinder strength for the G50 wall were higher than those for the S5 wall, even though the temperature difference between the wall core and atmosphere owing the hydration heat was greater in the S5 wall than in the G50 wall. This may be attributed to the higher value of E_i for G50 mixture. In addition, the GGBFS has more temperature sensitivity than SF, which showed that, under a hot temperature of 30–50 °C, a higher strength could be obtained in concrete with GGBFS than in concrete with SF at the same w/cm (Lee et al. 2013). The average ratios between the core strength and cylinder strengths were lower for the G65S5 wall than those of the other walls.

3.5 Comparison of Compressive Strength Development

Table 5 gives the average strength measured from the cores and field-cured cylinders at different ages. For the S5 mixture cured under the cold condition, the core strength was higher than the cylinder strength until an age of 28 days, beyond which a higher strength was obtained in the cylinder, showing the crossover behavior, that is, higher early-age temperatures result in higher early strength and lower long–term strength. As shown in Fig. 5, the temperature at the center portion of the S5 wall owing to the hydration heat was above 20 °C until an age of 4 days, and then remained slightly higher than the ambient temperature. The walls experienced higher early-age temperatures than the cylinders and concrete S5 is susceptible to the crossover effect. This indicates that a higher core strength was obtained until an age of 56 days, compared to the cylinder strength, and the core strength was also higher at the wall center than at its inner surface. It is noted that the threshold age for the crossover effect is commonly observed between 7 and 14 days in OPC concrete (Carino and Tank 1992; Neville 1995). However, the present S5 mixture showed a longer threshold age of around 56 days. The chemical composition of the modified cement used for the present mixture was close to that of moderate heat cement, giving lower C_3S and C_3A contents than OPC. The reduced C_3S and C_3A contents in the modified cement were unfavorable to strength development at an early age. Furthermore, the period when the temperature was maintained above 10 °C at the center region of a wall was the first 7 days only. For these reasons, a late threshold age for the crossover effect was obtained in the S5 mixture. The strength gain of the cores beyond 28 days was insignificant, revealing that the core strength ratios between 91 days and 28 days did not exceed 1.05. This implies that the high early-age curing temperature caused by the hydration heat reduced long-term strength gain of concrete S5. No

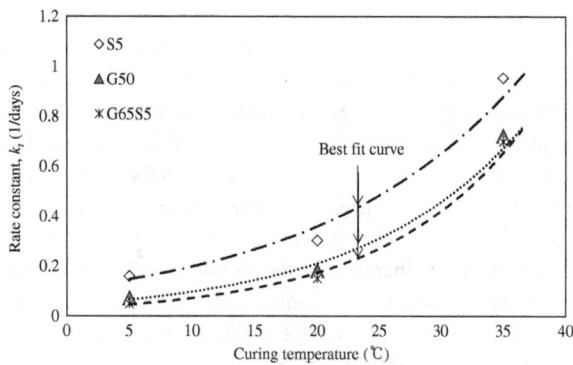

Fig. 7 Rate constant during initial hardening phase of concrete.

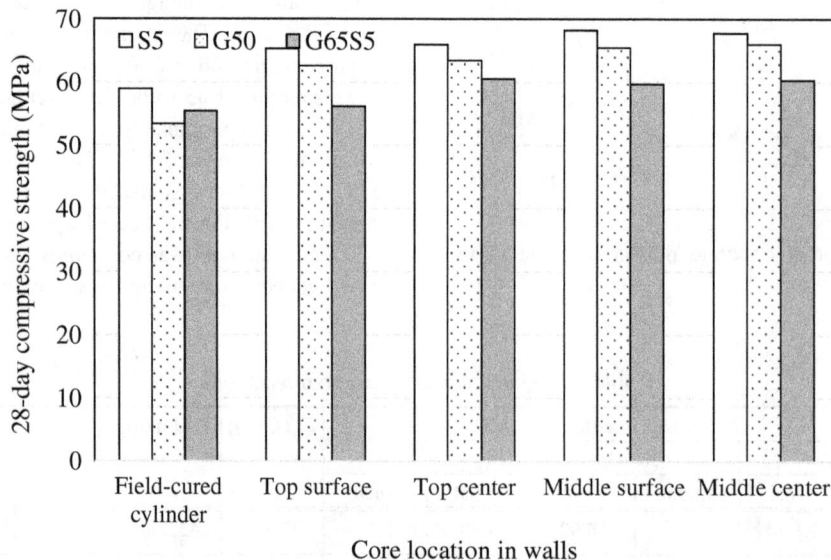

Fig. 8 Comparisons of field-cured cylinder and core strengths at age of 28 days.

Table 5 Compressive strength development of concrete mixtures.

Mixture	Specimens		Compressive strength (MPa) at different ages (days)					
			3	7	14	28	56	91
S5	Field-cured cylinder		21.5	37.9	50.9	58.9	67.5	75.8
	Core (location drilled from walls)	Top surface	32.4	49.2	61.9	65.3	65.5	66.7
		Top center	34.4	48.5	62.8	66.0	66.6	67.0
		Middle surface	36.6	51.4	64.7	68.3	68.3	69.7
		Middle center	44.9	52.3	61.4	67.8	69.8	71.0
G50	Field-cured cylinder		18.4	37.5	45.5	53.4	55.4	58.9
	Core (location drilled from walls)	Top surface	27.8	44.2	50.2	62.6	63.4	65.4
		Top center	35.9	50.0	56.0	63.5	64.6	65.7
		Middle surface	29.4	43.7	56.4	65.5	65.5	67.1
		Middle center	33.6	53.0	56.2	66.0	67.0	67.6
G65S5	Field-cured cylinder		30.7	45.3	52.2	55.4	55.1	58.3
	Core (location drilled from walls)	Top surface	32.7	48.4	54.5	56.2	56.8	58.3
		Top center	43.3	54.9	57.0	60.6	58.3	60.6
		Middle surface	35.8	48.3	55.8	59.7	61.3	62.8
		Middle center	41.3	55.6	56.0	60.4	63.3	63.7

crossover effect was observed in either the G50 or the G65S mixture. For the G50 mixture, which was stored at an ambient temperature of 20 °C, higher core strengths compared to cylinder strengths were obtained throughout the testing ages, showing the highest strength for the core drilled from the middle-center region of the wall, as given in Table 5. The G65S5 mixture, which was stored at 35 °C, showed smaller differences between the core and cylinder strengths, compared with the other mixtures.

4. Comparisons of Maturity Functions and Experiments

4.1 Review of Existing Models

Considering temperature variation by the heat of hydration in mass concrete or the change of external environment, Kim et al. (2002a) proposed a modified strength–maturity relationship for estimating strength development of concrete, as summarized in Table 6. Their modified maturity function assumed that the value of E_i is constant with aging because the activation energy can be regarded to be the characteristic property of concrete, whereas the limiting strength (S_u) is a function of age and temperature under variable curing temperature. In the maturity function, the parameters were based on regression analysis of test data. However, the value of S_u under variable temperature was not clearly identified, which would lead to additional hard task to obtain straightforwardly the strength development at a specified age.

Pinto and Schindler (2010) proposed an extended maturity approach to unify the distinctly different temperature sensitivities before setting and during the hardening period of

concrete, based on the Arrhenius maturity function. The effect of different activation energies on the strength–maturity relationship was taken into account, as summarized in Table 6. This maturity approach also demonstrated that the setting behavior needs to be taken into account in the strength–maturity relationship in order to improve the overall strength estimation of concrete at all ages. Hence, different activation energy values before and after the final setting of concrete were used in calculating equivalent age at the reference temperature. The offset time function (t_{sr}) was introduced to account for the effect of temperature on the setting time of the concrete. However, there are insufficient data for verification of the Pinto and Schindler model for estimating strength gain under variable temperature conditions.

Yang (2014) proposed a modification for the strength–maturity relationship proposed by Carino and Tank (1992) in order to explain that the maturity is related to the relative strength rather than the absolute strength. The maturity function computes the equivalent age (t_e) at the reference temperature (T_r), including the setting and hardening phases. Because the offset time (t_{0r}) at T_r is related to the setting behavior of concrete, the offset time (t_0) at a given temperature was assumed to be equal to the final setting time (t_s) of concrete at that temperature. This maturity function considered that during the hardening phase, increments of equivalent age are proportional to the affinity ratio of the rate constants using E_a, whereas the offset time at the setting phase is inversely proportional to the affinity ratio determined using E_s, because a higher temperature would result in a shorter setting time. Furthermore, the temperature–dependent hydration reaction also affects the value of the E_a at the hardening phase. Byfors (1980) showed that E_a decreases sharply beyond a certain age

Table 6 Summary of previous models for strength–maturity relationship.

Researcher	Formulation of the relationship
Kim et al. (2002a)	$$\frac{S}{S_u} = \left\{ 1 - \frac{1}{\sqrt{1 + \sum_{i=1}^{n} A \left[e^{-\frac{E_a(i)}{R[T_c(i)+273]}^{-\alpha t_i}} + e^{-\frac{E_a(i)}{R[T_c(i)+273]}^{-\alpha t_{i-1}}} \right] (t_i - t_{i-1})}} \right\}$$ where $A = 1 \times 10^7$ (experimental constant); $E_a(i) = 42830 - 43T_c(i)$ (in J/mol); $\alpha = 0.00017T_c(i)$; and $t_o = 0.66 - 0.011T_c(i) \geq 0$
Pinto and Schindler (2010)	$$\frac{S}{S_u} = \frac{k_r(t_e^* - t_{sr})}{1 + k_r(t_e^* - t_{sr})}$$ where $t_{sr} = t_s \exp\left[-\frac{E_s}{R} \left(\frac{1}{T_c(i)+273} - \frac{1}{T_r+273} \right) \right]$; and $$t_e^* = \sum_0^{t_{sr}} \exp\left[-\frac{E_s}{R} \left(\frac{1}{T_c(i)+273} - \frac{1}{T_r+273} \right) \right] t_i + \sum_0^t \exp\left[-\frac{E_i}{R} \left(\frac{1}{T_c(i)+273} - \frac{1}{T_r+273} \right) \right] t_i$$
Yang (2014)	$$\frac{S}{S_{28}} = \beta_1 \frac{k_r(t_e - t_{or})}{1 + k_r(t_e - t_{or})}$$ where $\beta_1 = \frac{S_u}{S_{28}} = 1 + \frac{1}{k_r \cdot 28}$; $k_T = k_r \exp\left[-\frac{E_a(i)}{R} \left(\frac{1}{T_c(i)+273} - \frac{1}{T_r+273} \right) \right]$; $t_{or} = t_{sr} \exp\left[-\frac{E_s}{R} \left(\frac{1}{T_c(i)+273} - \frac{1}{T_r+273} \right) \right]$; $$t_e = \sum_0^{t_{sr}} \exp\left[-\frac{E_s}{R} \left(\frac{1}{T_c(i)+273} - \frac{1}{T_r+273} \right) \right] \Delta t_i + \sum_{t_{sr}}^3 \exp\left[-\frac{E_a(i)}{R} \left(\frac{1}{T_c(i)+273} - \frac{1}{T_r+273} \right) \right] \Delta t_i + \sum_3^t \exp\left[-\frac{E_a(i)}{R} \left(\frac{1}{T_{A3}+273} - \frac{1}{T_r+273} \right) \right] \Delta t_i$$ $S_{28} = \left[\left(\frac{T_{A3}}{T_r} \right)^2 (w/cm)^4 + 0.97 \right] (S_{28})_{T_r}$; and $E_a(i) = E_i \cdot \exp(-0.00017T_c(i) \cdot t)$

that varies with curing temperature. Therefore, E_a somewhat depends on the curing temperature and age, which affects the value of equivalent age. However, the compressive strength development of concrete is insignificantly affected by the value of the apparent activation energy. For the three concrete mixtures tested, the estimates using the constant value of E_i produced the approximately same strength–age curves as the estimates using the variable value E_a given in Table 6. Hence, the parameter E_a in Yang's model needs to be replaced using the constant E_i for simpler calculation process. Yang (2014) showed that the strength development of HSC is independent of the curing temperature after an early critical age, and 3 days can be selected as this critical age. Considering the temperature-dependence of the setting and hardening phases and the early-age curing temperature effect until the critical period, t_e for HSC was calculated as the sum of three terms, as given in Table 6. To straightforwardly calculate the strength development at different ages, the relationship between S_{28} and $(S_{28})_{T_r}$ needs to be established for a given concrete. From the regression analysis of test data, Yang showed that the relationship between S_{28} and $(S_{28})_{T_r}$ is significantly affected by w/cm and the temperature until the early age of 3 days as a critical factor to represent the whole temperature history.

4.2 Comparisons with Test Results

Figure 9 shows some comparisons of estimated strength based on strength–maturity relationships in Table 6 with strength gain of concrete mixtures measured by the field-cured cylinders and cores drilled from the middle center regions of the walls. Table 7 shows the mean (γ_m), standard deviation (γ_s), and coefficient of variation (γ_v) of the ratios ($\gamma = S_{estimation}/S_{test}$) between estimated compressive strength of concrete and test results. In estimating the in situ strengths of walls using the reviewed maturity approaches, the temperature profile measured at each core location was used for the curing temperature. The ambient temperature of each chamber was also used for estimating the strength of field-cured cylinders. For the models proposed by Pinto and Schindler (2010) and Yang (2014), the experimental values given in Table 4 were used for key parameters such as t_{sr}, k_r, E_i, and E_s. For the limiting strength in the models of Kim et al. and Pinto and Schindler, 91-day strength was used. The model proposed by Kim et al. (2002a) underestimated the strength gain of the S5 mixture stored at 5 °C but tended to overestimate the strength gain of the G50 and G65S5 mixtures. This overestimation was greater for cylinder specimens than for cores, as shown in Fig. 9. The estimates obtained from the Pinto and Schindler model (2010) were also lower than the measured strength gain of the S5 mixture, but were in good agreement with the strength gain of cores drilled from the G50 and G665S5 walls. However, the Pinto and Schindler model still overestimated the field-cured cylinder strength of the G50 and G665S5 mixtures. Furthermore, this model gave no strength gain of cores at an age

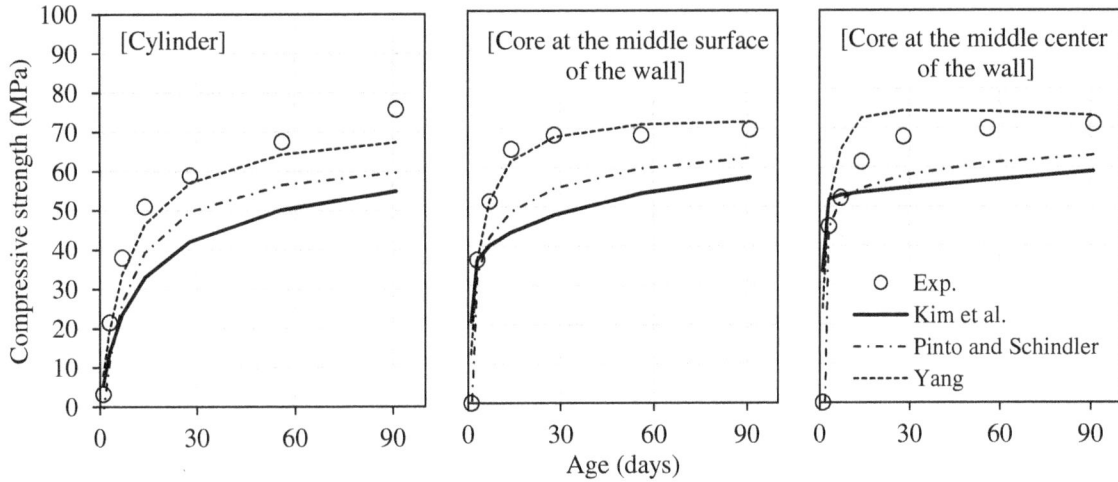

(a) S5 mixture under cold temperature condition

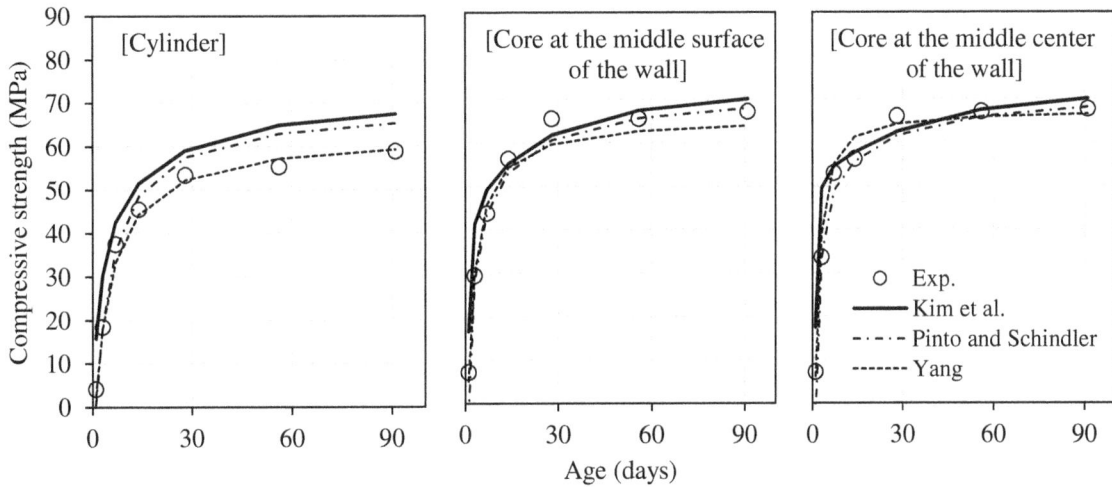

(b) G50 mixture under reference temperature condition

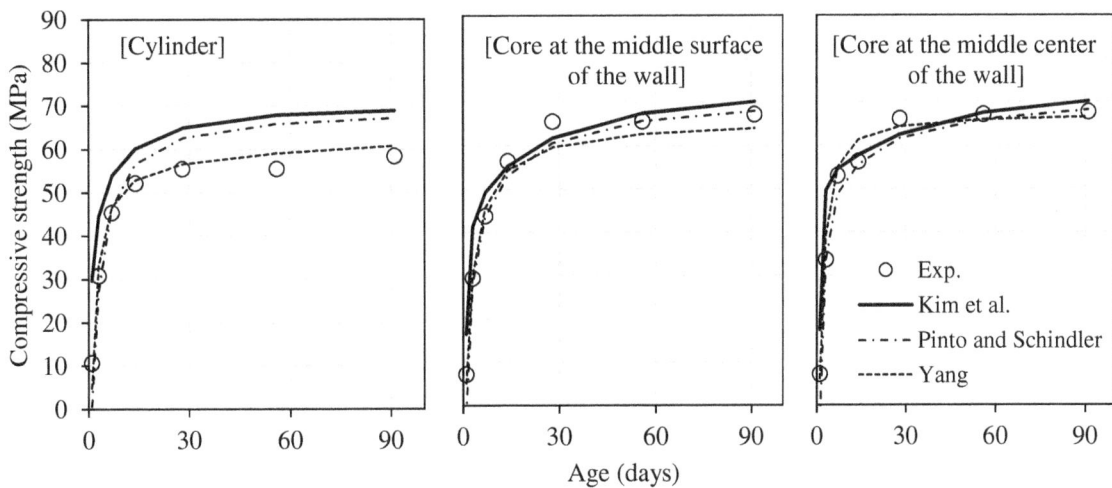

(c) G65S5 mixture under hot temperature condition

Fig. 9 Comparisons of measured strengths and estimates based on strength–maturity relationships given in Table 6.

Table 7 Statistical values of ratios between estimated and measured strengths.

Mixtures	Statistics	Kim et al. (2002a)					Pinto and Schindler (2010)					Yang (2014)				
		Age (days)				Total	Age (days)				Total	Age (days)				Total
		3	7	28	91		3	7	28	91		3	7	28	91	
S5	γ_m	1.084	0.866	0.757	0.825	0.883	0.942	0.888	0.843	0.890	0.890	1.130	1.101	1.044	1.022	1.074
	γ_s	0.312	0.188	0.057	0.060	0.213	0.273	0.144	0.028	0.062	0.150	0.217	0.177	0.066	0.080	0.144
	γ_v	0.288	0.218	0.076	0.073	0.241	0.290	0.162	0.033	0.070	0.168	0.192	0.161	0.061	0.078	0.134
G50	γ_m	1.475	1.098	0.992	1.072	1.159	0.980	0.969	0.974	1.040	0.991	1.063	1.021	0.965	0.986	1.009
	γ_s	0.103	0.042	0.066	0.043	0.201	0.035	0.030	0.060	0.040	0.049	0.072	0.079	0.037	0.024	0.065
	γ_v	0.069	0.038	0.067	0.040	0.173	0.036	0.031	0.062	0.0390	0.049	0.067	0.077	0.039	0.024	0.064
G65S5	γ_m	1.393	1.151	1.147	1.170	1.215	0.855	1.010	1.110	1.141	1.029	1.215	1.086	1.053	1.076	1.107
	γ_s	0.122	0.062	0.037	0.040	0.125	0.080	0.046	0.034	0.039	0.124	0.109	0.046	0.029	0.040	0.087
	γ_v	0.087	0.054	0.032	0.034	0.103	0.094	0.045	0.030	0.034	0.120	0.090	0.042	0.028	0.037	0.079

γ_m, γ_s, and γ_v indicate the mean, standard deviation, and coefficient of variation of the ratios ($\gamma = S_{estimation}/S_{test}$) between the estimates and measured strengths.

of 1 day. Pinto and Schindler model had lower values of γ_s and γ_v than Kim et al.'s model. The maturity approach proposed by Yang (2014) tended to overestimate the early strength until the age of 3 days for the S5 mixture and the G65S5 mixture, beyond which the overestimation gradually diminished. For the estimates based on Yang's model, no clear trend was observed for the differences in the γ values between the cylinders and cores. For the three mixtures at different ages of 3, 7, 14, 28, 56, and 91 days, the overall values of γ_m, γ_s, and γ_v determined from the reviewed models were 1.057, 0.219, and 0.207, respectively, for Kim et al., 0.979, 0.127, and 0.130 for Pinto and Schindler, and 1.058, 0.098, and 0.093 for Yang. This indicates that Yang's approach gives slightly more reliable estimates with less scatter, even at early ages.

5. Conclusions

To evaluate the effect of the hydration heat on the in situ strength development of three HSC mixtures in a massive wall stored under different ambient temperatures and examine the difference between field-cured cylinder strength and in situ strength measured from cores, 1200-mm thick mock-up walls were prepared and stored in approximately isothermal chambers at 5 °C (cold condition), 20 °C (reference condition), and 35 °C (hot condition). The HSC mixture proportions specifically determined for use in nuclear facility structures were identified as S5 (with 5 % SF) for storage under the cold condition, G50 (with 50 % GGBFS) for storage under the reference condition, and G65S5 (with 65 % GGBFS and 5 % SF) for storage under the hot condition. The in situ compressive strength development in the mock-up walls measured from core samples was compared with estimated strength based on three strength–maturity relationship models, considering the temperature rise owing to the hydration heat. The following conclusions may be drawn from this study:

1. The G50 and G65S5 mixtures had longer final setting times than the S5 mixture, regardless of the curing temperature. The differences in the setting times of the concrete mixtures increased with a decrease in the ambient curing temperature.
2. The initial apparent activation energies at the hardening phase were higher by 1.8, 2.2, and 2.6 times for the S5, G50 and G65S5 mixtures, respectively, than the apparent activation energies until the final setting, indicating that the temperature sensitivity of the strength development is more than that of setting time.
3. The differences in the 28-day strengths between the field-cured cylinders and cores were higher for the G50 wall than for the S5 wall. Meanwhile, the 28-day strength measured by cores from the G65S5 wall was similar to that of the companion field-cured cylinders.
4. For the S5 mixture stored at 5 °C, the core strength was commonly higher than the field-cured cylinder strength until an age of 28 days, beyond which a higher strength

was obtained in the cylinders, showing the crossover effect. Furthermore, the core strength at an early age was higher at the central region of the wall than at its inner surface region due to greater temperature rise from heat of hydration.
5. For the G65S5 mixture, a strength gain of more than approximately 80 % of design strength was achieved within the first 7 days, even for the field-cured strength. However, for the S5 mixtures stored at 5 °C, the cylinder strength at 7 days failed to meet the 80 % design strength level, whereas the core strengths achieved this value.
6. The strength–maturity model proposed by Yang provides better reliability in estimating in situ strength of concrete than that of Kim et al. and Pinto and Schindler, indicating that the internal curing effect owing to the hydration heat in a massive member needs to be considered to reasonably assess the early-age strength of concrete.

Acknowledgments

This work was supported by a Nuclear Research and Development program of the Korea Institute of Energy Technology Evaluation and Planning (KETEP) Grant funded by the Korea Government Ministry of Knowledge Economy (2011T100200161).

References

ACI Committee 214. (2010). *Guide for obtaining cores and interpreting compressive strength results (ACI 214.4R-10)*. American Concrete Institute, Farmington Hills, Michigan, USA.

ACI Committee 306. (2010). *Guide to cold weather concreting (ACI 306R-10)*. American Concrete Institute, Farmington Hills, Michigan, USA.

ASTM C42/C42 M, C150, C403/C403 M, C989, C1074. (2011). *Annual Book of ASTM Standards, V. 4.02*, ASTM International, West Conshohocken, PA. 2011.

Bamforth, P. B. (1980). In situ measurement of the effect of partial Portland cement replacement using either fly ash or ground granulated blast furnace slag on the performance of mass concrete. *Proceeding Institution of Civil Engineers, 69*(2), 777–800.

Bogue, R. H. (1955). *Chemistry of portland cement*. New York, NY: Reinhold Publisher.

Byfors, J. (1980). Plain concrete at early ages. CBI Research Report No. 3:80, *Cement and Concrete Research*.

Carino, N. J., & Tank, R. C. (1992). Maturity function for concretes made with various cements and admixtures. *ACI Materials Journal, 89*(2), 188–196.

Harris, D. W., Mohorovic, C. E., & Dolen, T. P. (2000). Dynamic properties of mass concrete obtained from dam cores. *ACI Materials Journal, 97*(3), 290–296.

Haug, A. K., & Jakobsen, B. (1990). In situ and design strength for concrete in offshore platforms. ACI SP 121-19 High-Strength Concrete. *Second International Symposium*, 369–397.

Hulshizer, A. J. (2001). The benefits of the maturity method for cold-weather concreting. *Concrete International, 23*(3), 68–72.

Kim, J. K., Han, S. J., & Park, S. K. (2002a). Effect of temperature and aging on the mechanical properties of concrete. Part II. Prediction model. *Cement and Concrete Research, 32*(7), 1095–1100.

Kim, J. K., Han, S. H., & Song, Y. C. (2002b). Effect of temperature and aging on the mechanical properties of concrete Part I. Experimental results. *Cement and Concrete Research, 32*(7), 1087–1094.

Lee, D. H., Kim, S. Y., Jeon, M. H., Kim, Y. H., & Lee, K. H. (2013). *Development of technology for the field application of blast-furnace slag powder concrete*. Daejeon, Korea: Land & Housing Institute.

Neville, A. M. (1995). *Properties of concrete*. New York: Addison Wesley Longman Limited.

Nili, M., & Salehi, A. M. (2010). Assessing the effectiveness of pozzolans in massive high-strength concrete. *Cement and Concrete Research, 24*(11), 2108–2116.

Parsons, T. J., & Naik, T. R. (1985). Early age concrete strength determination by maturity. *Concrete International, 7*(2), 37–43.

Pinto, R. C. A., & Schindler, A. K. (2010). Unified modeling of setting and strength development. *Cement and Concrete Research, 40*(1), 58–65.

Pucinotti, R. (2013). Assessment of in situ characteristic concrete strength. *Construction and Building Materials, 44*, 63–73.

Schrader, E. (2007). Statistical acceptance criteria for strength of mass concrete. *Concrete International, 29*(6), 57–61.

Sofi, M., Mendis, P. A., & Baweja, D. (2012). Estimating early-age in situ strength development of concrete slabs. *Construction and Building Materials, 29*, 659–666.

Uva, G., Porco, F., Fiore, A., & Mezzina, M. (2013). Proposal of a methodology for assessing the reliability of in situ concrete tests and improving the estimate of the compressive strength. *Construction and Building Materials, 38*, 72–83.

Vázquez-Herrero, C., Martínez-Lage, I., & Sánchez-Tembleque, F. (2012). A new procedure to ensure structural safety based on the maturity method and limit state theory. *Construction and Building Materials, 35*, 393–398.

Yang, K. H. (2014). High-strength concrete application technology for nuclear facilities. *Technical Report(1st)*, Department of Plant·Architectural Engineering, Kyonggi University, Suwon, Korea.

Zákoutský, J., Tydlitát, V., & Černý, R. (2012). Effect of temperature on the early-stage hydration characteristics of Portland cement: A large-volume calorimetric study. *Cement and Concrete Research, 36*, 969–976.

Permissions

The contributors of this book come from diverse backgrounds, making this book a truly international effort. This book will bring forth new frontiers with its revolutionizing research information and detailed analysis of the nascent developments around the world.

We would like to thank all the contributing authors for lending their expertise to make the book truly unique. They have played a crucial role in the development of this book. Without their invaluable contributions this book wouldn't have been possible. They have made vital efforts to compile up to date information on the varied aspects of this subject to make this book a valuable addition to the collection of many professionals and students.

This book was conceptualized with the vision of imparting up-to-date information and advanced data in this field. To ensure the same, a matchless editorial board was set up. Every individual on the board went through rigorous rounds of assessment to prove their worth. After which they invested a large part of their time researching and compiling the most relevant data for our readers.

The editorial board has been involved in producing this book since its inception. They have spent rigorous hours researching and exploring the diverse topics which have resulted in the successful publishing of this book. They have passed on their knowledge of decades through this book. To expedite this challenging task, the publisher supported the team at every step. A small team of assistant editors was also appointed to further simplify the editing procedure and attain best results for the readers.

Apart from the editorial board, the designing team has also invested a significant amount of their time in understanding the subject and creating the most relevant covers. They scrutinized every image to scout for the most suitable representation of the subject and create an appropriate cover for the book.

The publishing team has been an ardent support to the editorial, designing and production team. Their endless efforts to recruit the best for this project, has resulted in the accomplishment of this book. They are a veteran in the field of academics and their pool of knowledge is as vast as their experience in printing. Their expertise and guidance has proved useful at every step. Their uncompromising quality standards have made this book an exceptional effort. Their encouragement from time to time has been an inspiration for everyone.

The publisher and the editorial board hope that this book will prove to be a valuable piece of knowledge for researchers, students, practitioners and scholars across the globe.

List of Contributors

Floriana Petrone, Li Shan and Sashi K. Kunnath
Department of Civil and Environmental Engineering, University of California, Davis, CA 95616, USA

Dong-Kwan Kim
SEN Structural Engineers Co., R&D Team, Seoul 07229, Korea

Soo-yeon Seo
Department of Architectural Engineering, Korea National University of Transportation, Chungju, Korea

Ki-bong Choi
Department of Architectural Engineering, Gachon University, Seongnam, Korea

Young-sun Kwon
Jaeshin CTNG, Co., Ltd., Seoul, Korea

Kang-seok Lee
Department of Architectural Engineering, Cheonnam National University, Gwangju, Korea

Byung-Wan Jo, Sumit Chakraborty, Ji Sun Choi and Jun Ho Jo
Department of Civil and Environmental Engineering, Hanyang University, Seoul 133791, Korea

Yasser Sharifi, Iman Afshoon, Zeinab Firoozjaei and Amin Momeni
Department of Civil Engineering, Vali-e-Asr University of Rafsanjan, Rafsanjan, Iran

Shashi Ranjan Pandey and A. K. L. Srivastava
Department of Civil Engineering, National Institute of Technology, Jamshedpur, Jharkhand 831014, India

Shailendra Kumar
Department of Civil Engineering, Institute of Technology, Guru Ghasidas Vishwavidyalaya (A Central University), Bilaspur, CG 495009, India

S. Altoubat and M. Maalej
Department of Civil & Environmental Engineering, College of Engineering, University of Sharjah, Sharjah, UAE

F. U. A. Shaikh
Department of Civil Engineering, Curtin University, Perth, Australia

K. De Wilder and L. Vandewalle
Department of Civil Engineering, Building Materials and Building Technology Section, KU Leuven, 3001 Heverlee, Belgium

G. De Roeck
Department of Civil Engineering, Strucutral Mechanics Section, KU Leuven, Kasteelpark Arenberg 40, Box 2448, 3001 Heverlee, Belgium

Jinkoo Kim, Yong Jun, and Hyunkoo Kang
Department of Civil and Architectural Engineering, Sungkyunkwan University, Suwon, Korea

A. M. Hernández-Díaz
Department of Civil Engineering, Universidad Cato´lica de Murcia (UCAM), Campus de Los Jerónimos, 30107 Murcia, Spain

M. D. García-Román
Department of Civil Engineering, University of La Laguna, Campus de Anchieta, 38271 Santa Cruz de Tenerife, Spain

Hamdy M. Afefy
Department of Structural Engineering, Faculty of Engineering, Tanta University, Tanta 31511, Egypt

El-Tony M. El-Tony
Department of Structural Engineering, Faculty of Engineering, Alexandria University, Alexandria, Egypt

Jianzhuang Xiao
College of Civil Engineering, Tongji University, Shanghai 200092, China

Zhenping Sun
College of Civil Engineering, Tongji University, Shanghai 200092, China
College of Material Science and Engineering, Tongji University, Shanghai, China

Changqing Wang
College of Civil Engineering, Tongji University, Shanghai 200092, China
Postdoctoral Mobile Research Station, College of Material Science and Engineering, Tongji University, Shanghai, China
Nanyang Normal University, Nanyang 473000, China

Woo-Young Lim
Institute of Engineering Research, Seoul National University, Seoul 08826, Korea

Sung-Gul Hong
Department of Architecture and Architectural Engineering, Seoul National University, Seoul 08826, Korea

Doo-Yeol Yoo
Department of Architectural Engineering, Hanyang University, Seoul 133-791, Korea.

Young-Soo Yoon
School of Civil, Environmental and Architectural Engineering, Korea University, Seoul 136-713, Korea

Lan Chung, Jong-Jin Lim and Tae-Sung Eom
Department of Architectural Engineering, Dankook University, Yongin-si 448-701, Gyeonggi-do, Korea

Hyeon-Jong Hwang
College of Civil Engineering, Hunan University, Yuelu Mountain, Changsha 410082, Hunan, China

Keun-Hyeok Yang
Department of Plant Architectural Engineering, Kyonggi University, Suwon 443-760, Korea

Jae-Sung Mun
Department of Architectural Engineering, Graduate School, Kyonggi University, Suwon 443-760, Korea

Do-Gyeum Kim
Structural Engineering & Bridges Research Division, Korea Institute of Construction Technology, Goyang 411-712, Korea

Myung-Sug Cho
KHNP-Central Research Institute, Korea Hydro & Nuclear Power Co., LTD, Daejeon 305-343, Korea

Index

A
Alternate Path Method, 1-2
Anchorage Capacity, 28, 32, 34, 38
Aqueous Carbonation, 41-45, 47-48, 50, 52

B
Binary Blended Concrete, 55

C
Carbon Dioxide Emission, 41
Cementing Material, 55, 63-65, 69, 71
Cementitious Material, 41-44, 48-54, 56, 59, 63, 70, 182, 205, 207
Chloride-induced Steel Corrosion, 85
Co2 Sequestration, 41, 53
Column Removal Scenarios, 1, 8, 11
Composite Beams, 192-193, 196-197, 199-200, 202, 204
Composite Structure, 192, 195, 203-204
Concrete Fracture, 72, 83-84
Coordinate Measurement Machine, 94-95
Corrosion Damage, 85-86, 88, 92
Curing Temperature, 205-206, 210-212, 214, 217
Cyclic Material Models, 14
Cylinder Strengths, 205, 211-213

D
De-bonded Length, 28-30, 32, 34-35, 38
Deformation Capacity, 14, 28, 30, 32, 35, 39, 156, 172, 192
Diagonal Uniaxial Elements, 14-16, 25
Double Curvature Mode, 131-133, 135-137, 141-142
Double-k Fracture Parameters, 72, 74, 79, 83-84
Ductility Factors, 109, 120-121, 123-124

E
Effectiveness of Aqueous Carbonated Lime, 41
Equivalent Column Concept, 131, 133, 142

F
Failure Mode, 14, 19, 23, 39, 102-107, 167, 180, 188
Finite Element Model, 14, 145, 151
Flexural Strength, 28-29, 32, 35, 38, 59-60, 65, 70, 177, 189, 191-192, 204
Frame Structure, 1, 11, 145-147, 151-152, 154-157, 159, 161
Free Drying Shrinkage, 55, 60, 67

G
Ground Waste Glass (gwg), 55

H
Hardened Properties, 55-56
High-rate Loads (impacts And Blasts), 174
High-strength Concrete, 71, 108, 205-206, 218
Hydration Heat, 63, 205-207, 209, 212, 217
Hydrothermal Synthesis, 41-44, 48-51

L
Laboratory Simulation, 85
Load And Resistance Factor Method, 192
Load-carrying Capacity, 14, 19
Loadbearing Columns, 1
Loading Condition, 14, 76, 101, 146, 175

M
Macro Model, 14
Mock-up Wall Specimens, 206-208

O
Overstrength Factors, 109-110, 118, 120, 123-124

P
Partial Factor Method, 192-193
Partially De-bonded Nsm Frp Strip, 28, 38
Peak Ground Acceleration (pga), 24, 151
Prestressed Concrete, 94-95, 106, 108, 129-130, 184, 187
Progressive Collapse Analysis, 1, 4, 8, 12-13

R
Recycled Aggregate Concrete (rac), 145
Reinforced Concrete, 1, 12-16, 25-28, 39-40, 83, 85-86, 93, 107, 109-110, 123-125, 129-131, 133, 144, 146, 151, 157, 160, 162, 172-174, 178, 182, 186, 189-191
Reinforced Concrete (rc) Moment-frame Buildings, 1
Reinforced Concrete Wall Structure, 14
Resistance Factor, 192-193, 195, 197, 199-203
Resistance Factor Design Method (lrfd), 192
Response Factors, 109

S
Seismic Behavior Factors, 109-110
Seismic Response Analysis, 14, 160
Self-consolidating Concrete (scc), 55, 181

Shaking Table Test, 23-25, 145, 147, 152, 159-160

Shear Reinforcement, 16, 25, 95, 97, 102-106, 162-163, 165-167, 169-173

Simplified Design Procedure, 131

Single And Triple Lines Of Frp, 28

Single Curvature Mode, 131-132, 135-136, 140, 142

Slender Walls, 18

Span-to-depth Ratio, 14, 18, 25, 165, 170, 179, 188

Split-tension Cube Test, 72, 74, 82

Staggered Wall Systems, 109-110, 124

Steel Constitutive Model, 127, 129

Steel Fiber-reinforced Concrete (sfrc), 162

Stereo-vision Digital Image Correlation, 94, 97, 106

Strain Rate Effect, 145-147, 152, 154-157, 159-160

Strength-maturity Models, 205

Supplementary Cementitious Materials (scms), 206

T

Tension Stiffening Area, 127, 129

The In Situ Compressive Strengths, 205-206

The Strength-maturity Model, 205, 217

Threat-independent Approach, 1

Three Point Bend Test, 72, 74, 78, 80, 82

U

Ultra-high Performance Fiber-reinforced Concrete (uhpfrc), 162

Ultra-high-performance Fiber-reinforced Concrete (uhpfrc), 174

W

Waste Glass Micro-particles, 55

Weight Function, 72, 74, 77-79, 82-83